Computing Methodology for Switchgear

Mingzhe Rong · Yi Wu · Joseph Yan · Hao Sun

Computing Methodology for Switchgear

科学出版社

Springer

Mingzhe Rong
School of Electrical Engineering
Xi'an Jiaotong University
Xi'an, Shaanxi, China

Yi Wu
School of Electrical Engineering
Xi'an Jiaotong University
Xi'an, Shaanxi, China

Joseph Yan
Department of Electrical Engineering
and Electronics
University of Liverpool
Liverpool, UK

Hao Sun
School of Electrical Engineering
Xi'an Jiaotong University
Xi'an, Shaanxi, China

ISBN 978-981-95-2839-4 ISBN 978-981-95-2840-0 (eBook)
https://doi.org/10.1007/978-981-95-2840-0

Jointly published with Science Press
The print edition is not for sale in China (Mainland). Customers from China (Mainland) please order the print book from: Science Press.
ISBN of the Co-Publisher's edition: 978-703-05-8165-5

This Springer imprint is published by the registered company Springer Nature Singapore Pte Ltd.
The registered company address is: 152 Beach Road, #21-01/04 Gateway East, Singapore 189721, Singapore

Foreword

Switchgear is an important type of electrical equipment, acting as the enabling hardware for the protection, control, and regulation of power systems. The design of switching devices is a process requiring input from many disciplines, including heat transfer, mechanics, electrical insulation, and gas discharge. It has long relied on personal experience and expensive and time-consuming tests on prototype designs. This traditional approach leads to several challenges such as low design efficiency and difficulties in improving and optimising the product performance. Through the combination of basic theory, simulation technology, and calculation methods, key parameters can be predicted, bringing significant changes to the design practice of switching devices. This new approach helps revolutionise the traditional design practice which depends on experience and repeated tests while at the same time, greatly improving the design efficiency and providing strong guidance for the rapid development of new products and upgrading of existing ones.

This book is devoted to the calculation and analysis methods for arc discharges and electrodynamic and thermal stabilities encountered in the design of switching products. First, the basic theory of switching arc modelling is described with due consideration of the current interruption process. The arc modelling method is proposed according to two distinctive stages: the current-carrying arcing stage and the arc quenching stage. The calculation of gas material properties is described with the aim of providing baseline data for arc modelling. Optical diagnosis of important arc plasma parameters is presented as a method of verifying the results from arc modelling. Second, this book also deals with the electrodynamic and thermal stability design of switchgear products and describes methods for calculating the heating phenomenon and evaluating the effects of Lorentz forces associated with the current-carrying conductors. Mathematical model development is introduced for the operating mechanisms in both low-voltage and high-voltage circuit breakers. Finally, a number of case studies involving the current interruption processes in switching devices are provided to enable readers to apply the calculation methods to practical problem solving.

The content of this book draws from the authors' research on electrical apparatus technology in the past 20 years, covering theory, methods of calculation, and solutions to practical problems. It is engaging and should be used by academics, research students and design engineers as a reference book to guide research or product design. I would certainly recommend this book to relevant readers.

June 2018

Academician of the Chinese Academy of Science (CAS)

Preface

Switchgear as a group of electrical equipment performs the dual functions of control and protection in power systems. They are widely used in power plants, substations, switch stations and on load circuits. To meet their operational requirements, these switching devices should have reliable breaking ability for rated and short-circuit currents whilst also needing to exhibit electrodynamic and thermal stability not only at the rated and short-circuit currents but also with prolonged overload current.

In response to the trend of global development in power generation and delivery, there has been intense research and development (R&D) in recent years in switching technologies towards environmental friendliness, intelligence and miniaturisation, and for use with DC applications. Mechanical switching devices are accompanied by strong arc discharges during their closing and breaking operations. The physical processes involved are extremely complicated, posing stringent requirements on the design and manufacture of switching devices. Traditionally, switchgear design has been based on developmental tests but now there is a persistent desire to find a reasonably accurate calculation-based design method to shorten the product development cycle and lower R&D costs.

Introduction to Computing Methodology for Switchgear

To meet the challenges above, computing arises as an important branch in Electrical Engineering to aid switchgear design, which can be coined Computing Methodology for Switchgear. In short, it is the use of mathematical (rather than experimental) methods to predict, evaluate and optimise the arc behaviour, driving mechanism characteristics, and electrodynamic and thermal stability of switching equipment. This is supported by the following three considerations.

First, with the growing application of analytical and numerical methods in electromagnetics, heat transfer, structural mechanics, fluid mechanics, and atomic physics, knowledge and understanding of the complex physical processes involved in the operation of switching devices has been continuously advanced. Together with the rapid advance in computing power in the past 30 years, and the maturing of commercial computational software, this knowledge and understanding has enabled the modelling and simulation of physical phenomena in virtual environments. Therefore, integrating the different calculations related to switching devices into an interdisciplinary subject based on numerical simulation has become more relevant than ever.

Second, switching devices consist of several key components such as the arc interrupter, operating mechanism, tripping system, electrical contacts, etc. Describing and modelling the individual physical processes alone is not able to guide the design of switchgear products as a whole system. Thus, the establishment of Computing Methodology for Switchgear in Electrical Engineering focuses on solving the key scientific problems of a multi-physics nature such as those that are encountered in the design of switchgear, including the arcing and current zero phenomenon, mechanical stress and deformation, system dynamics with rapidly changing load, and thermal stability with strong electrical heating. By building up a complete set of mathematical models, numerical solution procedures, and a comprehensive knowledge pool, quantitative prediction of the performance of key elements in switching devices will be possible, eliminating the reliance of product development on expensive and time-consuming tests.

Finally, computing as a switchgear design tool is undergoing continual progress.

The development of efficient methods and high-performance computational software has greatly reduced the difficulties in numerical calculation, enabling wider applications in switchgear research and design. Numerical predictions are also steadily approaching reality in terms of microscopic and fundamental properties, allowing a more in-depth understanding of the complex physical processes in switching devices. Unquestionably, this will provide more accurate guidance to designers and be a tool for rapidly testing broader design ideas for switchgear products.

The Development of Computing Theory

Since the launch of switchgear, research on its technology and the method for its design has evolved significantly. From pure experimentation and testing, through semi-empirical design mixing experimental and theoretical understanding, to a modern state that uses advancing digital simulation tools.

Development of the Arc Model

Since the beginning of the 20th century, our research on arc models has undergone continual improvement. Most early models were based on empirical formulae for qualitative analysis. Among them, the most representative is the black-box arc model adopted in the 1930–1940s which treats the arc as a two-terminal component in an electrical circuit. Differential equations proposed by Mayr and Cassie describe the volt–ampere characteristics of the arc, which when coupled with other circuit parameters, represent the interruption process embedded in the arc–circuit interaction. However, this empirical model only considers the overall electrical characteristics and does not account for the complex physical and chemical processes which occur inside the arc. Furthermore, the model parameters, such as the arc power loss constant, steady-state arc conductance in steady state, and arc time constant are not universal and need to be determined by measurement under different interruption conditions for each type of design. The application of the black box model becomes more restricted in the modern era of rapidly advancing switching technology, including switching medium and diversified design.

In the 1940s, Barbu et al. first proposed the concept of magnetohydrodynamics, a subject that studies the motion of conductive fluids in a magnetic field. Building on the work in this field, methods of describing the arcing process in terms of the field distribution of flow quantities began to be developed in the 1970s. The arc models during this period were however relatively simple, i.e. based mainly on the one-dimensional formulation in space. This was a result of unavailable physiochemical properties of the high-temperature arcing gas and a lack of efficient numerical solution algorithms and computing power. This considerably limited the ability of arc modelling to aid switchgear design.

In the last two decades of the 20th century, full differential models for thermal plasmas including switching arcs were developed to calculate the arc and flow parameter distributions, such as gas temperature, pressure and flow field, and their change with time. Detailed information on enthalpy, specific heat capacity, viscosity, thermal conductivity, electrical conductivity, and radiation characteristics as a function of plasma state became available, which collectively brought about a breakthrough in arc modelling. Digital simulation of the arc behaviour under different interruption conditions for various structures of the interrupter was first realised during this period.

With high-performance computers becoming affordable, arc modelling based on the magnetohydrodynamic approach flourished in the early years of the 21st century. The work by our group at Xi'an Jiaotong University has been fruitful, focusing on a number of lines including arc root transfer and its mobile behaviour in low-voltage applications, arc splitting and ablation of splitter material at high current, instability of arcing in high-voltage circuit breakers, and non-equilibrium phenomenon in switching arcs.

Since 2010, switching arc-related activities have followed two directions: engineering applications and fundamental understanding, each supporting the other. For the former, an increasing number of switchgear manufacturers have been choosing to use commercially available coupled multi-physics solvers to perform switching arc simulation to aid the design and optimization of their products. For the latter, advanced plasma diagnosis has aided our understanding of the dynamic behaviour of arcs at a more fundamental and microscopic level. The outcome of this has been a more accurate and in-depth scientific knowledge that helps tackle the pressing challenges in switchgear technology, such as the dielectric recovery of non-equilibrium arced gases post current zero and the search for SF_6 alternative gases.

Structural Analysis and Study of Electrodynamic and Thermal Stability

The development of computing capabilities in electromagnetics, Newtonian mechanics, structural mechanics, and heat transfer has enabled the rapid growth of computing as a tool for designing switchgear operating mechanisms, which ensures the electrodynamic and thermal stability of switchgear products when subject to transient and stable electrical loads.

In the 20th century, the design of switchgear operating mechanisms and the determination of electrodynamic and thermal stability started with empirical descriptions. A typical example is the 'conductive bridge' model describing electrical contact resistance between the fixed and movable contacts in switching devices: the equivalent radius of the conducting spots on the surface is obtained using an analytical expression based on the surface interaction force and the hardness of the materials. From this, the equation representing the repulsive Holm force originating from the contraction of the current conducting cross-sectional area can be derived. Furthermore, different numerical methods for the magnetic circuit calculation of electromagnetic mechanisms, including analytical, graphical, and magnetic field partition-based, have all been developed to analyse the electric and magnetic fields. These analytical or semi-empirical methods were commonly used to characterise the electrodynamic and thermal stability of switchgear throughout the 20th century.

Moving into the 21st century, the structure of operating mechanisms and electrical contact systems in switching devices has become more diversified with increasing performance requirements. Alongside this, nonlinear ferromagnetic materials have also been used in electromagnetic operating mechanisms. As a result, the application of the above methods has become more restricted in product design. With computational software and computing power becoming more advanced, the finite element method (FEM) has seen more widespread applications to address the needs relating to different aspects of switchgear design, such as the electrodynamic stability of contacts and conductors, thermal stability, spring operating mechanisms, electromagnetic and permanent magnet operating mechanisms, and thermal trips in low-voltage switches. Comparison with experimental results has proven that the FEM has advantages in both calculation accuracy and applicability over the traditional methods.

However, the application of FEM analysis was hindered by the complexity of handling geometry and difficulties in meshing and setting up boundary and load conditions, limiting its widespread use in switchgear design. From around 2010, there has been rapid development in finite element analysis software. ANSYS Maxwell for electromagnetic field simulation has been an example of increasing user-friendliness in its interface and ease of use. The calculation accuracy of FEM-based solvers has continuously improved, and the number of functional modules has increased while reducing the difficulty in operation. FEM is now widely used in the analysis of operating mechanism characteristics and electrodynamic and thermal stability of switchgear product.

About This Book

Accurate representation and numerical modelling, supported by practical experiences, of the behaviour of key components and processes form the pillars to construct the theoretical framework for switchgear computing and is thus the most important issue to address. Since 2000, Xi'an Jiaotong University has conducted a comprehensive research program relating to switchgear computing, covering calculation of plasma properties, arc plasma diagnosis, non-equilibrium dielectric recovery after current zero in high-voltage circuit breakers,electrodynamic force analysis, high-speed electromagnetic mechanisms, arc splitting in low-voltage circuit breakers and rapid DC medium-voltage high-current breaking in air.

Based on our research findings, this book aims to summarise the relevant results and research trends in recent years. We look forward to receiving feedback and suggestions from readers to further improve its quality and readability.

Xi An, China
August 2024

Mingzhe Rong

Introduction

This book is based on the research conducted by the authors over a substantial period of time in the field of power-switching apparatuses, encompassing recent advances and future development trends. With due consideration of practical experiences, the focus of this book is on theoretical studies with the aim of establishing a framework for Computing Methodology for Switchgear through accurate modelling of the key functional components of switchgear products. Case studies closely related to practical engineering applications are provided. The book consists of seven chapters. Chapters 1–4 focus on the development of mathematical models, including the relevant theory and solution methods for arc simulation, with a particular interest in the post-arc period and the calculation and experimental verification of the material properties used directly in arc simulation. Chapters 5 and 6 introduce methods of calculation for the analysis of processes relating to mechanical stability due to Lorentz force, and operation of the operating mechanisms of switchgear. Chapter 7 provides case studies relating to switching arcs.

This book is intended for R&D engineers of power switching apparatuses, academics and postgraduate researchers in the fields of electrical discharges, thermal plasmas, gas-based switching technology and other relevant disciplines.

Chapter 1—Fundamentals and Computer Simulation of Switching Arcs
Chapter 2—Calculation and Analysis of the Post-Arc Dielectric Recovery Process
Chapter 3—Calculation of Material Properties of Arcing Gas
Chapter 4—Measurement of Plasma Parameters for Switching Arcs
Chapter 5—Calculation and Analysis of Electromotive Force
Chapter 6—Calculation and Analysis of Operating Mechanisms in Switchgear
Chapter 7—Simulation Examples for Switching Arcs.

Contents

Chapter 1
Fundamentals and Computer Simulation of Switching Arcs

1.1 Introduction

The electrical contacts in mechanical circuit breakers separate rapidly when they are called by the power system to interrupt a current. During this interruption process, the gaseous medium in the contact gap (air or SF_6) is ionised and turned into a cluster of high-temperature, bright and electrically conductive gas. This ionised gas is termed 'switching arc' and will be known as 'arc' for short in this chapter. In low-voltage air circuit breakers, the arc immediately stays immobile between the two contacts for a short while after its ignition before moving swiftly, under the action of Lorentz force and pressure gradient, along the guiding contacts towards the splitter plates. The arc is then forced by a strong gas blast and magnetic field to cut into the plates before cooling down and finally extinguishing to achieve current interruption. The effect of the arc is two-sided. On the one hand, it provides a pathway for dissipating the energy stored in the circuit and reducing the over-voltage generated in the interruption process, arc voltage also helps to limit the fault current. On the other hand, the arc causes contact erosion or even an explosion of the device in the worst-case scenario.

Switching arc research aims to understand the characteristics and mechanisms of relevant physical phenomena operating in the process from arc initiation to arc extinction, in order to maximise the desired effect of the arc and minimise any problems. There are two approaches to achieve this. The first one is to experimentally understand the dynamics of the arc through the measurement of macroscopic parameters such as the arc voltage, current, gas pressure, magnetic field, arc size and location, etc. The other approach is to make use of mathematical models to simulate and study the flow field, electric and magnetic fields, and temperature fields and their variation with time. These two approaches provide complementary information so are often used together.

The arcing process, following contact separation, is a crucial step in the operation of a circuit breaker. Knowledge and understanding of its characteristics bear theoretical significance to the design and optimisation of a new circuit breaker product.

M. Rong et al., *Computing Methodology for Switchgear*,
https://doi.org/10.1007/978-981-95-2840-0_1

Until recently, circuit breaker design has heavily relied on the traditional practice of making prototypes and performing evaluation tests, which does not meet the demand for modern equipment manufacturing. Computer simulation aided circuit breaker design and optimisation, however, can significantly improve the design efficiency and reduce development costs. For the latter, the establishment of arc models is a complex problem that requires interdisciplinary research. It can be difficult to directly measure many of the physical parameters of the arcs due to their characteristic short duration (<60 ms) and high temperature (>15,000 K). As an important part of the arc theory, many studies have been carried out to establish the dynamic model of gas arcs.

In this chapter, the simulation method for gas arcs is presented in three parts. First, the fluid model based on the Magnetohydrodynamics approach, radiation transfer, turbulence, and their relevant numerical solution methods are described in detail. Second, a three-dimensional arc model for the process when the arc cuts into the splitter plates of an air circuit breaker is developed and the effects of electrode sheath and contact erosion are also discussed. Finally, a two-dimensional axisymmetric arc model for SF_6 circuit breakers is established and modelling of steady-state arcs is carried out. This chapter aims to help the reader deepen their understanding of arc theory and modelling methods while mastering the essential skills for arc modelling.

1.2 Basic Physical Processes in Electric Arcs

When the current to be interrupted and the voltage across the contacts is higher than certain threshold values during the interruption process, there will be a cluster of high-temperature, bright and electrically conductive gas produced in the contact gap. Examples of arcing in low-voltage air circuit breakers and high-voltage SF_6 circuit breakers are illustrated in Fig. 1.1a, b, respectively. Research on switching arcs seeks to understand how the arc behaves and how it can be quenched under given conditions, a critical part to the design and optimisation of switchgear products. The theory on switching arcs still needs improvement, partly because of its coupled multi-physics nature and partly due to its interaction with the confining materials.

There are strong interactions between the flow field, thermal field, electric field and magnetic field in the arc plasma, as is shown in Fig. 1.2. When an arc is ignited, it starts to carry current. Strong joule heating raises the temperature and pressure of the conducting gas. Energy is then transferred from the arc to its surroundings through thermal conduction, convection and radiation. The high pressure initiates gas flow. As a result, the contacts (electrodes), walls and some of the gas in the arcing chamber are heated up. Subsequent changes in the gas temperature and pressure lead to varying electrical conductivity of the heated gas and redistribute the electric field and current density, which not only affects the joule heating (towards a different temperature) but also alters the Lorentz force (modifying the flow field). Thus the changes in temperature, velocity and pressure fields are closely coupled, evolving with time. In summary, arc simulation involves the calculation of gas flow field, thermal field,

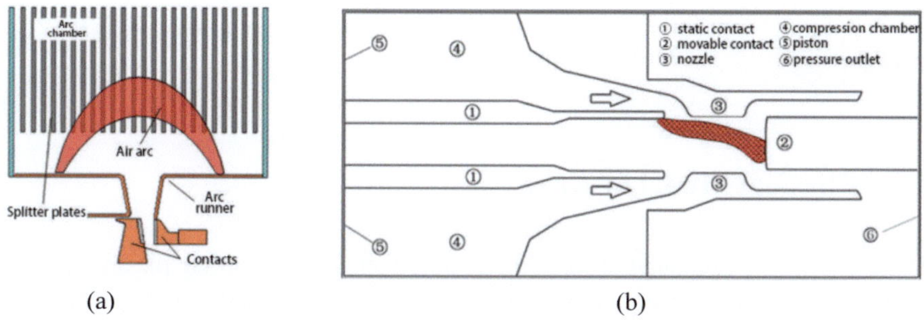

Fig. 1.1 Typical instants of arcing in gas circuit breakers. **a** air circuit breaker when the arc cuts into the splitter plates; **b** high voltage SF6 circuit breaker when the main nozzle is unblocked by the plug contact

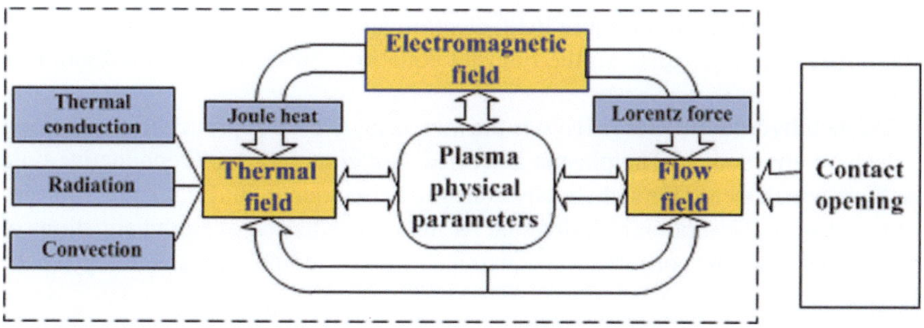

Fig. 1.2 Coupled physical processes in switching arcs. [This diagram can be improved to better represent the couling mechanisms]

electric and magnetic fields, and radiation transfer, coupled by momentum and energy sources and material properties. Establishing the model and solving the coupled equations is therefore key to the modelling of switching arcs.

Arc plasmas contain electrons, ions as well as neutral particles, where the degree of ionisation is largely controlled by the temperature. The density of different species in the plasma varies with temperature and pressure [1]. The plasma as a mixture is electrically neutral and its state can be assumed through local thermodynamic equilibrium (LTE). At atmospheric pressure and above, frequent collisions between the plasma particles warrant the treatment of the arcing gas as a continuum medium that is electrically conducting. In the presence of an electric field, collisions between charged particles and other plasma particles produce joule heating, or Ohmic heating. Magnetohydrodynamics (MHD) is a subject that studies the motion of a conductive fluid in an electromagnetic field, especially in a strong, externally applied magnetic field. The electrical conductivity of the fluid is expected to be very high, close to that of a fully ionised plasma; the externally applied magnetic field is expected to be strong so that the term $V \times B$ becomes important. The behaviour of the flow is therefore governed by the coupled fluid dynamics equations and Maxwell's equations. The

MHD approach is taken into account below in the development of the switching arc model.

1.3 Physical Model of Arcs

There is a strong exchange of mass, momentum and energy among plasma species, leading to thermal conduction, convection and radiation transfer at a macroscopic level. The composition of the plasma also changes as a result of dissociation, association, ionisation and recombination at the microscopic level. The collective effects of these processes can be described by a set of unified governing equations. The differences in the behaviour of various substances is characterised by their material properties [2]. Computer simulation of switching arcs includes the calculation of the flow field, thermal field, pressure field, electromagnetic field and radiation transfer, so the first step is to establish the relevant equations governing the distribution of quantities.

Magnetohydrodynamics (MHD) is the most widely used approach in recent years to develop arc models. The plasma is treated as a continuous and conductive fluid medium, which means the statistically averaged behaviour of fluid elements is considered instead of the motion of individual particles. The basic governing equations of the arc model are the conservation equations of mass, momentum and energy of the arc gas.

Due to the complex physical phenomena involving arc plasma, certain assumptions are made in order to make the problem solvable. simplifying their simulation. According to the established model, the following basic assumptions are adopted in this chapter [3]:

1. It is considered that the arc plasma is in the state of local thermal equilibrium (LTE) [4].
2. The plasma is electrically neutral without space charge.
3. The vaporisation of the electrodes or wall is not considered.

1.3.1 Governing Equations for Arc Flow

According to the theory of hydrodynamics, Reynolds number is used to determine whether the fluid is laminar or turbulent [5].

$$R_e = VD\rho/\mu \tag{1.1}$$

where V is the characteristic velocity, D is the characteristic length of the flow, ρ the density of the gas, and μ is the dynamic viscosity. When $R_e > 10^5$, the flow is normally turbulent. Although the gas velocity in the arc space may be high, flow can still be subsonic since the speed of sound increases with temperature. D is usually taken as the typical dimension of the flow confining channel. Taking the air circuit breaker as an example, the size of the flow channel is in the order of 10^{-2} m. In addition, the viscosity, μ, of the arcing gas at high temperature. In comparison, the gas density in the surrounding cooler region is $10^{-1} \sim 10^{-2}$ kg/m^3. Based on the analysis above, the R_e of the high temperature arc column in the air circuit breaker arc chamber is 10^2 and R_e in the lower temperature region is 10^4, so the flow in the chamber can be considered as laminar.

1. Conservation of mass

$$\frac{\partial \rho}{\partial t} + \nabla \cdot (\rho \vec{U}) = 0 \tag{1.2}$$

where t is time, ρ the gas density, and $\vec{U}$ is the velocity vector of gas flow.

2. Conservation of momentum

The solution of the velocity field is described by Navier–Stokes equations [6–8]:

$$\frac{\partial(\rho u)}{\partial t} + \nabla \cdot \left(\rho u \vec{U}\right) = \nabla \cdot (\eta \nabla u) + S_u \tag{1.3}$$

$$\frac{\partial(\rho v)}{\partial t} + \nabla \cdot \left(\rho v \vec{U}\right) = \nabla \cdot (\eta \nabla v) + S_v \tag{1.4}$$

$$\frac{\partial(\rho w)}{\partial t} + \nabla \cdot \left(\rho w \vec{U}\right) = \nabla \cdot (\eta \nabla w) + S_w \tag{1.5}$$

where u, v and w are the velocity components in x, y and z directions, respectively; η is the viscosity of the gas. The source terms in the momentum conservation equations are:

$$\begin{aligned} S_u = &-\frac{\partial p}{\partial x} + \frac{\partial}{\partial y}\left(\eta \frac{\partial u}{\partial x}\right) + \frac{\partial}{\partial z}\left(\eta \frac{\partial u}{\partial x}\right) + \frac{1}{3}\frac{\partial}{\partial x}\left(\eta \frac{\partial u}{\partial x}\right) \\ &- \frac{2}{3}\frac{\partial}{\partial x}\left(\eta\left(\frac{\partial v}{\partial y} + \frac{\partial w}{\partial z}\right)\right) + F_{mx} \end{aligned} \tag{1.6}$$

$$\begin{aligned} S_u = &-\frac{\partial p}{\partial y} + \frac{\partial}{\partial z}\left(\eta \frac{\partial v}{\partial y}\right) + \frac{\partial}{\partial x}\left(\eta \frac{\partial w}{\partial y}\right) + \frac{1}{3}\frac{\partial}{\partial y}\left(\eta \frac{\partial v}{\partial y}\right) \\ &- \frac{2}{3}\frac{\partial}{\partial y}\left(\eta\left(\frac{\partial w}{\partial z} + \frac{\partial u}{\partial x}\right)\right) + F_{my} \end{aligned} \tag{1.7}$$

$$\begin{aligned} S_w = &-\frac{\partial p}{\partial z} + \frac{\partial}{\partial x}\left(\eta \frac{\partial w}{\partial z}\right) + \frac{\partial}{\partial y}\left(\eta \frac{\partial v}{\partial z}\right) + \frac{1}{3}\frac{\partial}{\partial z}\left(\eta \frac{\partial w}{\partial z}\right) \\ &- \frac{2}{3}\frac{\partial}{\partial z}\left(\eta\left(\frac{\partial u}{\partial x} + \frac{\partial v}{\partial y}\right)\right) + F_{mz} \end{aligned} \tag{1.8}$$

In comparison with traditional hydrodynamic equations, the arc plasma is affected by electromagnetic forces. Since the plasma is electrically neutral, the electrical force acting on the gas is negligibly small and the volume force is mainly Lorentz force:

$$F_{mx} = j_y B_z - j_z B_y \tag{1.9}$$

$$F_{my} = j_z B_x - j_x B_z \tag{1.10}$$

$$F_{mz} = j_x B_y - j_y B_x \tag{1.11}$$

where F_{mx}, F_{my} and F_{mz} are, respectively, the components of magnetic force in the x, y and z directions of the Cartesian coordinate system, j_x, j_y and j_z, the corresponding components of current density and B_z, B_y and B_z the components of magnetic flux density.

3. Conservation of energy

$$\frac{\partial(\rho h)}{\partial t} + \nabla \cdot \left(\rho h \vec{U}\right) = \nabla \cdot \left(\frac{\lambda}{c_p} \nabla h\right) + S_h \tag{1.12}$$

with the source term being

$$S_h = S_\varphi + \sigma E^2 - q_{rad} \tag{1.13}$$

where h is specific enthalpy, λ the thermal conductivity, c_{p} the specific heat at constant pressure, and S_h is the source term of energy. S_φ is the viscous dissipation term in the flow, σ the electrical conductivity of the arc plasma, and E is the electric field. σE^2 is the joule heating produced by current flowing through an electrically resistive medium. It is the main cause of temperature rise in the arc column and can be obtained by solving the equation governing the electric field. q_{rad} represents the energy loss through radiation transfer. As the arc temperature is high, radiation is an important means of energy exchange in arc plasma. The calculation methods for electric field and radiation will be introduced shortly. The arc radiation model is delayed until Sect. 1.4.

4. Equation of state

Finally, the equation of state of gas, as given below, is required to solve the equations above

$$\rho = \rho(p, T) \tag{1.14}$$

For real gas, the equation of state has to be expressed in the form of a data table or by curves representing the density as a function of temperature and pressure and also species concentration in a mixture.

1.3.2 Governing Equations for Electric and Magnetic Fields

Arc plasma containing free-charged particles is a special type of fluid and the current flowing through it produces a magnetic field. The Joule heating and electromagnetic force in the source terms of the conservation equations need to be known. Maxwell equations are used to obtain the electromagnetic field in the arc.

The electric field is calculated by the following equation:

$$\nabla \cdot (\sigma \nabla \varphi) = 0 \tag{1.15}$$

$$\vec{E} = -\nabla \varphi \tag{1.16}$$

$$\vec{J} = \sigma \vec{E} \tag{1.17}$$

where φ is the electric potential, J is the current density vector and E is the electric field.

There are two methods to solve the magnetic field. The first one is to use Biot-Savart law and the other is to solve the magnetic vector potential equation. The advantage of the former is that the boundary problem for the magnetic field does not need to be considered in the calculation, but the number of iterations required in the calculation is proportional to the square of the number of grid elements. For a three-dimensional arc model, the number of elements is large and the calculation may take a long time. In addition, Biot-Savart's law is only suitable for uniform permeability within the calculation domain and is not applicable to problems with ferromagnetic material. Considering the complexity of three-dimensional magnetic field calculation, the magnetic vector potential method:

$$\nabla \cdot (\nabla A_i) = S_{Ai} \tag{1.18}$$

$$S_{Ai} = -\mu j_i \tag{1.19}$$

$$\vec{B} = \nabla \times \vec{A} \tag{1.20}$$

where A is the magnetic vector potential and μ the permeability of the gas. The subscript i represents the vector direction of current in three-dimensional space.

The material properties of the arc plasma appearing in the above equations can be calculated using fundamental theory, which includes density ρ, specific heat at constant pressure c_p, thermal conductivity λ, viscosity η and electrical conductivity σ. The composition and degree of ionisation of the plasma varies with temperature and pressure and so do the material properties. The curves of viscosity, thermal conductivity, specific heat at constant pressure and electrical conductivity of air plasma as a function of temperature and pressure are shown in Fig. 1.3 [9]. It can be

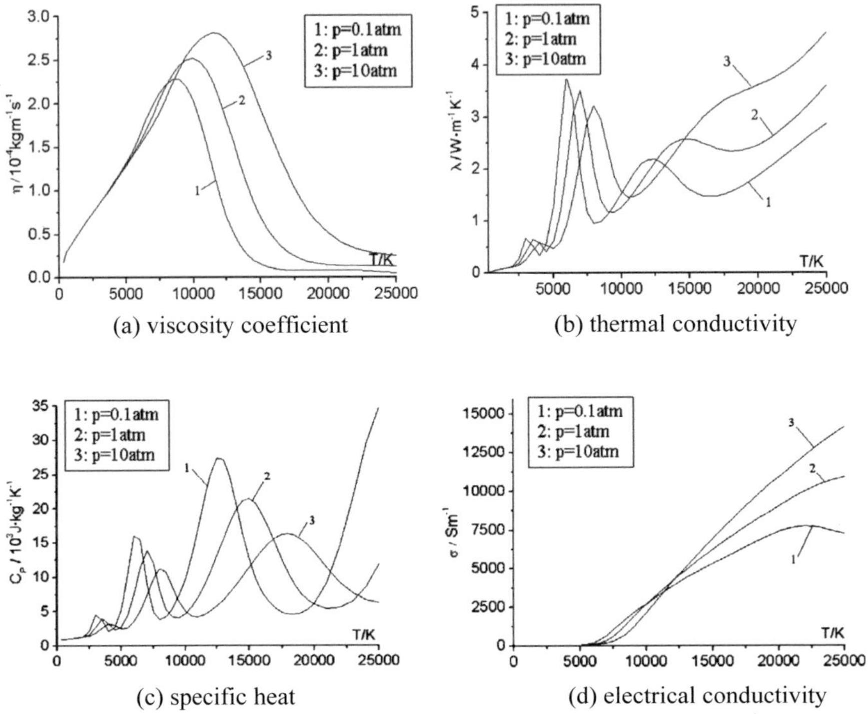

Fig. 1.3 Material properties of air plasma at different temperatures and pressures [9]

seen that the change in temperature and pressure has a great influence on the material properties of air. As a result, the solution to the governing equations will be affected by highly nonlinear material properties.

1.4 Model of Radiation Transfer

Radiation transfer is the fastest energy exchange process in an arc plasma and it is an essential part of the arcing process [2]. In the high temperature arc, vibration and excitation of the molecules and atoms and interaction of electrons with other particles lead to the emission of light. the spectrum in the wavelength range of 0.4–40 μm can be absorbed by the medium being projected on, turning into heat. Radiation is essentially an electromagnetic wave. As shown in Fig. 1.4, radiation transfer can be described by the general equation of radiation transfer for a single wavelength in directions [10]:

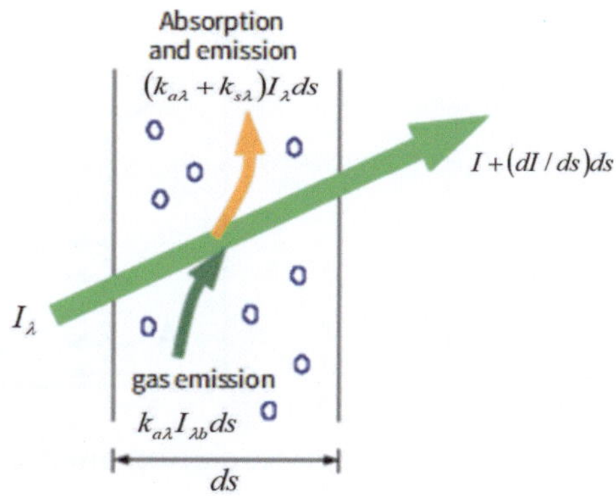

Fig. 1.4 Schematic diagram of radiation transfer

$$\frac{dI_\lambda}{ds} = k_{a\lambda} I_{\lambda b} - (k_{a\lambda} + k_{s\lambda}) I_\lambda + \frac{k_{s\lambda}}{4\pi} \int_0^{4\pi} I_\lambda(s, \Omega)\Phi_\lambda(\Omega) d\Omega \tag{1.21}$$

where I_λ is the monochromatic spectral radiation intensity, $k_{a\lambda}$ the spectral absorption coefficient, $k_{s\lambda}$ the spectral emission coefficient and Φ_λ is the phase function of spectrum scattering.

As the temperature in the column of a high current arc exceeds 20,000 K, it is generally considered that the radiation exchange of the arc with its surrounding region consists of two parts. As shown in Fig. 1.5, the high-temperature zone in Part A, the arc core, generally emits radiation along the radial direction; the low-temperature zone in Part B, located at the edge of the arc column, absorbs radiation. Arc radiation includes both continuum spectrum and line spectrum and the latter includes a large number of spectral lines, ranging from infrared to ultraviolet in wavelength. Therefore radiation transfer involving radiation emission and absorption is a very complex process. The spectral absorption coefficient at a temperature of 15,000 K and pressure of 1 atm is shown in Fig. 1.6 where the line absorption varies greatly [11]. Since it takes a huge amount of effort to calculate the emission and absorption of the line as well as the continuum spectra, the method of direct integration of the radiation spectra is not applicable to switching arc simulation [12]. In addition, the absorption coefficient heavily depends on temperature. Generally speaking, the higher the gas temperature, the greater the energy radiated out. Within an arc plasma, the temperature varies at different positions, leading to a difference in radiation emission and absorption. Radiation transfer calculation is one of the challenging problems in arc research.

In order to obtain the radiation term q_{rad} in Eq. 1.13, it is necessary to calculate the radiation field. A number of approximate and practical methods have been developed, such as the net emission coefficient method (NEC), simplified T4 method, method of partial characteristics (MPC), P-1 method, etc.

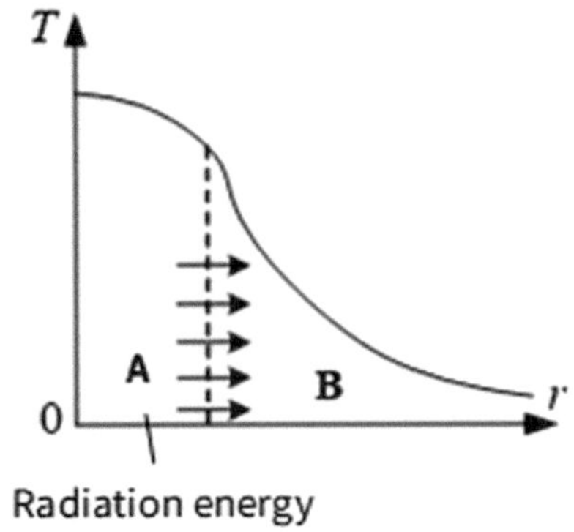

Fig. 1.5 Schematic diagram of radiation from an arc

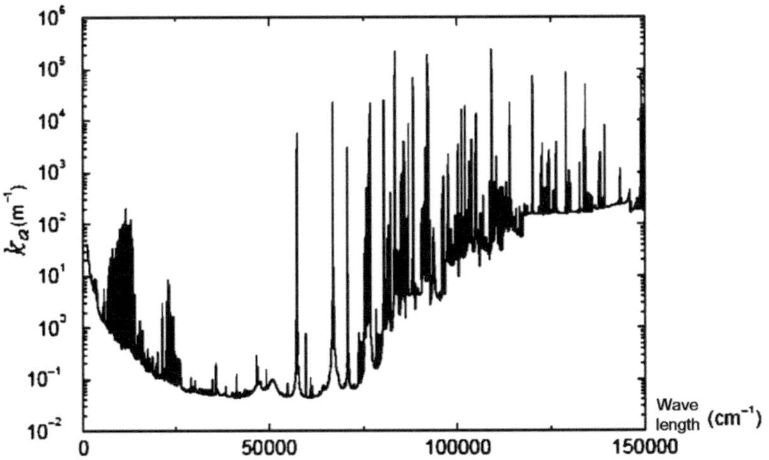

Fig. 1.6 Spectral absorption coefficient of air at 15,000 K and 1 atm

NEC refers to the difference between the energy emitted and absorbed at the centre of an isothermal sphere. Based on this definition, the NEC, ε_n, of isothermal spheres of different radii can be derived from Kirchhoff's law. The radiative energy source term can be calculated by the NEC as:

$$q_{rad} = 4\pi \varepsilon_n \tag{1.22}$$

The NEC of different gases have been calculated, e.g. in [13–15]. The simplified T4 method defines the total emission energy of a black body to be directly proportional to the fourth power of the temperature according to the Boltzmann law of black body radiation, and the net radiation from an arc is [16]:

$$q_{rad} = 4\alpha k\left(T^4 - T_0^4\right) \tag{1.23}$$

where α is the Stefan–Boltzmann constant, k the absorption coefficient and T_0 is the temperature of the ambient. These two methods are relatively simple and practical, without the requirement of solving equations. Hence they are widely used in the calculation of arc radiation. However, they only consider the emission of radiative energy in the high-temperature zone of the arc chamber, and ignore the absorption of radiation in the cooler region, thus introducing large error in the cold surrounding areas where radiation absorption takes place. In addition, these two methods cannot provide accurate radiation information for the calculation of vaporisation of the wall material as a result of absorption of radiation.

The partial characteristics method simplifies the radiation transfer calculation by introducing two integral functions determined by the distance between two points and their temperatures, i.e. the partial characteristic functions [17]. This method takes into account radiation emission and absorption at the same time, so it can more accurately calculate the absorption of radiation at the arc column edge. However, this method still consumes a large amount of computing time on integration and so far has only been applied to arcs in uniform pressure field. It is generally only used for two-dimensional arc simulation.

In addition, a semi-empirical model was proposed by M. T. C. Fang at the University of Liverpool for fast and approximate radiation calculation in axially blown SF_6 arcs [18, 19]. Net emission is considered in the arc core ($T > 0.83\ T_{max}$ for a monotonic radial temperature profile) and net radiation absorption takes place in the region from $T < 0.83T_{max}$ to $T = 4{,}000$ K. According to this model, 80% of the radiative energy from the arc core is absorbed in the re-absorption region. This method is an improvement of the NEC method and brings the results closer to the actual situation; however Fang's method is only applicable to simple nozzle arcs. As for the transient arc in low voltage circuit breakers, the arc shape is irregular and fast changing so this method may not be applicable.

With advances in radiation transfer calculation techniques, the P1 model has attracted more and more attention in recent years for arc modelling. Its calculation accuracy has been well-verified. Fang and Eby [20, 21] used the 5-band P1 model for SF_6 arcs and the results show agreement with the measured temperature; Khokan [22] also used the five-band P1 model to carry out a two-dimensional radiation transfer calculation for a DC mercury arc lamp. This model is commonly used in the solution of two-dimensional problems, its use in three-dimensional arcs of low-voltage circuit breaker has not been explored.

1.5 Model of Turbulence

Since Reynolds discovered the turbulence phenomenon in circular tube flow in 1883, research has not been stopped. However, due to the complexity of turbulence, there is no generally accepted definition that can fully describe the characteristics of turbulence. Turbulence is a random, multi-scale phenomenon. Research has shown that the minimum turbulence length scale, the Kolmogrov scale, is much larger than the mean free path of gas molecules under normal conditions, so the fluid in turbulent flow can still be regarded as a continuous medium. Thus, a mathematical model for turbulent flow can be established in a similar way to the description of fluid flow and the Navier–Stokes equations can still be used to describe time-averaged turbulent flow.

According to the number of turbulence scales that can be accurately solved in the multi-scale system, the existing methods for turbulence modelling can be divided into three types: direct numerical simulation (DNS), large eddy simulation (LES) and Reynolds average N-S equation (RANS). The first two methods need huge computational effort and available computing power cannot meet the demand for engineering problems. So they are only used in fundamental research. Methods of their type have to be used for turbulence modelling in engineering applications.

The RANS method divides the turbulent flow into two parts: the mean velocity and the pulsation. The velocity of the flow can be calculated from Eq. 1.35,

$$u_i = \overline{u}_i + u_i' \tag{1.24}$$

where u_i is instantaneous velocity component, $\overline{u}_i$ the time average velocity components, i = 1,2,3 representing the direction of x, y and z of the Cartesian coordinate system, $u_i\prime$ is the offset of average velocity.

Substituting the equation above into the N-S equation for a time-averaged velocity, it can be obtained as:

$$\frac{\partial}{\partial t}(\rho\overline{u}_i) + \frac{\partial}{\partial x_j}\left[\rho\overline{u}_i\overline{u}_j + p\delta_{ij} - \overline{\overline{\tau}} - \tau_{ij}^{\mathrm{R}}\right] = F_{\mathrm{m}i} \tag{1.25}$$

$$\tau_{ij}^{\mathrm{R}} = -\rho\overline{u_i'u_j'} \tag{1.26}$$

Equation 1.25 represents the Reynolds-averaged Navier–Stokes equation. Compared with the Navier–Stokes equation for laminar flow, the additional $\tau_{ij}{}^{\mathrm{R}}$ term is called the Reynolds stress, which is an unknown term describing the momentum of the turbulent pulsation.

The RANS method aims at the establishment of an additional relationship to close the turbulent flow equations. This is fulfilled by the development of an empirical or semi-empirical relationship between the high-order and low-order statistical terms according to the very nature of turbulence, experimental observation or DNS results. Up to date, methods based on this approach fall into two categories: first-order moment model (eddy viscosity model) and second-order moment model. This chapter mainly introduces the first-order moment model.

The first-order moment model is the most widely used turbulence model in engineering problems due to its simple form, high applicability and its acceptable accuracy. It was originally proposed by Boussinesq [23] following the idea of molecular viscosity, assuming that the Reynolds stress satisfies the following linear relationship:

$$\tau_{ij}^{\mathrm{R}} = \mu_{\mathrm{t}}\left(\frac{\partial u_i}{\partial x_j} + \frac{\partial u_j}{\partial x_i}\right) - \frac{2}{3}\left(\rho k + \mu_{\mathrm{t}}\frac{\partial u_k}{\partial x_k}\right)\delta_{ij}, \delta_{ij} = \begin{cases} 1, i = j \\ 0, i \neq j \end{cases} \tag{1.27}$$

where μ_{t} is the turbulent viscosity coefficient or eddy viscosity coefficient, and k is the turbulent kinetic energy per unit mass. By dimensional analysis, it can be found that:

$$\mu_t \propto ul \tag{1.28}$$

where u is the characteristic velocity scale and l is the characteristic length scale. In terms of expressing the two scales, the eddy viscosity model can be further divided into zero equation, single equation and two equation models.

Semi-empirical turbulence models developed in the 1930s, such as Prandtl's mixing length model [24] and those based on Taylor's vorticity transport theory [25] and Von Karman's similarity theory [26], are essentially zero equation models. Fang has shown that the Prandtl mixing length model can produce satisfactory predictions for SF_6 switching arcs under certain conditions, which works based on nozzle arcs in strong axial flow bear similarities to boundary layer problems [27].

The single equation model is used to solve the transport equation for turbulent kinetic energy or eddy viscosity coefficient with good accuracy and robustness. Among them, the B-B model and S-A model are excellent representatives of the single equation model, particularly the S-A model [28]. The transport equation for eddy viscosity is constructed with the aid of experience and dimensional analysis. Many experimental results are used to calibrate the model coefficient, which has good robustness and calculation accuracy. At present, it is implemented in numerous commercial computational fluid dynamics (CFD) software packages and has been applied widely in the field of aerospace [29].

The standard k-epsilon model was developed by Lauder et al. [30] in the 1970s. Two equations, one for the turbulence kinetic energy k and the other for turbulence energy dissipation rate ε, are used to represent the transport and dynamic change of turbulence parameters (at least to a certain extent). It is the foundation of the two-equation model. Following this, the re-normalised group k-epsilon model (RNG k-epsilon) and realisable k-epsilon model were developed. These two models exhibit superior performance in specific applications [31].

The k-epsilon model is more widely used compared with the other variants because it can provide reasonable accuracy and is relatively more economical. Equations 1.29–1.31 give the mathematical details.

$$\mu_t = C_\mu \frac{k^2}{\varepsilon} \tag{1.29}$$

$$\frac{\partial}{\partial t}(\rho k) + \frac{\partial}{\partial x_i}(\rho k u_i) = \frac{\partial}{\partial x_j}\left[\left(\mu + \frac{\mu_t}{\sigma_k}\right)\frac{\partial k}{\partial x_j}\right] + G_k + G_b - \rho\varepsilon - Y_M \tag{1.30}$$

$$\frac{\partial}{\partial t}(\rho \varepsilon) + \frac{\partial}{\partial x_i}(\rho \varepsilon u_i) = \frac{\partial}{\partial x_j}\left[\left(\mu + \frac{\mu_t}{\sigma_\varepsilon}\right)\frac{\partial \varepsilon}{\partial x_j}\right] + C_{1\varepsilon}\frac{\varepsilon}{k}(G_k + C_{3\varepsilon}G_b) - C_{2\varepsilon}\rho\frac{\varepsilon^2}{k} \tag{1.31}$$

where G_k is the turbulence kinetic energy generation rate that is related ot the gradient of the time averaged velocity, G_b the turbulence kinetic energy generation rate due to the buoyancy, and Y_M the contribution of the fluid compressibility towards the turbulence kinetic energy in high Mach number flow. $C_{1\varepsilon}$, $C_{2\varepsilon}$, $C_{3\varepsilon}$ and C_μ are empirical constants.

In engineering applications, it is often necessary to select an appropriate turbulence model according to the characteristics of the problems to be tackled. The k-epsilon model, among the two-equation models, is widely used in the simulation of SF_6 nozzle arcs and air arcs. There is an indication that the standard k-epsilon model overestimates the turbulent effect and the model constants ($C_{1\varepsilon}$, $C_{2\varepsilon}$, $C_{3\varepsilon}$) need to be changed to bring the prediction closer to reality [32].

Another group of two-equation turbulence models are the k-ω models which has two representatives: the standard k-ω model and the k-ω SST model. It is generally accepted that the standard k-ω model is preferred for modelling near wall turbulence while the k-ω SST model is more accurate in predicting flow separation in air in the presence of strong adverse pressure gradient.

Turbulence in arc flow is a phenomenon with more complexity in comparison with flow near room temperature. Research effort so far has been on the suitability of existing turbulence models to predict the macroscopic effect of turbulence mass, momentum and energy transport rather than the physical mechanisms responsible for the onset, transition and development process in arc flow.

1.6 Solution of Arc Model

According to the above analysis, it can be seen that a general form of the governing equations needs to be established in order to simplify the analysis and solution to the governing equations. The governing equations for arc flow introduced above can be written in the following general form:

Table 1.1 Description of solved-for variables in (1.32)

Equation	a_t	Φ	Γ_Φ	S_Φ
Mass	1	1	0	0
Momentum	1	u	η	S_u
	1	v	η	S_v
	1	w	η	S_w
Energy	1	h	λ/c_p	S_h
Electric field	0	φ	σ	0
Magnetic field	0	A_x	1	S_{Ax}
	0	A_y	1	S_{Ay}
	0	A_z	1	S_{Az}
Radiation field equation	0	G_i	$1/(3k_{ami})$	S_{Gi}

$$a_t \frac{\partial(\rho\Phi)}{\partial t} + \nabla \cdot \left(\rho\Phi\vec{U}\right) = div(\Gamma_\Phi \nabla\Phi) + S_\Phi \qquad (1.32)$$

where Φ represents the solved-for variables including u, v, w, h, φ, A_x, A_y, A_z, and G_i. Their meanings as well as the material property coefficients in (1.32) are listed in Table 1.1.

1.6.1 Discretisation of Governing Equations

There are three main discretisation methods for the system of governing equations, each based on a different principle: finite difference method (FDM), finite element method (FEM), and finite volume method (FVM) [33]. The finite difference method was the first method to be used for such calculations. It divides the solution domain into a system of difference grids. The continuous solution domain is replaced by a finite number of grid nodes and the derivative in the partial differential equations is replaced by difference quotient to obtain the difference equations of finite unknowns on the discrete points. The finite element method is a popular numerical solution method that combines the basic method of discretisation in the finite difference method and the more accurate method of using approximate function to perform integration over the finite elements. However its solution is slower than those of FDM and FVM.

The basic principle of the finite volume method is to divide the computational domain into finite grids and the finite volumes around a grid do not overlap (ΔV is used to represent a finite volume, or control volume). A set of discrete equations is obtained by integrating the governing partial differential equations over each control volume. The unknowns are the solved-for variables on the grid points, such as gas speed, density, and temperature. The value of a variable in each volume, φ, is represented by the value at the centre of mass of the control volume. The discretised equations

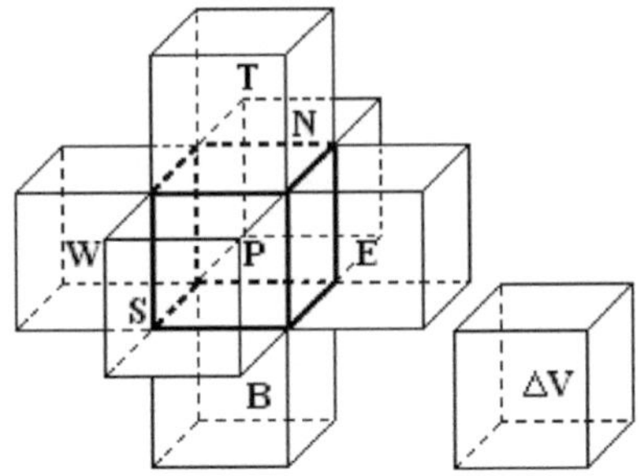

Fig. 1.7 3-D mesh control volume

derived in the finite volume method ensures the conservation of physical quantities and the relevant coefficients has clear physical meanings. The computational effort required to obtain the solution is relatively less than the finite element method. This method is commonly used to solve the problems involving arc plasma systems.

As shown in Fig. 1.7, the nodes of control volumes adjacent to control volume P are W, E, S, N, B and T. By discretisation of the flow equations, the variables on control volume P can be obtained as follows [34]:

$$\alpha_p \varphi_p = \alpha_W \varphi_W + \alpha_E \varphi_E + \alpha_S \varphi_S + \alpha_N \varphi_N + \alpha_B \varphi_B + \alpha_T \varphi_T + b \tag{1.33}$$

where the coefficients of discretised equations are

$$b = \alpha_P^0 \varphi_P^0 + S_C \Delta V \tag{1.34}$$

$$\alpha_P = \sum \alpha_{nb} + \Delta F + \alpha_P^0 - S_P \Delta V \tag{1.35}$$

$$\alpha_P^0 = \frac{\rho_P^0 \Delta V}{\Delta t} \tag{1.36}$$

and α_P^0 is the initial value of α_P, S_C is the constant part of the discretised source term, and S_p is the part of the discretised source term that changes with time and φ_p. ΔF is a coefficient related to the control volume interface, and α_{np} is the α coefficient linked to an adjacent control volume.

1.6.2 *Use of Fluent Software Package for Simulation*

It can be seen from the above equations relating to arc simulation that the mathematical model is described by a set of coupled non-linear partial differential equations. Obviously, an analytical solution method is not applicable to such problems and a numerical solution method is needed. There are two main approaches to obtain the numerical solution, one is to use self-programming whose code is transparent and can be changed. This approach has however its limitations, such as that different codes are produced for different arcing chambers; codes developed by different authors may differ in their solution; model set-up requires heavy human effort; and post-processing may be difficult. With the development of computer technology, another practical way to obtain a numerical solution is to use commercial computational fluid dynamics (CFD) software packages. Existing work has shown that it is feasible to use a commercial CFD package to perform arc simulation [35]. In this book, Fluent (version 18.1 released in 2018) is used as the equation solver.

Fluent is a general purpose CFD package and application specific coding and settings are needed. The following steps are needed for arc modelling based on Fluent:

1. Definition of computational domain and grid generation. Gambit is used to create the geometry of the arcing chamber, generate the mesh, and export the mesh, through a file, into the Fluent solver.
2. Implementation of the arc model. This includes two aspects. Firstly, the continuity, momentum, species concentration in the case of a gas mixture, and energy equations need to be implemented into the default equations provided by Fluent. The material properties and source terms need to be fed into the solver using the interface provided. Secondly, equations for user-defined solved-for variables need to be set up to obtain solutions to other governing equations, such as those for the electric and magnetic fields and radiation transfer. User written codes are often required for the calculation of material properties and source terms. All governing equations are coupled to obtain the arc solutions. This is the key step in arc simulation based on Fluent.
3. Generation of the Case file that contains all necessary information for the simulation and producing the executable by compiling and linking the solver code with the dynamic link library (.dll file). Parameters relating to the solution process may need to be set to specific values according to the application before the simulation is run to obtain the results.
4. Finally, process the output data.

Switching arcs involve high-speed gas flow in the arc chamber and are driven by Lorentz force. The governing equations are strongly coupled. Use of the coupled solver may be preferred over the segregated solver to promote convergence in the solution process. The energy and flow equations are directly coupled during the solution process, leading to faster convergence than the Segregated solver. However, it needs a larger amount of computer memory, about 1.5–2 times that of the segregated method.

The implicit scheme is commonly used to discretise the nonlinear arc governing equations into algebraic equations for each control volume, containing variables defined on the control volume and its neighbours. If there is a total of N coupled governing equations and a total of M volumes in the system, then N linear equations are generated for each control volume, and the system has a total of $N \times M$ equations. The Gauss–Seidel method is used to solve the equation, as in (1.37).

$$\varphi_k^n = \left(\sum_{l=1}^{k-1} \alpha_{kl} \varphi_l^n + \sum_{l=k+1}^{M \times N} \alpha_{kl} \varphi_l^{n-1} + b_k \right) \Bigg/ \alpha_{kk} \tag{1.37}$$

where k and l are the number of rows and columns of the coefficient matrix of the equations, α and b are the coefficients in the equations, and n is the number of iteration. By solving these equations values of the solved-for variables in all control volumes will be obtained.

Solution of the arc governing equations is achieved with the execution of the user-written code, as shown in Fig. 1.8. Initialisation is carried out at the very beginning before the first iteration cycle of the first time step begins. Within each iteration, all variables are updated first according to the results from the previous iteration or time step, as the start point. The first step is to update the material properties and derived variables in each cell (control volume) based on the most recent temperature and pressure fields, for example the electrical conductivity, viscosity coefficient and thermal conductivity. The electric field, current density and magnetic flux density are then obtained.

The second step is to update the boundary conditions and source terms (e.g. Lorentz force, joule heating, etc.) for each governing equation. The new set of equations, as a result of changes in the material properties and source terms, are solved. The mass, momentum and energy conservation equations are solved concurrently, which differs from the segregated solver. This is followed by the solution to the other governing equations.

The convergence of the solution process is reflected in the residuals resulting from the iterative process. Threshold values for the residuals or a fixed number of iteration can be specified to end the iterative process. Parameters that may affect the

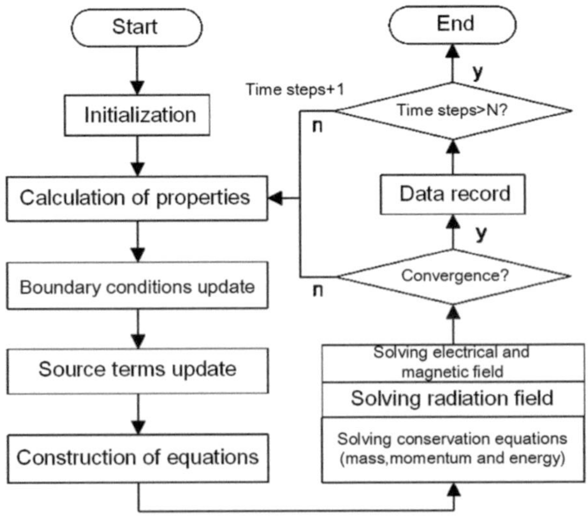

Fig. 1.8 Procedure flow of coupled secondary development

convergence of the solution process include the size of the time step, the number of iterations, and the relaxation factors, etc.

User-defined code in Fluent appears in the form of user-defined functions (UDFs). This includes functions to calculate material properties and source terms, control the computation, perform initialisation and define boundary conditions. These functions are integrated with the solvers through macros in standard C language.

1) Functions for initialisation

Initial conditions form the start point of a transient simulation. each solved-for variable requires its own initial condition in the whole computational domain, such as the initial gas temperature and pressure fields in the arc chamber. For the solution of a steady-state arc case, a cylindrical initial arc column is assumed in this chapter for easy convergence, the temperature and pressure of which is 15,000 K and 1 atm, respectively, while the other field values are 0. For a transient arcing process, the steady-state solution may be used as its initial conditions.

The macro function, DEFINE_INIT (), is created by Fluent to initialise the soled-for variables. This type of function can only be performed once in each simulation, and they will be executed before all other calculation.

2) Functions to calculate material properties

It can be seen from Fig. 1.3 that the material properties (thermal conductivity, viscosity, specific heat, etc.) of arc plasma are non-linear functions of temperature and pressure. In the iterative solution process, these material properties need to be updated continuously until the end of the iteration.

The macro function of DEFINE_PROPERTY () is used to pass known property value to the solver. For properties like thermal conductivity, viscosity, electrical conductivity, specific heat, and radiation absorption coefficient, two-dimensional linear surface interpolation is used. For those properties relating to electric, magnetic and radiation fields, macro functions such as DEFINE_ DIFFUSIVITY () can be used to define them.

3) Functions for source terms

Some source terms in the arc governing equations, such as joule heating and Lorentz force, are not built in source terms and have to be defined by the user and pass into the solver. Most source terms are related to the solved-for variables, Φ. In order to improve the convergence of the solution process, source terms are often linearised, as shown in Fig. 1.9 as

$$S = S_c + S_p \Phi_p \tag{1.38}$$

where S_c is the constant part of the source term, and S_p is the slope. Before defining the source terms, the user will need to perform the linearization to obtain S_c and S_p.

The macro function DEFINE_SOURCE () is provided in Fluent for the user to define the S_c and S_p of the source terms. It is called by the solver to fetch the source term for each cell before the discretised equations are solved.

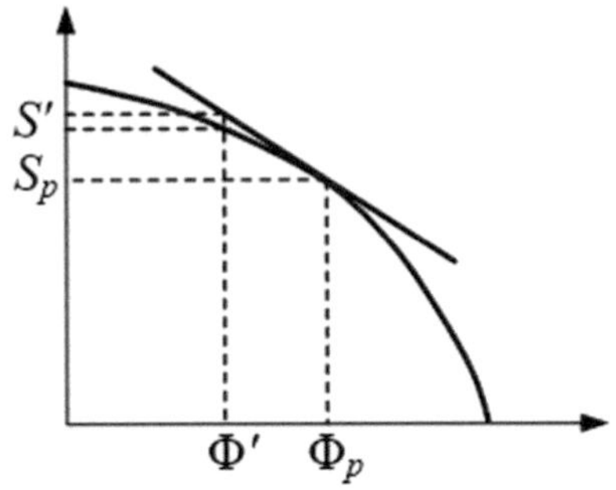

Fig. 1.9 Linearisation of source terms. Φ_P is the solved-for variable in cell P and Φ' is the value in the neighbouring cells

4) Functions for boundary conditions

Boundary conditions represent the interaction between a system and its environment. The governing equations and their initial and boundary conditions together form a complete mathematical description of a physical process. The boundary conditions in an arc simulation mainly involve mass flux, velocity, temperature, electromagnetic field, radiation field and pressure at the boundary. The boundary types in the examples in this chapter are mainly wall and outlet. The built-in macro function DEFINE_PROFILE () can be used to define boundary conditions. The first type of boundary condition and the second type of boundary condition are mainly used in the arc simulation. For example, current density distribution is used as the boundary condition for electric potential equation (electrical field). Thermal flux or fixed temperature is usually used as the boundary conditions for the energy equation. Boundary conditions are updated for each iteration.

5) Functions for control and adjustment

Over the course of simulation, it may be necessary to check, record, adjust or assign data items, such as the instantaneous current, current density, and arc induced magnetic field. At certain points in an iteration loop, user operation can be invoked using control macro-functions such as DEFINE_ADJUST (), DEFINE_EXECUTE_AT_END (), and DEFINE_ON_DEMAND () to perform adjustment at the beginning or to prepare for the next iteration.

The macro function DEFINE_ADJUST () can be used to obtain the relevant gradients on each cell and to calculate the current density and magnetic flux density. This function is executed once at the beginning of each iteration. A macro-function executed at the end of the iteration, DEFINE_EXECUTE_AT_END (), is also available to prepare for the next iteration or time step, such as the activation of the change in arc root position, value of instantaneous current or electrode velocity. The macro function DEFINE_ON_DEMAND () can be used for data inspection and modification only when it is necessary, for example to force a change to a variable or to test the convergence of an equation, such as by changing the value of the arc drive magnetic field.

1.7 Simulation of Air Arc Splitting Process

For air circuit breakers, the most popular method to quench the arc is to use the metal splitter plates. The governing principle is that the original single arc column is cut into several short arc columns located between the metal plates such that multiple pairs of electrode region voltage drops are developed on the metal plates. This not only rapidly increases the arc voltage but also cools the arc at the same time, producing enhanced current limiting effect and completing the fault current breaking [36]. Therefore, the characteristic behaviour of the arc in the chamber is key to the performance of the breaker, and the arc cutting process is one of the most important aspects. In addition, when the current is high and the arc stays on the metal plates for long, the temperature of the plate surface will reach its melting point. This will inevitably lead to the diffusion of metal vapour due to contact (electrode) erosion into the arc column and modify the arc characteristics. The influence of the ablated metal vapour is particularly important for medium-voltage, which around 10 kV to 70 kV. DC arc interruption which requires a large number of metal plates.

In low-voltage air circuit breakers, especially medium-voltage air circuit breakers with a higher number of metal splitters, the splitters occupy the majority of the arc chamber space and interact with the arc over a period that is longer than those when the arc is between the two contacts or between the arc runners. In other words, when ablation of the plates is significant, the influence of the metal vapour from the metal splitters is more important than that of the vapour from contact and arc runner erosion. The influence of metal vapour on the arc has always been a difficult issue to understand. In recent years, with advances in arc mathematical models, strong research focus has been placed on this problem. Gonzalez and Gleizes considered the effect of metal evaporation in free-burning arc burning in atmospheric argon gas [37]; Murphy et al. identified the role of metal vapour in welding arc [38]; Zhang et al. studied the effect of copper electrode vapour on SF_6 nozzle arcs [39]. Rong also carried out research on the influence of copper vapour on the opening process of air circuit breakers [40].

In air circuit breakers, the most complex stage of the arcing process is the splitting of the arc by the metal plates, for the following three reasons. First, the complex structure of the splitters poses difficulties to the establishment of the model. Second, most of the metal plates are made of nonlinear ferromagnetic materials whose nonlinear characteristics must be considered in the analysis. Third, it is necessary to consider the complex interaction between metallic contact and arc plasma, including the influence of the metal vapour.

Therefore, in order to study the influence and provide a more accurate description of the role of the metal vapour in the arc cutting process by the splitters in air circuit breakers, a three-dimensional mathematical model on the motion and splitter-cutting of the air arc is established following the MHD approach. The extremely complex physical and chemical processes have to be simplified with convincing arguments for the establishment of the mathematical model, as have been reported in various research work.

1.7.1 Mathematical Models and Boundary Conditions

In the movement and cutting process of the arc, strong exchange of mass, momentum and energy takes place between various plasma species under the influence of the flow field, thermal field, electromagnetic field and other physical and chemical processes This includes the energy exchange at macroscopic level due to thermal conduction, convection and radiation transfer, as well as microscopic processes such as dissociation, association, ionisation and recombination. However, the mathematical description of these complex processes can be unified through the equations based on the theory of MHD [41]. In addition, the plasma sheath at the interface of the arc column and the contact surface plays an important role in arc rooting and arc voltage during the arc-cutting process. Without considering the role of the sheath, the bending and confinement of the arc column (which are very important phenomena in the cutting process before the formation of the new arc root) will not be represented in the simulation [42]. The sheath region is therefore taken into account in the modelling process in this chapter and the whole computational domain is divided into the arc column region, the sheath region and the solid metal region.

1. Establishment of geometric model and division of calculation region

Figure 1.10 shows the structure of the arc chamber used in this chapter to study the arc motion and cutting process. This is a simplified model for an air circuit breaker that contains only one iron splitter. In order to reduce the amount of calculation, only half of the symmetric model is adopted in this chapter considering the symmetry of various fields. Apart from the extended region for the solution of the magnetic field, the computational domain in Fig. 1.10 is 120 × 24.2 × 7 mm in x, y and z directions, respectively. The origin of the coordinate system is on the symmetric surface (x–y plane z = 0). The distance between the two electrodes is 17 mm, the thickness of the metal splitter (iron plate) is 2 mm, and each electrode (anode, cathode and splitter) is surrounded by a sheath region of 0.1 mm in thickness. Arc chambers are wrapped in insulating shells.

Figure 1.11 is a schematic diagram of the symmetric surface in the geometric structure. The whole domain can be divided into the following areas: arc column,

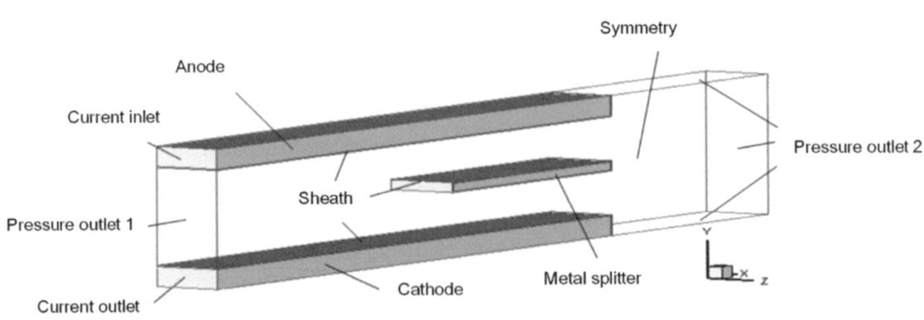

Fig. 1.10 Simplified structure of arc chamber with metal plate

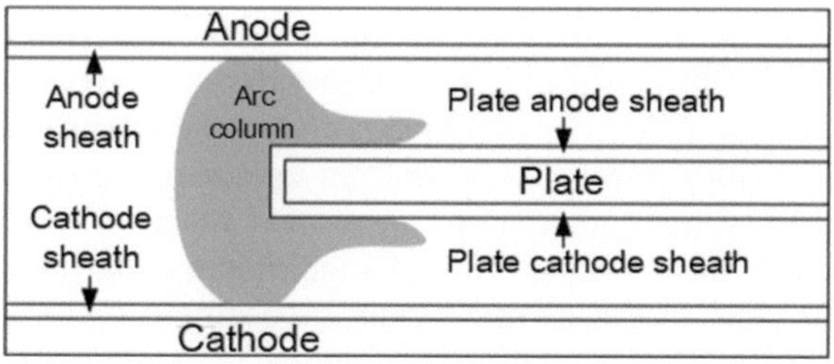

Fig. 1.11 Schematic diagram of the constituting components of the computational domain on the symmetrical plane (z = 0)

solid metal (including the contacts and splitter), and anode and cathode sheathes adjacent to the metallic surfaces. There are also extended areas not shown in the diagram that are used to solve the equation for magnetic field.

1) Arc column and sheath region

In the mathematical model, mechanisms represented by the mass, momentum and energy conservation equations as well as the electric and magnetic field equations are coupled to describe the state of the arc plasma. In addition, to obtain the distribution of iron vapour concentration in the presence of splitter plate erosion and the influence of the vapour on arc behaviour, a mass concentration equation for the metal vapour is also solved. The specific equations are as follows:

(1) Mass conservation equation

$$\frac{\partial \rho}{\partial t} + \nabla \cdot (\rho \overline{V}) = S_m \tag{1.39}$$

$$S_m = \begin{cases} \dot{m}(\Delta S/\Delta V) & \text{Sheath} \\ 0 & \text{Fluid} \end{cases} \tag{1.40}$$

$$\dot{m} = q_{vap}/h_v \tag{1.41}$$

$$q_{rad} = 4\pi \varepsilon_N \tag{1.42}$$

(2) Momentum conservation equation

$$\frac{\partial (\rho v_i)}{\partial t} + \nabla \cdot \left(\rho v_i \overline{V}\right) = \nabla \cdot (\eta \nabla v_i) - \frac{\partial p}{\partial x_i} + \left(\overline{J} \times \overline{B}\right)_i \tag{1.43}$$

(3) Energy conservation equation

$$\frac{\partial (\rho H)}{\partial t} + \nabla \cdot \left(\rho H \overline{V}\right) = \nabla \cdot \left(\frac{\lambda}{c_p} \nabla H\right) + \sigma E^2 - q_{rad} + q_\eta + S_m h_t \tag{1.44}$$

$$h_t = h_{vg} + |j_e| \frac{V_c}{\dot{m}} \tag{1.45}$$

In the mass conservation equation above, S_m is the volumetric mass source term in the sheath region, which is related to the mass flux, $\dot{m}$, at the metal surface (the energy flux q_{vap} for erosion of the metal divided by the latent heat of melting and evaporation h_v) with a conversion factor from surface source to volume source where ΔV is the mesh volume adjacent to the erosion surface and ΔS is the area of the element face on the metal surface. The radiation energy source term q_{rad} in the energy equation is obtained by the use of net emission coefficient [43]. The concept of equivalent total enthalpy, h_t, is introduced to represent the energy that is assigned to the metal vapour coming into the flow from the metal surface where in addition to the energy carried by the metal vapour, the energy input as a result of the cathode emitted electrons accelerated in the sheath is also included. h_{vg} is the enthalpy of evaporation of the metal vapour at the boiling point, j_e is the electron current density at the cathode surface, and V_c is the sheath voltage at the cathode. V_c is set to be zero on the anode surface.

(4) Electric and magnetic field equations

The method of magnetic vector potential in Eqs. 1.18–1.20 is adopted to solve the magnetic field in the whole domain including the extended domain. The B-H curve of iron is used to take into account the influence of the non-linear ferromagnetic material on magnetic field.

(5) Mass concentration equation

The mass concentration equation of the metal vapour is introduced to describe the convection and diffusion effects in the mixture.

$$\frac{\partial(\rho c_m)}{\partial t} + \nabla \cdot \left(\rho c_m \overline{V}\right) - \nabla \cdot (\Gamma_c \nabla c_m) = 0 \tag{1.46}$$

where, c_m is the mass concentration of the metal vapour, which ranges from 0 to 1 (dimensionless), c_m is the diffusion coefficient of the vapour.

2) Solid electrode and splitters area

An energy conservation equation in the contact (electrode) and splitter regions is also established. Compared with the arc, the conductivity of the electrodes and the splitters is very high, and the joule heating generated by the metal's resistance can be ignored, hence:

$$\frac{\partial}{\partial t}\left(\rho_s c_p T_s\right) - \nabla \cdot (k_s \nabla T_s) = 0 \tag{1.47}$$

2. Thermodynamic and transport properties of the mixture of air and iron vapour

To study the influence of metal vapour on the arc behaviour, the thermodynamic properties and transport coefficients of the air-iron vapour mixture should first be obtained. This chapter adopts the method used in [9] with the assumption that the plasma is in LTE state. The chemical composition of the gas is obtained by minimising the Gibbs free energy of the gas system and following this the thermodynamic properties, such as the density and specific heat capacity, can be directly calculated. The Chapman-Enskog approximation is then used to solve the Boltzmann equation to obtain the transport coefficients of the gas or gas mixture, such as the electrical conductivity, viscosity and thermal conductivity.

3. Mathematical model of the electrode sheath

The term 'sheath' refers to a thin layer that bridges the electrode surface with the high temperature arc column. LTE does not hold in this region and a special mathematical model needs to be established. The sheath plays an important role in the formation of new arc roots and arc voltage during arc-cutting by the splitters. It can be a cathode sheath and anode sheath where complex physical processes operate. There has been much research to understand and model the sheath layers [44–47]. This section aims at large-scale three-dimensional air arc simulation. The influence of the gas state deviating from LTE is considered in an approach as described below.

1) Cathode sheath

The temperature of the metal surface outside the arc root is lower than the melting point of the metallic material. According to the LTE property of the gas, this arc root surrounding region will not be conductive. In fact, the non-LTE state of the sheath allows it possess a high electrical conductivity [47]. Therefore, the concept of effective electrical conductivity is introduced in this chapter to take this into account.

$$\sigma_{eff} = j\frac{\Delta y}{U_s} \tag{1.48}$$

where $\Delta y = 0.1$ mm is the cathode sheath thickness, taken from the largest sheath thickness found from experimental observations, j is the current density, and U_s is the cathode sheath voltage drop. For iron, which is a non-refractory material, it is difficult to obtain the accurate voltage drop of the cathode sheath, especially when significant contact erosion exists. Based on results reported in existing work, a voltage drop of 14.5 V will be used when the current density exceeds 10^8 Am^{-2} [48]. Research also shows that for circuit breaker arcs, the sheath voltage drop tends to be higher when the current density is relatively low during the formation of the arc root [49]. So a maximum cathode sheath voltage drop of 22.6 V is chosen when the current density is less than 10^4 Am^{-2} [50].

2) Anode sheath area

The study in [51] showed that the electrical conductivity in the anode sheath layer is the result of ambipolar diffusion of charged particles. To accurately describe the non-LTE effect of the anode sheath layer, processes involving micro-particles need to be considered. However, the application of this approach to a three-dimensional arc model leads to excessive computing effort. In [52], an LTE-diffusion model was proposed where the thickness of the anode sheath is set to 0.1 mm which is larger than the dimension of the region dominated by electron diffusion, and it is further assumed that its effective electrical conductivity is equal to that of the arcing gas immediately adjacent. The results of this model were compared with those of the full-diffusion model that explicitly considered the role of electrons. The rationality of the simplified model is demonstrated by the close agreement between its results and those from the full diffusion model. The simplified LTE-diffusion model is therefore used in this chapter to describe the anode sheath.

4. Boundary conditions

Boundary conditions are mathematical representation of the interaction between a system in the context of computer simulation and its environment. They are essential for a unique and correct solution to the governing partial differential equations. The following boundary conditions are specified for the arc governing equations.

1) Temperature (thermal) boundary conditions

For the arcing chamber shown in Fig. 1.10, thermal boundary conditions must be given in order to solve the energy equation and obtain the temperature distribution in the domain. For the problem considered in this chapter, the boundary conditions are associated with metal and solid insulating walls.

(1) The metal surface

When the arc cuts into the splitters, energy is injected from the arc column into the iron plates. When the surface temperature reaches the melting/boiling point, the iron plates will begin to melt and evaporate in the arc root region. For iron, its latent heat of melting is 274 Jg^{-1}, far below its latent heat of evaporation which is 6365 Jg^{-1}, giving rise to an evaporation dominated mode. The liquid layer on the metal surface is thin and the molten material quickly evaporates on reaching the boiling point. The liquid layer on the metal surface can be ignored in the calculation, and this simplifies the calculation of the ablation rate of the metallic material. At the same time, since the duration of erosion at a fixed position is very short during the transient process where the arc roots move continuously, the dimensional change of the metal plates can be ignored. Based on energy balance consideration at the metal surface, a thermal flux boundary condition on the surface is given. The erosion rate of the metal is obtained through the balance equation, as described below.

(2) Treatment of cathode surface

The interaction between arc and cathode is a complex physical process. In this chapter, from the perspective of energy conservation and based on LTE in the arc column, an energy balance equation between the arc and metal is constructed on the surface of the cathode. The energy received by the cathode spot consists of the conduction heat flux $q_{con\ arc}$ from the arc, radiation $q_{rad\ arc}$ and the heat generated by the ion bombardment q_{ion}. The outgoing energy consists of the heat flux $q_{con\ c}$, the surface radiation $q_{rad\ c}$, the cathode electron emission energy q_e and the evaporation energy q_{vap} of the cathode material when the temperature reaches the melting point. The principle of energy conservation requires:

$$q_{vap} + q_e + q_{conc} + q_{radc} = q_{conarc} + q_{ion} + q_{radarc} \tag{1.49}$$

In Eq. 1.49, the expressions of each term are as follows:

$$q_{conarc} = -\lambda_a \frac{\partial T}{\partial y} \tag{1.50}$$

$$q_{conc} = -\lambda_c \frac{\partial T}{\partial y} \tag{1.51}$$

$$q_{ion} = j_i \left(V_c + V_i - \Phi_c + \frac{5k_B}{2e} T_c \right) \tag{1.52}$$

$$q_e = |j_e| \Phi_c \tag{1.53}$$

$$q_{vap} = h_v \dot{m} \tag{1.54}$$

The meanings of variables in Eqs. 1.50–1.54 are as follows:

λ_a is the thermal conductivity of the arc near the cathode surface, λ_c is the thermal conductivity of the cathode material, j_i is the ionic current, V_c is the voltage drop of the cathode sheath, Φ_c is the work function of the cathode material, $V_i = 7.9$ V is the plasma ionisation energy, k_B is the Boltzmann constant, e is the electron charge, T_c is the temperature of the cathode spot, j_e is the electron current density, and h_v is the sum of the latent heat of melting and evaporation of the cathode material.

The total current density of the cathode spot is the sum of ion current density and electron current density, that is $|J| = j_i + j_e$. Therefore, the relative relationship between j_i and j_e needs to be given in the calculation. An approximate value $j_e/|J| = 0.78$ given in [39] is adopted in this chapter.

In addition, in the energy balance equation, the radiation energy from the electrode surface can be ignored due to the low temperature of the cathode spot. Electrode heating due to arc radiation is small compared to other energies and can also be ignored [51]. After the temperature of the cathode surface reaches its melting point, the temperature will no longer rise and the additional incident energy, q_{vap}, will be used for melting and evaporating the cathode material. Through Eq. 1.54, the ablation rate of the metal surface can be obtained, and based on this, the boundary conditions at the cathode surface for the metal vapour concentration equation can be obtained. Outside the cathode spot, the energy exchange between the metal and the gas is only through thermal conduction.

(3) Treatment of anode surface effect

Similar to the case of the cathode, the energy balance equation was established on the surface of the anode and the erosion rate of the metal is derived from this eqution:

$$q_{conarc} + q_{radarc} + q_{ie} = q_{cona} + q_{rada} + q_{vap} \tag{1.55}$$

In Eq. 1.55, the meaning of each term is given below:

$$q_{conarc} = -\lambda_a \frac{\partial T}{\partial y} \tag{1.56}$$

$$q_{conc} = -\lambda_c \frac{\partial T}{\partial y} \tag{1.57}$$

$$q_{ie} = |j_n| \left[\Phi_a + V_a + \frac{5k_B}{2e}(T_g - T_a) \right] \tag{1.58}$$

$$q_{vap} = h_v \dot{m} \tag{1.59}$$

The meanings of the variables in Eqs. 1.55–1.59 are as follows:

q_{cona} is the thermal conduction through the anode electrode, q_{vap} is the energy loss from the anode due to evaporation, q_{ie} is the energy flux brought to the anode surface by electrons, j_n is the current density component perpendicular to the anode surface, T_a is the anode surface temperature, T_g is the fluid temperature next to the anode spot, V_a is the anode voltage drop, 2.1 V according to [53], and Φ_a is the anode material work function, 4.5 V.

The radiation terms q_{radarc} and q_{rada} are similar to the cathode surface and negligible compared to other energy fluxes. Similarly, when the metal surface temperature exceeds the melting point, the erosion rate of the anode can be obtained through Eq. 1.54. the boundary conditions for the energy equation and iron vapour concentration equation can thus be set based on the mass and energy fluxes.

(4) Insulation wall

The atmosphere and the hot arcing gas is separated by the insulation wall. Energy will be lost via the plastic wall in the form of thermal conduction. The small thickness and low thermal conductivity of the plastic wall makes the normal conduction flux dominant. Therefore, the heat flux at the inner surface of the wall can be approximated by the one-dimensional heat conduction equation.

$$Q = -k(T - T_0)/d \tag{1.60}$$

where, k is the thermal conductivity of the wall material, T is the surface temperature of the wall, and d is the thickness of the wall. T_0 is the environmental temperature.

2) Velocity boundary conditions

The velocity boundary condition on the wall is the non-slip boundary condition commonly used [54].

3) Pressure boundary conditions

The arcing chamber is connected with the environment through two air outlets. In the calculation, the fixed pressure boundary condition is set at the outlet (1 atm).

4) Electromagnetic field boundary conditions

A common practice in arc modelling is that, at the interface between the arc and the electrode, Neumann type boundary condition is set for the electric potential equation, which is in fact the current density. An average current density or an exponential distribution descending from the centre of the arc root is often used for steady state arcs. It is however difficult to apply this approach to a transient arcing process in which the arc root moves rapidly. With the establishment of the mechanisms of current conduction in the cathode and anode sheaths, the solution domain now includes the electrodes, sheathes and arc region. The boundary conditions for the potential equation can be conveniently set at the surface of the electrodes (away from the arc roots) where electric current flows in and out. At power frequency, skin effect is negligible and an average current density can be used over the area of incoming current. The surface of outgoing current is set to zero potential. On the insulating wall, the normal gradient of potential is set to zero according to the usual method. At the current incoming surface,

$$j = I/S \tag{1.61}$$

where I is the total current, and S is the cross-sectional area of the current incoming surface.

In this chapter, the magnetic vector potential method is used to obtain the magnetic field in the arc chamber. On the symmetric surface of this half symmetric geometry, the following is required:

$$\overline{n} \times \overline{A} = 0 \tag{1.62}$$

$$\upsilon \nabla \cdot \overline{A} = 0 \tag{1.63}$$

For the boundary far away from the arcing chamber (the extended domain only for magnetic field), there is

$$\upsilon \nabla \times \overline{A} \times \overline{n} = 0 \tag{1.64}$$

$$\overline{n} \times \overline{A} = 0 \tag{1.65}$$

where υ is the magnetic reluctivity of the material.

5) Mass concentration boundary conditions

In the arcing chamber, during the formation of the arc root and its subsequent movement on the plates, the distribution of metal vapour in the arcing chamber will constantly change due to the convection and diffusion effects. Chapter 1.6 describes this process through the concentration Eq. 1.46. The boundary conditions are determined by the mass of metal vapour and air contained in the boundary cells and updated in each iteration.

$$c_m = \frac{m_m}{m_m + m_{air}} \tag{1.66}$$

$$S_m = \dot{m} \Delta S \Delta t \tag{1.67}$$

where, c_m is the vapour mass concentration of wall material, S_m is the vapour mass source of wall material contained in a boundary cell, m_{air} is the mass of the air contained in the cell, $\dot{m}$ is the erosion rate of wall material, ΔS is the area of the cell interfacing the electrode, and Δt is the time step size of simulation.

1.7.2 Method for Solving Arc Model

The three-dimensional mathematical model of air arc is a set of coupled non-linear partial differential equations with proper boundary conditions. It requires parallel computing in the solution process. The MpCCI interface software is a vendor neutral solution for co-simulation and file-based data transfer. In this chapter, the computational fluid dynamics (CFD) software, Fluent, and the electromagnetic field analysis module of Ansys are used together for coupled calculation to solve the three-dimensional arc model under the control of MpCCI, which serves as an interface for coupled simulation. Ansys, which is also suitable for large-scale parallel computing, can deal with non-linear electromagnetic field problems well. In this chapter, through the interface MpCCI, the advantages of two commercial solvers are taken of. With the aid of user defined code in the solvers and MpCCI, it enables the solution of the mathematical model that governs the arc motion and arc cutting process in the presence of contact erosion.

1.7.3 Consideration of Simulation Results for Arc Cutting Process with Contact Erosion

In the work presented in this section, the time step size for the coupled simulation was set to 2 μs. The calculation does not include the arc ignition process, instead a steady state arc solution between the two electrodes was used as the initial state of the calculation [40].

Figure 1.12 shows the gas temperature distribution during the arc cutting process. Figure 1.13 shows the distribution of iron vapour concentration at the corresponding times. It can be seen that from t ≈ 0.1 ms to 0.5 ms, the arc begins to expand and starts to move along the electrodes under the influence of the Lorentz force and the flow field. With the increase in current and the attraction effect of the ferromagnetic splitters, the velocity of the arcing gas also increases, and the arc is continuously stretched in the direction of the ferromagnetic plates. The differences in treating the

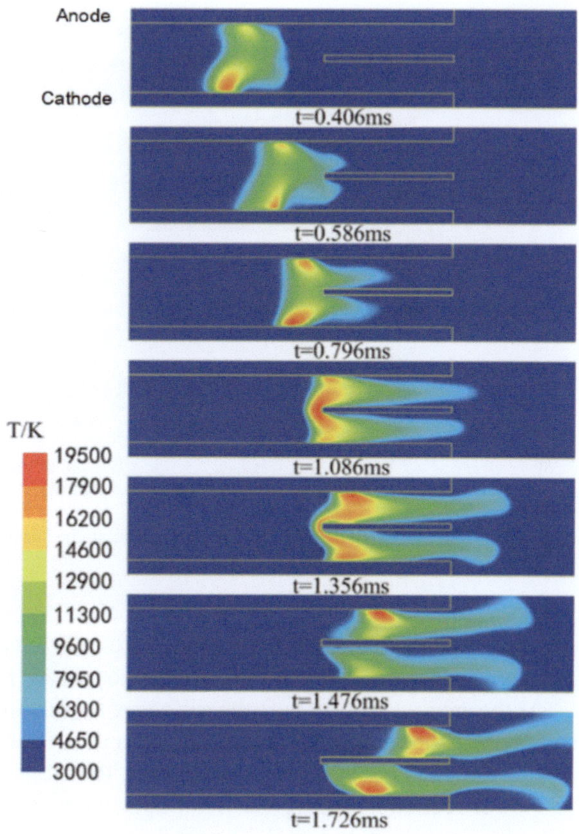

Fig. 1.12 Predicted temperature distribution on the symmetrical plane (z = 0) of the arcing chamber in the presence of contact erosion

cathode and anode leads to the anode arc root travels further inside the splitters than the cathode arc root, as can be seen from Fig. 1.12 (t = 0.586 ms).

Figure 1.14 shows the velocity field in the arcing chamber at two instants. It is worth noting that in Fig. 1.14a, a vortex appears before the cathode arc root, where A and B are the positions of the cathode and anode arc roots, respectively, and C is the position of the vortex. At this time, the flow is mainly directed toward the upper part of the arcing chamber. The energy in front of the cathode is taken away by the high-temperature gas along the direction of the flow, and the gas temperature in the area in front of the cathode rises relatively slowly. This means that the conductivity of this area is relatively low, which directly leads to the slow movement of the cathode arc root at this stage.

When t > 0.7 ms, the front of the arc column starts to touch the splitters while the arc column continues to elongate. At this stage, as shown in Fig. 1.14b, the velocity near the arc roots increases continuously, and the flow direction gradually becomes mainly horizontal. This makes it easier to heat the area in front of the two arc roots, leading to rapidly increasing electrical conductivity and accelerated movement of the arc roots. However, the arc column starts to be blocked by the plate, stopping it

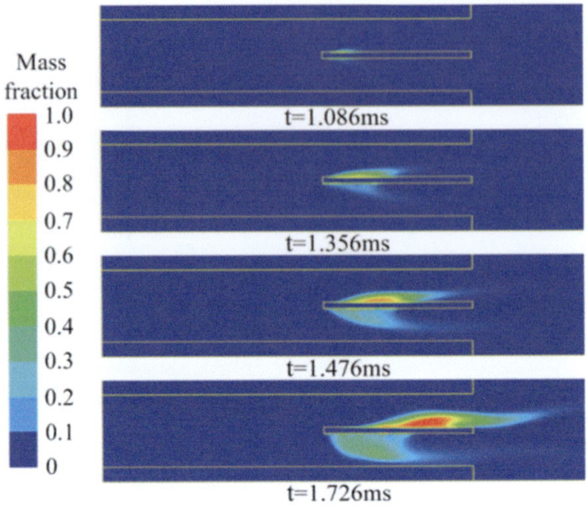

Fig. 1.13 Distribution of iron vapour concentration on the symmetrical plane of the arcing chamber (z = 0)

from move forward quickly. The cathode arc root, which falls behind the arc column, now moves faster and catches up with the anode arc root.

As the arc root moves along the surface of the electrodes, the arc column gradually bends and squeezes by the cold plate. As a result the arc column starts to surround part of the plate surface. At t = 1.086 ms, the arc is cut by the plate into an upper and lower section. As energy is continuously transferred from the arc column to the iron plate, as shown in Fig. 1.13, metal vapour begins to appear near the surface of the iron plate. However, at this stage, most of the current still flows around the plate, as shown in Fig. 1.15a.

When t > 1.2 ms, two new arc roots begin to form on the plate surface. As is shown in Fig. 1.15b, the current density in the plate also increases rapidly. It can be seen in Fig. 1.13 that the contact erosion becomes more intense at this stage. Under the influence of strong convection, the iron vapour moves forward but mainly concentrates in the area near the arc root.

It can also be seen from Fig. 1.13 that the metal vapour generated in the cathode arc root region is more intense than that in the anode arc root region. This is due to the higher current density, or stronger energy flux, in the cathode arc root. From Fig. 1.3 it can be seen from the properties of the arcing gas that the electrical conductivity of iron vapour is much higher than that of air below 15,000 K. This explains why the formation position of the cathode arc root is in front of the position of the anode arc root, as shown in Fig. 1.15b.

With the arc column moving into the splitter space, two distinct arc columns are formed and confined in the gaps between the metal plate and the electrodes while continuously moving forward. In this final stage, the velocity of the lower arc column is significantly lower than that of the upper column. This is due to the difference in

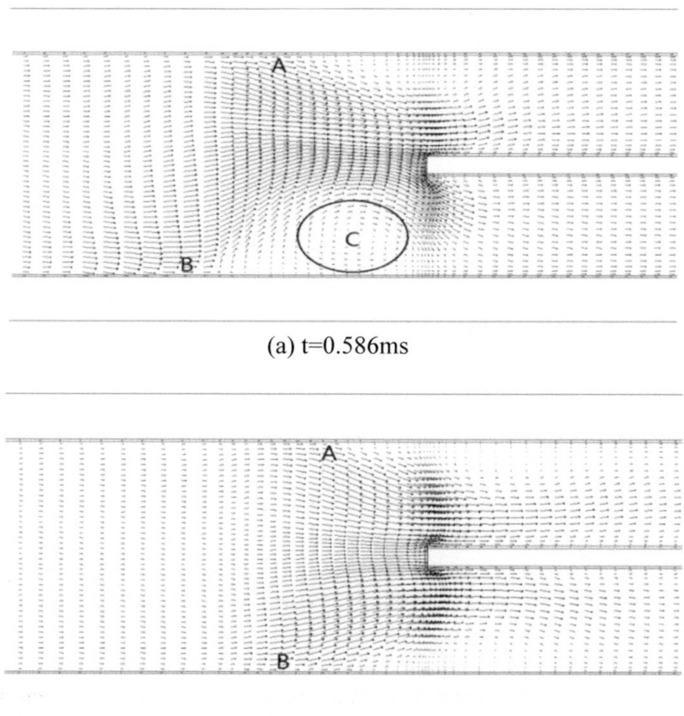

(a) t=0.586ms

(b) t=0.796ms

Fig. 1.14 Velocity field on the symmetry plane. The maximum velocity is 256 ms^{-1} and 284 ms^{-1}, where A and B are arc roots on anode and cathode, respectively, and C is the centre of the vortex

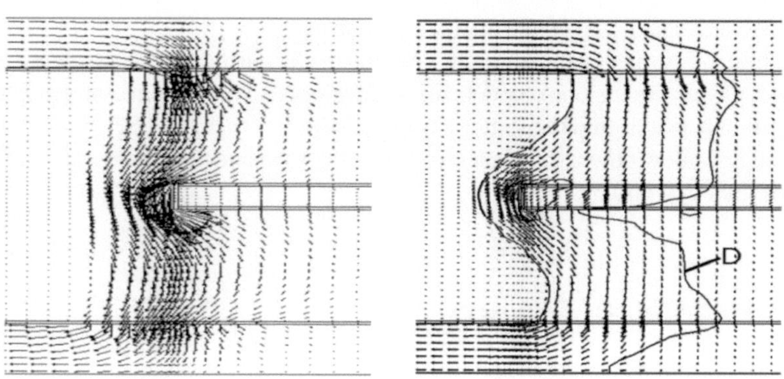

(a) t = 1.086 ms (b) t = 1.356 ms

Fig. 1.15 The current density distribution on the symmetrical plane, the maximum current density is 1.1×10^8 Am^{-2}, where D is the isoline of current density 10^7 Am^{-2}

the positions of the arc roots on both sides of the plate, resulting in a transverse current density component in the metal plate. As can be seen from the structure in Fig. 1.11, when the arc root at the cathode side of the plate is in front of the anode side root (lower face of the plate), the current in the plate flows from right to left. The resultant Lorentz force is opposite to the arc motion and thus slows down its motion. In contrast, the Lorentz force in the space above the plate accelerates the motion of the arc.

1.8 Nozzle Arc Simulation

SF_6 circuit breakers have been widely used in power delivery networks and in recent years there is a need for compact and rapid design with enhanced performance and reliability. Therefore, the study of SF_6 arcs has become one of the focuses for manufacturers and research institutes. Supersonic nozzles (convergent-divergent nozzles) are widely used in pressurised high voltage SF_6 circuit breakers. The development of high voltage circuit breakers is closely related to the study of SF_6 arcs in supersonic nozzles. When the ratio of inlet to outlet pressures exceeds a certain value, supersonic flow can be formed in the expansion section of the nozzle. With high voltage circuit breaker design, generation of a shock wave is inevitable in the interruption process. In theory, a shock wave represents a discontinuity of flow, resulting in sudden changes of density, velocity, pressure and other properties, so it can strongly affect the arc characteristics. At the same time, SF_6 arc is usually turbulent and unstable, which affects the distribution of velocity and temperature and also the formation of shock waves. To improve the short-circuit current breaking ability of high voltage circuit breakers, it is necessary to understand the arc characteristics under the influence of turbulence and shock waves in supersonic nozzles.

LTE is also assumed in this section. Treating the SF_6 arc as a compressible conductive fluid, a two-dimensional model can be developed. The presence of shock waves in a supersonic nozzle and their interaction with the arc in a model SF_6 circuit breaker containing a hollow downstream electrode are analysed and discussed under different ratios of inlet and outlet pressures.

1.8.1 Modelling of SF6 Arcs

1. Governing equations

SF_6 arc in a supersonic nozzle is axisymmetric and can be described by Navier–Stokes equations with specific source terms and addition equations governing the electric and magnetic fields. The following equations are written in the cylindrical polar coordinate system.

1) Mass conservation equation

The mass conservation equation is expressed in Eq. 1.2.

2) Momentum conservation equation

The radial and axial momentum equations are respectively

$$\frac{\partial(\rho v)}{\partial t}+\nabla\cdot(\rho\vec{u}v)=-\frac{\partial p}{\partial r}+\nabla\cdot(\mu_{eff}\nabla v)+\frac{1}{r}\frac{\partial}{\partial r}\left(r\mu_{eff}\frac{\partial v}{\partial r}\right)+\frac{\partial}{\partial z}\left(r\mu_{eff}\frac{\partial w}{\partial r}\right)$$
$$-\frac{2}{3}\frac{\partial}{\partial r}\left[\mu_{eff}\frac{1}{r}\frac{\partial}{\partial r}(rv)+\mu_{eff}\frac{\partial w}{\partial z}\right]-\frac{2\mu_{eff}v}{r^2}-j_zB_\theta+\rho g \tag{1.68}$$

$$\frac{\partial(\rho w)}{\partial t}+\nabla\cdot(\rho\vec{u}w)=-\frac{\partial p}{\partial z}+\nabla\cdot(\mu_{eff}\nabla w)+\frac{1}{r}\frac{\partial}{\partial r}\left(r\mu_{eff}\frac{\partial v}{\partial z}\right)+\frac{\partial}{\partial z}\left(r\mu_{eff}\frac{\partial w}{\partial z}\right)$$
$$-\frac{2}{3}\frac{\partial}{\partial z}\left(\mu_{eff}\frac{1}{r}\frac{\partial}{\partial r}(rv)+\mu_{eff}\frac{\partial w}{\partial z}\right)-\frac{2\mu_{eff}v}{r^2}+j_rB_\theta \tag{1.69}$$

$$\mu_{eff}=\mu_l+\mu_t \tag{1.70}$$

where p is the gas pressure, v and w are the radial and axial velocity components, respectively, g is the gravitational acceleration, μ_{eff} is the effective viscosity, μ_l is laminar viscosity, μ_t is the turbulence viscosity, J_r is the radial electric current density, J_z is the axial current density, and B_θ is the magnetic flux density in the azimuthal direction.

3) Energy conservation equation

$$\frac{\partial(\rho h)}{\partial t}+\nabla\cdot(\rho h\vec{u})=\frac{\partial}{\partial z}\left(k_{eff}\frac{\partial T}{\partial z}\right)+\frac{1}{r}\frac{\partial}{\partial r}\left(k_{eff}r\frac{\partial T}{\partial r}\right)$$
$$-Q_{rad}+\vec{j}\cdot\vec{E}+\frac{5k_B}{2e}\left(j_z\frac{\partial T}{\partial z}+j_r\frac{\partial T}{\partial r}\right) \tag{1.71}$$

$$k_{eff}=k_l+k_t \tag{1.72}$$

where k_{eff} is the effective thermal conductivity, k_l is the laminar thermal conductivity, k_t is the turbulent counterpart thermal conductivity, k_B is the Boltzmann constant, h is the specific enthalpy, and e is the electron charge, $e = 1.602 \times 10^{-19}$ C.

The source term, $\vec{j}\cdot\vec{E}$, is the joule heating. The radiative energy loss, Q_{rad}, is obtained by radiation transfer calculation. The last term represents a source term associated with the drift of electrons in the plasma.

4) Equation of state

The equation of state of the gas is expressed in Eq. 1.14.

5) Electric and magnetic fields

Since the arc column is axisymmetric, its magnetic field is dominated by its azimuthal component. For the sake of simplicity, only the azimuth component generated by the arc current is considered in this section. Assuming that the arc medium has a uniform permeability μ_0, the value of the magnetic flux density can be calculated by Ampere's law, as shown in Eq. 1.73.

$$B_\theta = \frac{\mu_0 \int_0^r J_z 2\pi\xi d\xi}{2\pi r} \tag{1.73}$$

where Jz is the axial current density and r is the radial coordinate.

2. Mathematical model of turbulence and radiation

1) Turbulence model

Turbulence occurs in the supersonic nozzle of high voltage circuit breakers [19] and has an important influence on the momentum and energy transport in the arc region. The closure of the governing equations of the system requires additional relationships between the time averaged flow quantities and the turbulence enhanced momentum and energy exchange. The practical approach in engineering applications is to establish the turbulence equation(s) based on empirical or semi-empirical hypothesis. Since the zero-dimensional Prandtl mixing length model is superior over other more complex turbulence models [20], it is used in this section. The viscous effect of turbulence is related to the turbulence length scale and the velocity gradient of the main flow:

$$\mu_t = \rho l_m^2 \left| \frac{\partial w}{\partial r} \right| \tag{1.74}$$

$$l_m = c\delta \tag{1.75}$$

$$\delta = (\theta_\delta / \pi)^{0.5} \tag{1.76}$$

$$\theta_\delta = \int_0^\infty (1 - \rho/\rho_\infty) 2\pi r dr \tag{1.77}$$

where δ is the thermal radius of the arc column and c is the turbulence parameter to be calibrated according to experiment. Previous studies have shown that the Prandtl number of supersonic nozzle arcs is usually around 1 [20]. Therefore, only the turbulence constant c needs to be determined based on a set of experimental results for the same nozzle. In this chapter, the turbulence constant is set to 0.20.

2) Radiation model

Excitation of plasma particles in high-temperature arc leads to the emission of electromagnetic waves, or radiation. The internal energy of particles increases when they absorb radiation. This process is called radiation transfer which is an important energy exchange mechanism in the arcing gas. Radiation is also the main energy source for nozzle ablation in high voltage gas blast circuit breakers.

Both theoretical calculations and experimental measurements show that the reabsorption of arc radiation by cold gas at the arc edge is an important source of arc heating [130] (并没有130文献). When arc burns in a molecular gas, intense reabsorption of radiation takes place at the arc edge through photo-excitation, photo-decomposition and photo-ionisation in the ultraviolet spectrum. In order to obtain the radiative source term in the energy equation, it is necessary to perform radiation transfer calculation. The radiation of arc plasma covers a very wide wavelength range containing hundreds of spectral lines. Since the calculation of the continuum and line spectral absorption coefficient requires a huge amount of time, radiation transfer calculation based on direct integration is not practical. In many cases, the net emission coefficient (NEC) method is used to estimate the radiative energy loss from the arc centre [21]. However, this method is based on the assumption of uniform temperature distribution to calculate the net radiation loss. It cannot estimate the reabsorption of radiation in the cold gas surrounding the arc. In the numerical simulation of the arc characteristics of circuit breakers, the temperature of the arc edge is very important for the estimation of the breaking characteristics.

Many practical approximation methods have been proposed for radiation energy in arc plasma simulation by scholars. Of these, the semi-empirical computing and modified net emission coefficient model (MNEC method) [55] is widely used in high voltage circuit breaker arc simulation, and the experimental measurement results show good consistency. According to this model, the high temperature region ($T > 0.83T_{max}$) of the arc column is the net radiation region, and the reabsorption of arc radiation energy is ignored. The region from 4,000 K $\leq T < 0.83T_{max}$ to is considered to be the radiation energy reabsorption region, and a certain proportion of the outward radiation energy in the high temperature region is considered to be absorbed in this region for heating the low temperature part of the arc edge. In the temperature range below 4,000 K, since the conductivity of the gas is almost zero, radiation and absorption of energy by the arc are very weak, and so it is not considered. The distribution of reabsorption power density in the reabsorption region shows a parabolic trend. The proportion of radiation energy reabsorbed from the arc centre is fixed. Based on previous experience results, it is assumed that 80% of arc energy is reabsorbed in this section.

This method is an improvement on the radiation model of the net radiation coefficient, focusing on the strong radiation reabsorption at the edge of the SF_6 arc, which is closer to the actual situation. In this section, the model is further improved. In the reabsorption region, not only the reabsorption but also the radiation energy loss of the arc itself is taken into account. The arc radius is set as the radial radius with

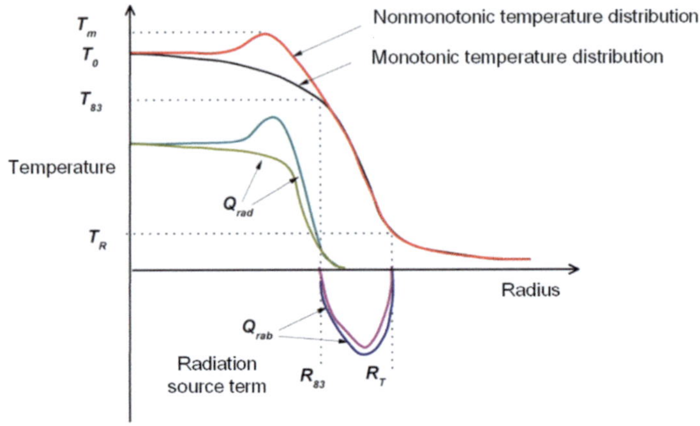

Fig. 1.16 Radiation model diagram, T_{max}, T_O, T_{83}, T_R respectively represent the maximum temperature of the arc, temperature at the centre of the arc, 0.83 times the maximum temperature, and the temperature corresponding to the arc radius (R); R_{max}, R_O, R_{83}, R respectively represent the radial position of the arc corresponding to the above temperatures

the conductivity of the arc edge equal to 1. The detailed model diagram is shown in Fig. 1.16.

3. Input of SF_6 arc properties

The numerical simulation of SF_6 arc plasma at LTE requires the thermal properties of arc plasma at different temperature and pressure conditions (such as mass density, specific heat at constant pressure, sound velocity, specific enthalpy), transport coefficient (such as thermal conductivity, conductivity, viscosity coefficient) and net radiation coefficient. The change of temperature and pressure has a great influence on these properties, which is obviously nonlinear, and the change of these properties will have a strong influence on the thermal field, electromagnetic field, airflow field and radiation field. Therefore, the governing equations of the arc model are non-linear.

The thermodynamic properties and transport coefficients of an SF_6 arc plasma in a LTE state are obtained by the calculations in some papers. The net radiation coefficient, calculated by Lowke et al. [56] is a function of temperature, pressure, and arc radius. All these parameters are prepared as inputs in the form of interpolation lists before calculation and applied to the cyclic iteration of arc CFD simulation.

For the convenience of calculation, it is set to a smaller value of 10^{-6} in the area where the conductivity is zero, 10^6 in the solid metal electrode, and 10^{-6} for insulating materials such as nozzle materials.

4. Geometric modelling and mesh generation for SF_6 nozzle arc

The work in this section mainly focuses on the Laval nozzle with a downstream hollow electrode. The geometric diagram of this nozzle is shown in Fig. 1.17. The total length of the nozzle is 257.41 mm, the outlet diameter of the nozzle is 46.36 mm,

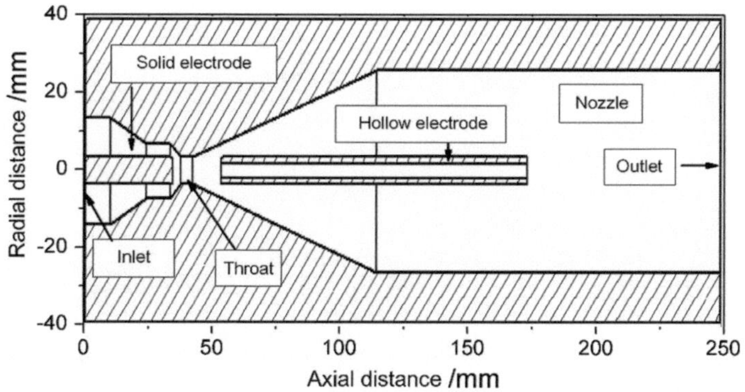

Fig. 1.17 Geometry diagram of nozzle

and the throat diameter of the nozzle is 6.33 mm. The upstream and downstream electrodes are of the same size as the nozzle throat and are located 35.36 mm and 55.01 mm away from the nozzle inlet respectively. The length of the downstream electrode is 124.10 mm, which is a hollow electrode with an air passage of diameter 3.56 mm. In previous studies [57, 58], the downstream electrodes were often ignored and assumed to be collectors of current, while in this section the effect of its actual size on the arc characteristics has been considered.

The quality of mesh generation directly affects the accuracy of simulation results. If the mesh number is too small, the accuracy of the calculation results will be reduced. If the mesh number is too large, the computing time required to achieve the convergence precision is too long. Therefore, considering the calculation accuracy and computing time, it is very important to generate the appropriate mesh number for the calculation model. Mesh selection is completed through repeated tests. That is to say, if the number of mesh sections continues to increase without changing the calculation result, the critical value of the mesh number is the final number of mesh sections. Compared with triangular grid units, quadrilateral structured grid units improve calculation accuracy and convergence. Therefore, this chapter adopts quadrilateral structured grid generation for models of circuit breakers containing supersonic nozzles, and the mesh number is 20,000.

5. Setting of boundary conditions

Boundary conditions are necessary in order to obtain the definite solution of the equations. Reasonable boundary conditions will provide good convergence and accuracy for the solution of the equations. For the simulation of SF_6 arc plasma in supersonic nozzle, the following boundary conditions are mainly used in this section.

1) Pressure boundary conditions

At the upstream inlet of the nozzle, the stagnation point pressure P_0 is 1.36 MPa, and the stagnation point temperature T_0 is 300 K. The mass, momentum and energy flux at the inlet are determined by isentropic conditions.

$$w_{in} = \sqrt{\frac{2\gamma}{\gamma - 1} RT_0 \left[1 - \left(\frac{P_{\text{cells}}}{P_0} \right)^{\frac{\gamma-1}{\gamma}} \right]} \tag{1.78}$$

$$\rho_{in} = \rho_0 \left(\frac{P_{\text{cells}}}{P_0} \right)^{\frac{1}{\gamma}} \tag{1.79}$$

$$T_{in} = T_0 \left(\frac{P_{\text{cells}}}{P_0} \right)^{\frac{\gamma-1}{\gamma}} \tag{1.80}$$

where w_{in} is the inlet velocity, ρ_{in} is the inlet mass density, T_{in} is the inlet temperature, P_{cells} is the pressure of the boundary layer, γ is specific heat ratio, R is ideal gas constant. The pressure ratios at the inlet and outlet were set to be 0.50, 0.33, 0.20, 0.10, and 0.01, corresponding to the outlet pressures of 0.680 MPa, 0.453 MPa, 0.272 MPa, 0.136 MPa, and 0.0136 MPa, respectively.

2) Thermal boundary conditions

Considering the interaction between the electrode and the arc, the effective average thermal conductivity method is adopted to deal with the large temperature gradient on the electrode surface [8]. Since the PTFE nozzle has a low thermal conductivity and a low temperature layer between the the arc and the surface of the nozzle, separating the high temperature arc region from the nozzle surface. An adiabatic boundary is set in this section at the nozzle surface, ignoring the thermal conduction of the nozzle material.

3) Velocity boundary conditions

The axisymmetric boundary condition is applied to the arc simulation, that is, the radial velocity on the axis of symmetry is zero.

1.8.2 Characteristics of Steady-State SF6 Arcs in Supersonic Nozzle

This section is mainly about SF_6 steady-state arc in circuit breaker, focusing on the internal mechanism of shock wave generation within a supersonic nozzle, and revealing the influence of the presence of a downstream electrode on the arc characteristics (e.g. airflow field, temperature field and pressure field, etc.).

Previous studies have shown that with of shock waves in the arc region, the presence of arc can significantly affect the distribution of airflow field [59]. Due to the axisymmetric characteristic of the solution, the pressure and temperature contour plots are presented simultaneously in the same figure in Fig. 1.18 when the steady-state current is 600 A and the outlet pressure is 0.68 MPa and 0.272 MPa respectively.

The pressure drop mainly occurs in the arc region between the electrodes. The highest arc temperature occurs upstream of the downstream electrode and moves

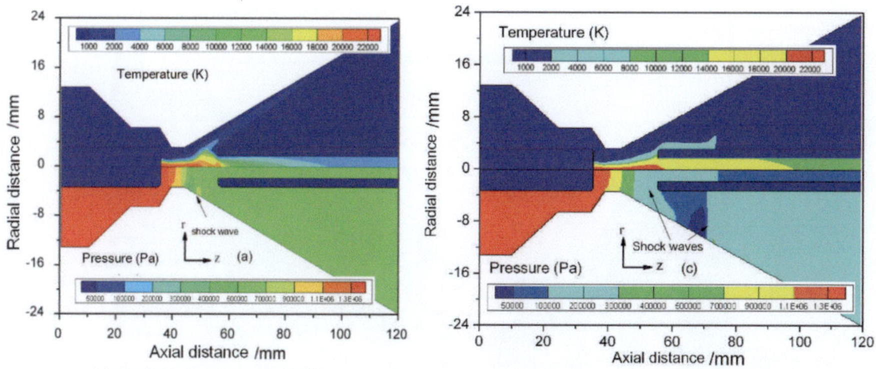

(a) Outlet pressure: 0.680MPa (b) Outlet pressure: 0.272MPa

Fig. 1.18 Temperature and pressure distribution of SF_6 nozzle arc, I = 600 A. Top: temperature; Bottom: pressure

upstream with an increase of outlet pressure. The arc temperature inside the hollow electrode downstream decreases as the axial position moves downstream due to the loss of the ohmic heating effect.

When the outlet pressure is 0.272 MPa, if the downstream hollow electrode is ignored, the calculated shock wave position occurs in the downstream of the actual hollow electrode electrode, which has less influence on the arc burning area. Although there is a flow channel inside the hollow electrode, the shock waves generated by the blocking effect of the downstream electrode reduce the velocity of the flow along the axis of symmetry. The position of the shock wave moves slowly upstream as the outlet pressure rises. When the outlet pressure is lower than 0.272 MPa, the temperature field and airflow field in the upstream region of the hollow electrode and the arc characteristics between the upstream and downstream electrodes are not affected by the outlet pressure. When the outlet pressure is 0.688 MPa, the high pressure fluid encroaches into the combustion arc region upstream of the hollow electrode, bringing an opposite pressure gradient and generating subsonic airflow with relatively low velocity in the thermal boundary layer between the arc centre and the cold airflow.

As shown in Fig. 1.19, since the temperature and velocity in the arc centre are higher, the encroaching high air pressure mainly affects the thermal boundary layer with lower velocity, resulting in a vortex, i.e. the stratified flow of airflow. In the thermal boundary layer, the hot arc centre and the cold flow strongly interact, affecting the distribution of the pressure field and the flow field. It can be concluded that both the high pressure fluid invading the thermal boundary layer and the blocking effect of the downstream electrode can generate the low-speed fluid which invades the centre of the arc, forming shock waves and causing the expansion of the arc column.

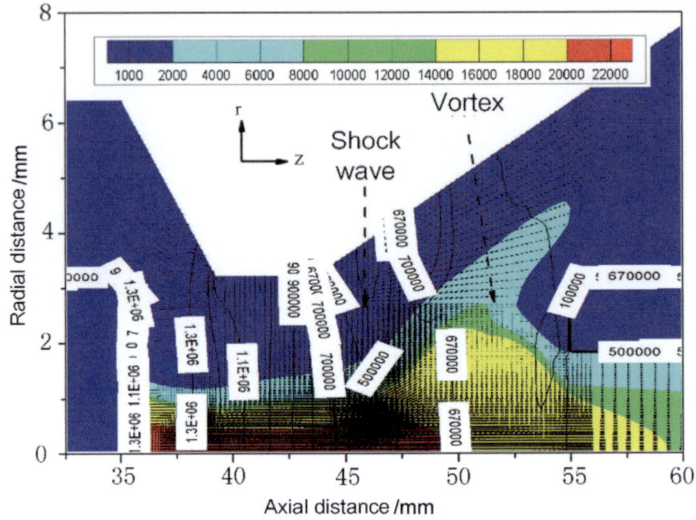

Fig. 1.19 Arc column broadening effect of SF_6 nozzle arc, I = 600 A, outlet pressure: 0.68 MPa

1.9 Summary

As the safety guard of a power system, a circuit breaker is responsible for switching on and interrupting the circuit in the system. One of its most important functions is to quickly interrupt the fault current to protect other equipment in the system from damage. During the breaking process, when the breaker electrodes are separated, an arc plasma with a temperature of more than 10,000 K will be generated. Whether the arc can be extinguished as soon as possible is crucial to the system protection and to the life of the circuit breaker itself. This chapter focuses on the basic theory and method of gas medium arc calculation. Through the study of this chapter, readers can deepen their understanding of arc theory and modelling methods, and preliminarily master the methods and skills of arc calculation. The three following key points can be taken from this chapter.

1. The basic theory, physical model and solution method of arc simulation technology based on MHD method are introduced in detail. First, based on the gas dynamics principle and Navier–Stokes equations, the mass conservation, momentum conservation and energy conservation equations describing the dynamic characteristics of an arc are given according to the calculation model of compressible laminar flow coupled with the calculation of electrode temperature field. The change of physical properties of an arc with variation in temperature and pressure are also considered. At the same time, the arc flow field is coupled with the electromagnetic field, and the magnetic field is solved by using the method of magnetic vector potential. Fluent, a commercial software package for fluid computing, is adopted as the solution platform for arc simulation. Based on

the coupled calculation method, the user-defined program development process of Fluent, which is used to achieve multi-field coupling calculation of low-voltage arcs, is introduced.

2. In view of the air arc, a three-dimensional mathematical model of air circuit breaker is established in consideration of splitters ablation and arc cutting process. By introducing the mass concentration equation of iron vapour, the distribution of metal vapour is described. The influence of metal vapour on properties of gas mixture and arc dynamic process is taken into consideration. By establishing the plasma sheath model and the energy balance equation of the metal surface, the influence of the non-equilibrium effect of the electrode surface on the arc root motion is considered. Through the simulation study, the inner mechanism of the metal vapour's importance to arc behaviour in the cutting process is revealed.
3. Taking an SF_6 arc as an example, the influence on the breaking performance of the circuit breaker is studied by simulation method, where the interaction of energy, turbulence and shock waves at the current zero area in the high-voltage SF_6 circuit breaker is taken into consideration. The two-dimensional SF_6 arc steady state model is established based on LTE hypothesis and MHD theory. The internal mechanism of a shock wave in the supersonic nozzle and its behaviour with changes in pressure is discussed in detail. The influence of the existence of a downstream electrode on arc characteristics (airflow field, temperature field and pressure field, etc.) is also introduced.

References

1. Naghizadeh K Y, Cressault Y, Gleizes A. Net emission coefficient of air thermal plasmas [J]. Journal of Physics D: Applied Physics, 2002, 35(22): 2925–2934.
2. Wang Q P, Hu W P, Liu Y F. Dynamic characteristics and application of arc in SF6 and its mixture gas [M]. Xi'an: Xi'an Jiaotong University Press, 1997: 20–25.
3. Gleizes A, Swierczynski B, Gonzalez J J. Contribution to a 3D modeling of a switching arc device [C]. In: Proc. 14th Int. Conf. on Gas Discharges and Their Applications (Liverpool), 2002: 7–150.
4. Trepanier J Y, Reggio M, Camarero R. LTE computation of axisymmetric arc-flow interaction in circuit breakers [J]. IEEE Transactions on Plasma Science, 1991, 19(4): 80–89.
5. Sun H T. Study on three-dimensional dynamic MHD mathematical model of switching arc for low-voltage circuit breakers and the factors influencing the switching characteristics [D]. Xi'an: Xi'an Jiaotong University, 2002: 25–50.
6. Chen M Z. Fundamentals of viscous fluid dynamics [M]. Beijing: Higher Education Press, 1993: 31–38.
7. Chen H P. Computational fluid dynamics [M]. Beijing: Hydraulic and Electric Power Press, 1995: 20–45.
8. Shi A M. Fundamentals of gas dynamics [M]. Beijing: Science Press, 1988: 35–40.
9. Murphy A. Transport coefficients of air, argon-air, nitrogen-air, and oxygen-air plasmas [J]. Plasma Chemistry and Plasma Processing, 1995, 15(2): 279–307.
10. Yu Q Z. Radiation heat transfer theory [M]. Harbin: Harbin Institute of Technology Press, 2000: 108–113.

11. Chauveau S, Deron C, Perrin M Y, et al. Radiative transfer in LTE air plasma for temperature up to 15000 K [J]. Journal of Quantitative Spectroscopy & Radiative Transfer, 2003, 77(2): 29–45.
12. Secevast'yanenko VG. Radiation transfer in a real spectrum. Integration over frequency [J]. Journal of Engineering Physics, 1979, 36(2): 138–148.
13. Shayler P J, Fang M T C. Radiation transport in wall-stabilised nitrogen arcs [J]. Journal of Physics D: Applied Physics, 1978, 11(12): 1741–1756.
14. Gleizes A, Rahmani B, Gonzalez JJ, et al. Calculation of net emission coefficient in N2, SF6 and SF6-N2 arc plasmas [J]. Journal of Physics D: Applied Physics, 1991, 24(8): 1300–1309.
15. Naghizadeh-Kashani Y, Cressault Y, Gleizes A. Net emission coefficient of air thermal plasmas [J]. Journal of Physics D: Applied Physics, 2002, 35(22): 2925–2934.
16. Karetta F, Lindmayer M. Simulation of the gasdynamic and plastic processes in low voltage switching arcs [J]. IEEE Transactions on Components, Packaging and Manufacturing Technology, 1998, Part A, 21(1): 96–103.
17. Aubrecht V, Lowke J. Calculations of radiation transfer in SF6 plasmas using the method of partial characteristics [J]. Journal of Physics D: Applied Physics, 1994, 27(10): 2066–2073.
18. Yan J D, Fang M T C. Visualization of arcing process in an auto expansion circuit breaker [J]. IEEE Transactions on Plasma Science, 1999, 27(1): 40–41.
19. Zhang J H, Fang M T C, Newland DB. Theoretical study of a 2 kA arc in a supersonic nozzle [J]. Journal of Physics D: Applied Physics, 2000, 20(3): 368–379.
20. Eby S D, Trepanier J Y, Zhang X D, et al. Modeling radiative transfer in SF6 circuit breaker arcs with the p-1 approximation [J]. Journal of Physics D: Applied Physics, 1998, 31(13): 1578–1588.
21. Dixon C M, Yan J D, Fang M T C. A comparison of three radiation models for the calculation of nozzle arcs [J]. Journal of Physics D: Applied Physics, 2004, 37(23): 3309–3318.
22. Paul K C, Takemura T, Matsuno H, et al. Predicted results of a HID DC current lamp considering a P1 radiation model [J]. IEEE Transactions on Plasma Science, 2004, 32(1): 118–126.
23. Boussinesq J. Théorie de l'écoulement tourbillonnant et tumultueux des liquides dans les lits rectilignes à grande section [M]. Paris: Gauthier-Villars et fils, 1897.
24. Obermeier F. Prandtl's mixing length model-revisited [J]. PAMM: Proceedings in Applied Mathematics and Mechanics, 2006, 6(1): 65–68.
25. Taylor G I. The transport of vorticity and heat through fluids in turbulent motion [J]. Proceedings of the Royal Society of London. Series A, Containing Papers of a Mathematical and Physical Character, 1932, 135(828): 685–702.
26. Kármán T V. The fundamentals of the statistical theory of turbulence [J]. Journal of the Aeronautical Sciences, 1937, 4(4): 131–138.
27. Herbert M V, Herd R J. Boundary-layer separation in supersonic propelling nozzles [C]. 1964.
28. Rahman M M, Agarwal R K, Siikonen T. A modified one-equation turbulence model based on turbulent kinetic energy equation [C]. 54th AIAA Aerospace Sciences Meeting, 2016.
29. Sharp D C. Challenges and solutions for embedded and networked aerospace software systems [J]. Proceedings of the IEEE, 2010, 98(4): 621–634.
30. Launder B E, Spalding D B. The numerical computation of turbulent flows [J]. Computer Methods in Applied Mechanics and Engineering, 1974, 3(2): 269–289.
31. Papageorgakis G C, Assanis D N. Comparison of linear and nonlinear RNG-based k-epsilon models for incompressible turbulent flows [J]. Numerical Heat Transfer: Part B: Fundamentals, 1999, 35(1): 1–22.
32. Mashhadimoslem H. A new simplified calculation model of geometric thermal features of a vertical propane jet fire based on experimental and computational studies [J]. Process Safety and Environmental Protection, 2020, 135: 301–314.
33. Wang F J. Computational fluid dynamics analysis [M]. Beijing: Tsinghua University Press, 2005.
34. Tao W Q. Numerical heat transfer [M]. 2nd Ed. Xi'an: Xi'an Jiaotong University Press, 2001.
35. Lindmayer M, Springstubbe M. Three-dimensional simulation of arc motion between arc runners including the influence of ferromagnetic material [J]. IEEE Transactions on Components and Packaging Technologies, 2002, 25(3): 409–414.

36. Zhang G S. Electrical Theory Basics[M]. Xi'an: Xi'an Jiaotong University Press, 1996.
37. Gonzalez J J, Gleizes A, Proulx P, et al. The modeling of a free-burning arc in the presence of metal vapor [J]. Journal of Applied Physics, 1993, 74(5): 3065–3070.
38. Murphy A B, Tanaka M, Yamamoto K, et al. Modeling of thermal plasmas for arc welding: The role of the shielding gas, the properties and of the metal vapor [J]. Journal of Physics D: Applied Physics, 2009, 42(19): 194006.
39. Zhang J L, Yan J D, Fang M T C. Electrode evaporation and its effects on thermal arc behaviors [J]. IEEE Transactions on Plasma Science, 2004, 32(3): 1351–1361.
40. Wu Y, Rong M Z, Sun Z Q, et al. Numerical analysis of arc plasma behavior during electrode opening process in low-voltage switching device [J]. Journal of Physics D: Applied Physics, 2007, 40(3): 795–802.
41. Wu Y. Simulation and experimental study on multi-field coupling process of low-pressure air arc [D]. Xi'an: Xi'an Jiaotong University, 2006.
42. Lindmayer M, Marzahn E, Mutzke A, et al. The process of arc splitting between metal plates in low voltage arc chutes [J]. IEEE Transactions on Components and Packaging Technologies, 2006, 29(2): 310–317.
43. Cressault Y, Hannachi R, Teulet P, et al. Influence of arc vapors on the properties of air thermal plasmas [J]. Plasma Sources Science & Technology, 2008, 17(036016): 1–9.
44. Benilov M S, Marotta A. A model of the cathode region of atmospheric pressure arcs [J]. Journal of Physics D: Applied Physics, 1995, 28(9): 1869.
45. Cayla F, Freton P, Gonzalez J J. Arc/cathode interaction model [J]. IEEE Transactions on Plasma Science, 2008, 36(4): 1944–1954.
46. Coulombe S, Meunier J L. Arc-cold cathode interactions: Parametric dependence on local pressure [J]. Plasma Sources Science & Technology, 1997, 6(4): 508–517.
47. Lowke J J, Morrow R, Haidar J. A simplified unified theory of arcs and their electrodes [J]. Journal of Physics D: Applied Physics, 1997, 30(14): 2033.
48. Yokomizu Y. Total voltage drops in electrode fall regions of argon and air arcs in current range from 10 to 20 000 A [J]. Journal of Physics D: Applied Physics, 1996, 29(5): 1260.
49. Mutzke A, Rüther T, Kurrat M, et al. Modeling the arc splitting process in low-voltage arc chutes [C]. Pittsburgh, USA, 2007.
50. Schmitz H, Riemann K U. Analysis of the cathodic region of atmospheric pressure discharges [J]. Journal of Physics D: Applied Physics, 2002, 35(14): 1727–1735.
51. Morrow R, Lowke J J. A one-dimensional theory for the electrode sheaths of electric arcs [J]. Journal of Physics D: Applied Physics, 1993, 26(4): 634.
52. Lowke J J, Tanaka M. LTE-diffusion approximation for arc calculations [J]. Journal of Physics D: Applied Physics, 2006, 39(16): 3634–3643.
53. Yokomizu Y, Matsumura T, Henmi R, et al. Total voltage drops in electrode fall regions of SF6, argon and air arcs in current range from 10 to 20 000 A [J]. Journal of Physics D: Applied Physics, 1996, 29(5): 1260–1267.
54. Rong M Z, Ma Q, Wu Y, et al. The influence of electrode erosion on the air arc in a low-voltage circuit breaker [J]. Journal of Applied Physics, 2009, 106(2): 023308.
55. Zhang J F, Fang M T C, Newland D B. Theoretical investigation of a 2 kA arc in a supersonic nozzle [J]. Journal of Physics D: Applied Physics, 1987, 20(3): 368–379.
56. Liebermann R W, Lowke J J. Radiation emission coefficients for sulfur hexafluoride arc plasmas [J]. Journal of Quantitative Spectroscopy & Radiative Transfer, 1976, 16(3): 251–264.
57. Yan J D, Fang M, Jones C. Electrical and aerodynamic behavior of arcs under shock conditions [J]. IEEE Transactions on Plasma Science, 1997, 25(5): 840–845.
58. Fang M T C, Kwan S, Hall W. Arc-shock interaction inside a supersonic nozzle [J]. IEEE Transactions on Plasma Science, 1996, 24(1): 85–86.
59. Liu R. Effect of air distribution on aerodynamic field and coal combustion in an arch-fired furnace [J]. Energy & Fuels, 2010, 24(10): 5514–5523.

Chapter 2
Calculation and Analysis of the Post-arc Dielectric Recovery Process

2.1 Introduction

High temperature and current conducting arc plasma is produced between the contacts of a circuit breaker in the current interruption process. As shown in Fig. 2.1, this process includes two key stages. The first stage is from arc initiation to a point before the final current zero (when arc is still burning in a relatively stable manner) and the other is the decay stage around zero current (from the onset of steep change of arc voltage to the end of the post-arc current) [1]. A successful breaking operation of the circuit breaker critically depends on the arc quenching process in the current zero period. The extinguishing principle of the arc is shown in Figs. 2.1 and 2.2.

In the current zero period, there are two concurrent and interrelated processes in the arc: one is the dielectric recovery process, namely u_{if} in Fig. 2.2. In this process, the ionised gas in the arc zone changes rapidly from the conductive state to the insulating state, so that the contact gap can withstand the applied voltage without re-ignition. The other is the voltage recovery process, namely u_{hf} in Fig. 2.2, where the system applies a recovery voltage across the contact gap following the breaking of the circuit at current zero.

During the current zero period, both processes take place at the same time. Whether a short-circuit current can be successfully interrupted depends on the competition between the two processes. If the dielectric recovery strength of the contact gap, u_{if}, is higher than the value of the recovery voltage, u_{hf}, the gas in the contact gap will not re-ignite and the current will then be successfully interrupted (shown in the red dotted line in Fig. 2.1. In contrast, if the u_{if} is lower than u_{hf} over a length of time, the contact gap will re-ignite (shown by the black dotted line in Fig. 2.1. It is vitally important to have a clear understanding of the dielectric recovery behaviour of the gas in the contact gap over the current zero period for different types of circuit breakers with different voltage ratings: it not only helps avoid interruption failure of circuit breakers due to contact gap re-ignition, but also minimises the potential hazards of circuit breaker explosion as a result of interruption failure.

M. Rong et al., *Computing Methodology for Switchgear*,
https://doi.org/10.1007/978-981-95-2840-0_2

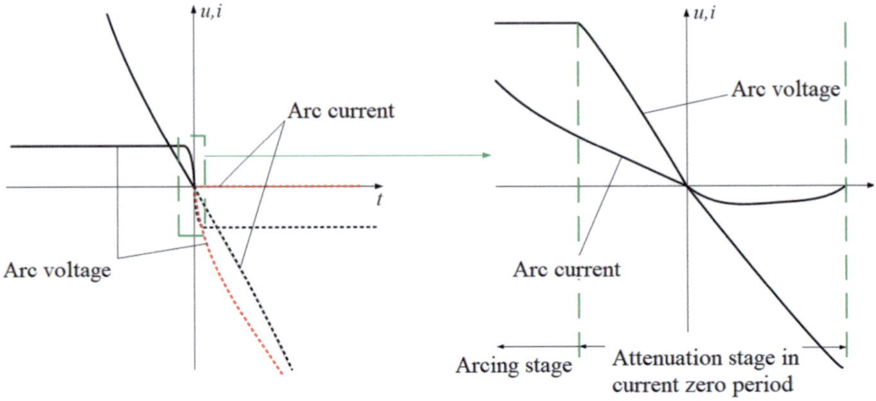

Fig. 2.1 Dielectric recovery of the contact gap of a circuit breaker following the arcing stage

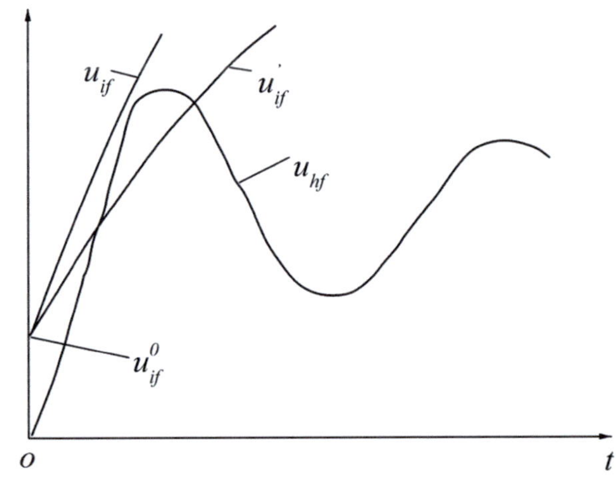

Fig. 2.2 Typical scenarios in the race of recovered dielectric strength of the contact gap and the system recovery voltage during post-arc phase of the interruption process

The simulation study of arc plasma produced in circuit breakers can not only reveal the dynamic behaviour of arc plasma more directly, but also provide important reference knowledge for the development and design of circuit breakers as well as the improvement of their breaking capability. Therefore, simulation study has long been the focus of switching arc research.

As mentioned in the previous chapter, the study of arc plasma involves complex physical and chemical processes, including the coupling of fluid mechanics, turbulent transport, thermal radiation, chemical reaction, nozzle ablation and contact erosion. Therefore, the simulation of the above processes has been a long-standing challenge. In recent years, significant progress has been made in the modelling of arc plasma with high-performance computers made available. It is generally accepted that the arc can be modelled using the local thermal equilibrium (LTE) assumption because the arcing process lasts for a sufficiently long duration (milliseconds) in comparison with the characteristic times of the strong physical and chemical reactions involved. With this assumption, modelling of equilibrium arc based on the magnetohydrodynamics (MHD) approach has focused on the macroscopic parameters of the arc, including arc voltage, temperature field, flow field and pressure field. Since the end of the last century, with the maturity of the arc model, simulation studies on switching arc have been carried out with the LTE assumption and the accuracy of simulation has been verified by various experiments.

The plasma in the post-arc phase has its unique behaviour at microscopic level, which is different from the arcing process. The following three aspects during the current zero phase needs to be considered in the numerical modelling of the dielectric recovery process:

(1) Thermally non-equilibrium behaviour
(2) Chemically non-equilibrium behaviour
(3) Non-Maxwellian distribution of electron energy.

The research on the non-equilibrium behaviours of plasma in the current zero phase above make it possible to improve the conformity with experiment results for arc calculation and simulation. However, due to the complexity of the non-equilibrium, there is no practical method in the field of arc numerical calculation to realise the simulation model considering the three effects at the same time. Even if the non-equilibrium behaviours of thermodynamics and chemistry or the non-Maxwellian distribution of electron energy are considered separately, there are still much challenge in simulation and experiment, which means the relevant research work is still in the development stage. Among them, there is least research on chemical non-equilibrium. Only preliminary attempts have been made and there is still a long way towards maturity in knowledge and understanding.

In view of the gaps described above, the physical mechanism and numerical calculation method of the thermal non-equilibrium behaviour, chemical non-equilibrium behaviour and the non-Maxwellian distribution of the electronic energy of arc plasma are first introduced in this chapter. Important results and conclusions from relevant research are then given to present the research status and development trend of the post-arc simulation work, which serves as a reference for the calculation method of dielectric recovery behaviour.

2.2 Thermal Non-Equilibrium Behaviour

2.2.1 Causes of Thermal Non-Equilibrium Behaviour

Irrespective of the media used in a gas circuit breaker (air, nitrogen, hydrogen, SF_6, CO_2, etc.), there are violent chemical reactions in the arc due to sharp temperature changes during the breaking process, including excitation, decomposition, ionisation, charge transfer and their respective reverse processes. In these chemical reactions,

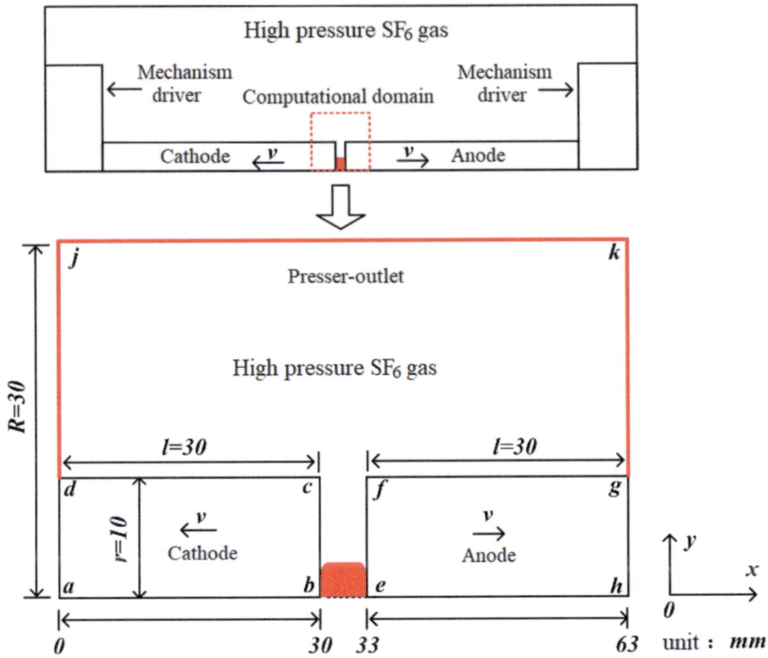

Fig. 2.3 Simple electrode model

the arc is no longer composed of a single type of molecule or atom, but a large number of molecules, atoms, ions and electrons. Depending on their mass, these particles can be divided into heavy particles (including atoms, molecules and ions, as shown in Fig. 2.3 and electrons.

Due to the large difference in mass, there is a great difference between the behaviour of heavy particles and that of electrons inside the arc. As electrons have a relatively small mass, the motion characteristics of electrons are more easily changed than that of heavier particles under the action of electric field. In the arcing period, there are frequent collisions between electrons and heavy particles within the typical time scale of the arcing process, making their kinetic energy of random movement approach each other. At the macroscopic level, the temperature of electrons and heavy particles tend to be equal, which is the basis of the local thermal equilibrium hypotheses in the arcing period.

In general, electrons absorb the electric energy that sustains the plasma, and then transfer the energy to the heavy particles through collisions, mostly through elastic collisions. The energy transferred from an electron in a single elastic collision is very small. For example, the mass of an SF_6 molecule, m_h, is about 2.7×10^5 times that of an electron, m_e, in an SF_6 plasma. The energy transferred in a single elastic collision is only $2m_e/m_h = 1/(1.8 \times 10^6)$ of that of the electron. Therefore, if the energy exchange between an electron and the heavy particles is not sufficient, it is difficult for the heavy particle to attain an energy equal to the electron purely through elastic collisions.

In the current zero phase, there are regions with large temperature gradient or regions with low electron number density inside the plasma zone, due to the rapid attenuation of temperature. In these cases, the local collisional energy exchange between electrons and heavy particles is insufficient, resulting in a difference between the thermal kinetic energy of electrons and that of heavy particles, and hence thermal non-equilibrium is developed [2].

In order to study the thermal non-equilibrium behaviour during the arc extinguishing process, the thermal non-equilibrium model (also known as the two-temperature model) is used to study the thermodynamic deviation of an arc [2]. In this model, the electrons have much smaller mass and their rate of energy change is higher than that of the heavier particles. As a result electrons and heavy particles have different temperatures. The difference in mass is small between heavy particles and they usually have the same temperature (gas temperature or plasma temperature).

2.2.2 Present Status of Research on Thermal Non-Equilibrium

The material properties under thermal non-equilibrium conditions are the basis of the two-temperature arc model. A relatively mature theoretical framework has been established for the determination of the material properties under thermal non-equilibrium conditions, and the research work involves almost all common gas media.

In the field of arc research, compared with the rapid development in the calculation of thermal non-equilibrium properties, there are only a few studies on the two-temperature arc model. The main research teams include Xi'an Jiaotong University, China, the University of Liverpool, UK, the University of Toulouse, France and Kanazawa University, Japan, etc. For example, Hsu et al. [3] first studied the two-temperature model for a free-burning arc. It was found that thermal non-equilibrium effect is relatively obvious in the low-temperature region of the arc boundary. The spatial distribution of electrical conductivity changes due to the thermal non-equilibrium effect, and this leads to the redistribution of current density. Girard et al. [2] developed a two-temperature model for SF_6 in the post-arc phase and found that the thermal non-equilibrium effect slows down the decay rate of arc temperature in the current zero phase. On this basis, Gonzalez et al. developed a simplified two-dimensional gas-blast SF_6 arc model and found that due to insufficient elastic collisions between electrons and heavy particles in the current zero phase, the temperature of electrons is higher than that of heavy particles. With a given recovery voltage, the breakdown between electrodes is more likely to occur in such a non-equilibrium situation [4]. Wang et al. [5] focused on the current zero phase characteristics of SF_6

nozzle arc by using a two-temperature model. It was shown that the degree of deviation from the thermodynamically balanced state increases with the current attenuation rate, and the strong convection effect in the nozzle causes insufficient collisions and energy exchange between electrons and heavy particles, indicating the potential importance of using the two-temperature model to evaluate the breaking capacity of high-voltage SF_6 circuit breakers.

At present, research on the two temperature arc model is mainly focused on the nozzle arc of the high voltage circuit breakers. Because of the symmetry of the nozzle, the simulation model of the arc generally adopts a 2.D axisymmetric structure. In the field of switching arc, no research has been reported on the study of the two-temperature model for 3-D applications.

2.2.3 *Two-Temperature Arc Modelling*

1. Assumptions

In the two-temperature arc model, the energy obtained by electrons from the electric field cannot be transferred to the heavy particles efficiently and rapidly in the form of elastic collisions and this leads to electrons and heavy particles having different temperature. Therefore, electrons and heavy particles follow different energy conservation equations. At the same time, electron conservation needs to be considered in the mass density of arc plasma that attains quasi-neutrality. Ignoring the effects of diffusion, electrons and heavy particles will have the same velocity, respectively.

In order to simplify the calculation, the following assumptions are adopted in the two-temperature arc model:

(1) The arc plasma is in a chemical equilibrium state (detailed content of chemical equilibrium will be further introduced in Sect. 2.2). Electrons and heavy particles move together with the arc, ignoring the influence of particle diffusion on the arc characteristics.
(2) The heavy particles and electrons follow the Maxwell-Boltzmann distribution but have different temperatures and equations of state.

2. Governing equations

The arc plasma can be described by the Navier–Stokes equations and corresponding electromagnetic field equations. The governing equations of the two-temperature arc model are described as follows, in which the energy conservation equations of electrons and heavy particles are described separately.

(1) Mass conservation equation

The mass conservation equation under the two-temperature model is the same as that under the equilibrium hypothesis.

$$\frac{\partial \rho}{\partial t} + \nabla \cdot (\rho \vec{u}) = 0 \tag{2.1}$$

where ρ is the density and **V** is the velocity vector of the flow.

(2) Momentum conservation equation

The velocity field is described by Navier–Stokes equations:

$$\frac{\partial \rho \vec{u}}{\partial t} + \nabla \cdot (\rho \vec{u} \vec{u}) = -\nabla \cdot \boldsymbol{\tau} - \nabla \mathrm{p} + \mathbf{J} \times \mathbf{B} \tag{2.2}$$

where p is pressure; $\boldsymbol{\tau}$ is pressure tensor [3]; $\mathbf{J} \times \mathbf{B}$ is the Lorentz force; **J** is the current density; and **B** is the magnetic flux density. The gravity term is ignored in most cases.

(3) Energy equation

In the two-temperature arc model, the energy conservation equations of the electrons and the heavy particles are considered separately to solve the temperature of the heavy particle, T_h, and the electron, T_e [3, 6, 7].

Energy conservation equation for heavy particles:

$$\frac{\partial (\rho h_h)}{\partial t} + \nabla \cdot (\rho h_h \vec{u}) = \nabla(\lambda_h \nabla T_h) + Q_{rad}^h + Q_{eh} + \Phi + \frac{\partial p_h}{\partial t} \tag{2.3}$$

Energy conservation equation for electrons:

$$\frac{\partial (\rho h_e)}{\partial t} + \nabla \cdot (\rho h_e \vec{u}) = \nabla(\lambda_e \nabla T_e) + Q_J + Q_{rad}^e - Q_{eh} + \frac{\partial p_e}{\partial t} + \frac{5k_B}{2|e|}\mathbf{J} \cdot \nabla T_e \tag{2.4}$$

The above energy conservation equations are coupled, and all of the physical parameters of arc plasmas are related to three independent variables: gas pressure, electron temperature, and heavy particle temperature. In addition, enthalpy is used as the solved-for variable for the energy conservation equations in this chapter. In the process of modelling, temperature can also be selected as the solved-for variable according to the characteristics of the numerical solution method, and the corresponding equation then needs to be rearranged according to the solved-for variable.

When $T_e = T_h$ is the energy conservation equation of heavy particles plus the energy conservation equation of electrons recovers to the total energy conservation equation, which is the most useful way to confirm the accuracy of the equation of the two-temperaturemodel.

h is the total enthalpy, $h = \sum_{i=1}^{n} h_i$, with n is species in the plasma. That is, the total enthalpy is equal to the sum of the enthalpy of all particles (including electrons).

h_e is the electron enthalpy, $h_e = \frac{5k_B}{2\rho} n_e T_e$, which means that only the enthalpy of electrons due to translation is considered in the electron energy equation. k_B is the Boltzmann constant and n_e is electron number density. For the ideal gas model, there is $\rho h_e = \frac{5}{2} k_B n_e T_e$, $T_e = T_h$.

The enthalpy of heavy particles, h_h, is much more complicated than that of electrons, h_e, and there are still some disputes about how best to express it [8]. Generally, it can be considered in two cases: monatomic and polyatomic.

The effects of translation, reaction and electron transition on the enthalpy of heavy particles are considered in the enthalpy of a single atom as follows:

$$h_i = \frac{5k_B}{2\rho} n_i T_i + \frac{1}{\rho} n_i E_i + \frac{k_B}{\rho} n_i \left(T_e^2 \frac{\partial \ln Q_i^{el}}{\partial T_e} + T_h^2 \frac{\partial \ln\left(Q_i^{rot} Q_i^{vib}\right)}{\partial T_e} \right) \tag{2.5}$$

where E_i is the change in reaction enthalpy to generate particle i from the reference state.

As for the enthalpy of polyatomic aspecies, the changes of enthalpy caused by vibration and rotation are introduced on the basis of single atoms as follows:

$$h_i = \frac{5k_B}{2\rho} n_i T_i + \frac{1}{\rho} n_i E_i + \frac{k_B}{\rho} n_i \left(T_e^2 \frac{\partial \ln Q_i^{el}}{\partial T_e} + T_h^2 \frac{\partial \ln\left(Q_i^{rot} Q_i^{vib}\right)}{\partial T_e} \right) \tag{2.6}$$

where Q_i^{el} is the energy level partition function of electron, Q_i^{rot} is the rotational energy level partition function of molecule, and Q_i^{vib} is the vibrational energy level partition function of molecule.

For the sake of simplicity in many cases, whether monatomic or polyatomic, only the translational and reaction effects are considered for the enthalpy of heavy particles, and the temperature of all heavy particles is considered to be the same T_h, which is

$$h_h = \sum_{i \neq e}^{n} \frac{5k_B}{2\rho} n_i T_h + \frac{1}{\rho} n_i E_i \tag{2.7}$$

In the diffusion terms of the energy equation, λ_h、λ_e、T_h and T_e represent the thermal conductivity and temperature of the heavy particles and electrons, respectively.

The joule heating term, Q_J, is the main energy source of plasma. In arc plasma, current is considered to be mainly caused by electron flow, and the current caused by ion flow is negligible, so the joule heat source term is attributed to the electron energy equation here [3, 9]. It is important to note that the joule heating term here is different from the joule heat in the LTE model [7, 9].

$$Q_J = \sigma \mathbf{E_p} \left(\mathbf{E_p} - \frac{\nabla p_e}{|e| n_e} \right) \tag{2.8}$$

where $\mathbf{E_p}$ is the electric field strength obtained by solving the electrostatic potential equation, p_e is the electron pressure, and e is the electronic charge. If the gas is assumed to be ideal, it can be written as

$$Q_J = \sigma \mathbf{E_p} \left(\mathbf{E_p} - \frac{\nabla (k_B n_e T_e)}{|e| n_e} \right) \tag{2.9}$$

Φ is the viscous dissipation term. This term is small and is ignored in most of the literature.

The enthalpy transport term of the current carrier part is written as

$$\frac{5k_B}{2e} \mathbf{J} \cdot \nabla T_e \tag{2.10}$$

(4) Equation of State for the gas

The equation of state of two-temperature arc plasma is related not only to the temperature of heavy particles but also to the temperature of electrons as

$$p = p_e + p_h \tag{2.11}$$

For the sake of simplicity, the ideal gas equation is adopted [4]

$$p_e = n_e k_B T_e \tag{2.12}$$

$$p_h = \sum_{i \neq e} n_i k_B T_h \tag{2.13}$$

(5) Electric and Magnetic field equations

The electric field is solved by the electrostatic potential equation as follows [9]

$$\nabla \cdot (\sigma \nabla V_p) = 0 \tag{2.14}$$

where σ is the electrical conductivity, V_p is the electrostatic potential, the electric field strength is obtained as $\mathbf{E_p} = -\nabla V_p$ and the current density is $\mathbf{J} = \sigma \mathbf{E_P}$.

In the calculation of arc plasma, the magnetic field is usually solved by solving the magnetic vector equation

$$\nabla (\nabla \mathbf{A}) = -\mu_0 \mathbf{J} \tag{2.15}$$

where μ_0 is the vacuum permeability and magnetic flux density $\mathbf{B} = \nabla \times \mathbf{A}$.

3. Mathematical description of arc energy transfer process

Q_{eh} is the energy exchange term between electrons and heavy particles, through which the energy conservation equation of heavy particles and electrons is coupled. From a microscopic point of view, electrons move faster than heavy particles, and electrons transfer energy to heavy particles through collisions. It is calculated by:

$$Q_{eh} = \sum_{i \neq e} \frac{3}{2} k_B (T_e - T_h) n_e \frac{2 m_i m_e}{(m_i + m_e)^2} \nu_{ei} \delta_{ei} \tag{2.16}$$

In general, we can simplify this to

$$Q_{eh} = 3 k_B (T_e - T_h) \rho_e \sum_{i \neq e} \frac{\nu_{ei} \delta_{ei}}{m_i} \tag{2.17}$$

where $\nu_{ei} = \sqrt{\frac{8 k_B T_e}{\pi m_e}} n_i Q_{ei}$ is the collision frequency between an electron and a heavy particle i. $\sqrt{\frac{8 k_B T_e}{\pi m_e}}$ is the electron mean velocity, Q_{ei} is the collision cross section of electrons and heavy particles. Q_{ei} is a function of electron temperature and electron number density [10]. The collision cross-section of an electron and an ion is calculated by coulomb potential, and the collision cross section of an electron and a neutral particle can be obtained from relevant literature. δ_{ei} is an inelastic coefficient which is adjusted to account for energy loss due to inelastic collisions in the real application.

4. Input of material property data of the arcing gas

In Chap. 4, the calculation method of material properties of the arcing gas is introduced in detail. Similar to the equilibrium state, the material properties of thermal non-equilibrium arc can be summarised as functions of electron temperature, T_e, heavy particle temperature, T_h, and pressure, p. In the process of solving the thermal non-equilibrium fluid equations, the material properties of the gas can be obtained by multi-level interpolations according to the local electron temperature, T_e, the temperature of heavy particles, T_h, and the pressure, p in the iterative solution process.

2.2.4 Exemplary Calculation

(1) Geometric structure and boundary conditions

Figure 2.3 is the geometric model of the two-temperature arc calculation in a simplified electrode model. As shown in the figure, the electrode gap and its surrounding area were selected as the calculation domain, and the fluid boundary connected with the external area was set as the pressure outlet to link to the external SF_6 environment. In this simulation, the SF_6 filling pressure is 4 atm. In addition, the post-arc process in this example is in the free recovery mode, that is, the influence of the recovery voltage on the thermal non-equilibrium behaviour of the arc is not considered.

Compared with the boundary conditions of the equilibrium model, the two-temperature model uses the electron energy conservation equation and the heavy particle energy conservation equation instead of the total energy conservation equation, so the boundary conditions of the electron and heavy particle temperatures need to be treated separately.

Under the equilibrium assumption, the electron temperature is equal to the temperature of the heavy particles, and the temperature of the electrode surface should not exceed the melting point of the electrode material at that point. Under the constraint of such temperature boundary conditions, the temperature calculated by the equilibrium model near the electrode is usually very low (no more than 3,000 K), resulting in very low electrical conductivity in the gas adjacent to the electrode (< 0.01 S/m), which behaves like an insulation medium. Considering the physics, an insulation gas layer adjacent to the electrode surface is against the experimental evidence As for the conditions at the pressure outlet, the simplest way to deal with it is to assume that $T_e = T_h$ [4]. When the gas flow outlet is sufficiently far away from the arc region, the local gas will be in equilibrium from a physical point of view. Generally speaking, with two-temperature arc modelling attention should be paid to ensuring that the pressure outlet boundary is far away from the arc.

As for the external wall of the electrode, i.e. *ad*, *dc*, *fg* and *gh* in Fig. 2.3, the electron temperature gradient here is generally set to be 0 [2]. There is not a very clear theory on how to set the electron temperature here.

(2) Governing equations

Figure 2.4 shows the arc temperature distribution calculated based on the equilibrium model and the two-temperature model at an instant after current zero. It can be seen that in the case of 2,000 A, the two models have a similar temperature distribution in the central region of the arc columns. Due to the higher calculated temperature of electrons and heavy particles in the two-temperature model on the edge of the arc column, the conductivity at the edge of the arc is higher than that at the edge of LTE. As a result, the current density distribution of the two-temperature model is not as concentrated as that in LTE, the temperature in the centre of the arc column is lower, and the whole arc column is wider than the case with LTE, the temperature gradient at the edge is also significantly smaller than that in the LTE model.

In the two-temperature model, the value of T_e/T_h is termed the 'degree of non-equilibrium' in the example here, which can be used to measure the degree of deviation from thermodynamic equilibrium in a two-temperature system: the larger the ratio, the more serious the deviation from thermodynamic equilibrium is.

Figure 2.5 shows the distribution of degree of non-equilibrium in the arc region of the two-temperature model at an instant after current zero. In the central region of the arc, the current density is large and the electron number density is high. Energy can be efficiently exchanged between electrons and heavy particles through elastic collisions. Therefore, the degree of non-equilibrium is essentially 1, which can be considered as being in thermodynamic equilibrium. At the edge of the arc, the current density is low and the collision frequency is reduced. The the electron temperature

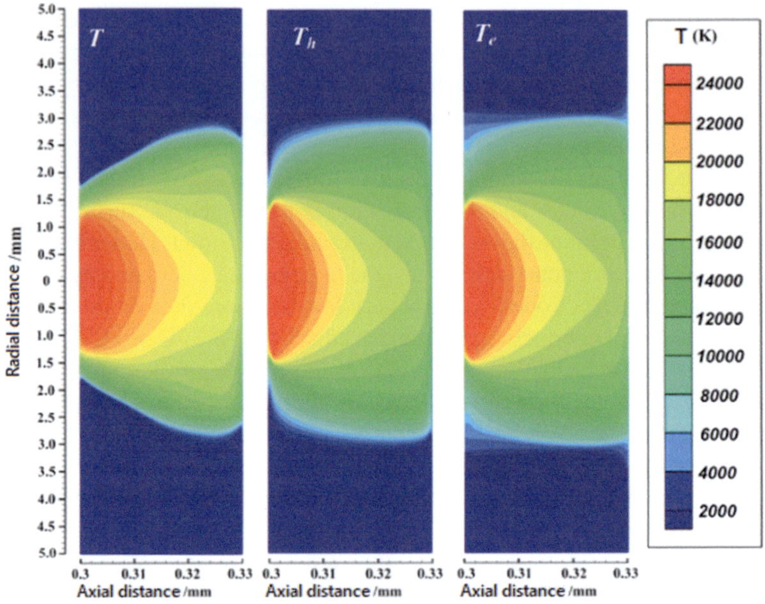

Fig. 2.4 Arc temperature distribution calculated based on equilibrium model and two-temperature model

is higher than the heavy particle temperature, and the degree of non-equilibrium is about 2.5. Meanwhile, the degree of non-equilibrium is highest in the near-electrode region, at about 4. The main reason is that the electrode imposes a strong cooling effect on the plasma, causing the arc to deviate from thermodynamic equilibrium.

Fig. 2.5 Distribution of unbalance degree in arc region of two-temperature model

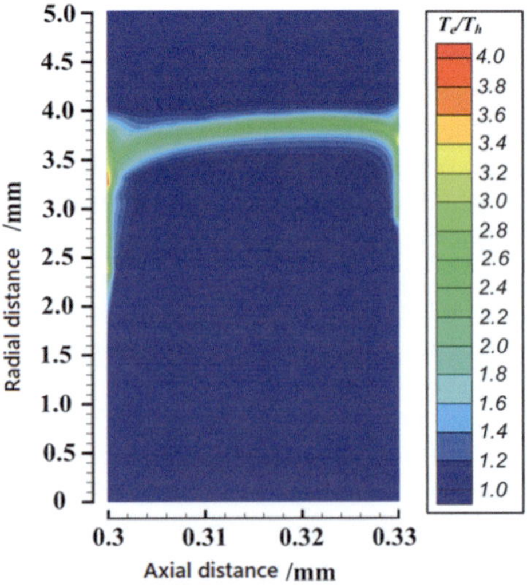

Figure 2.6 shows the velocity distribution of the flow in the arc region in the equilibrium case and the two-temperature model at the same instant as in Fig. 2.7. With both models, there is a strong jet flow from the cathode to the anode, and the highest velocity is concentrated on the axis of symmetry. For the equilibrium model, the high-speed axial flow is closer to the anode, forming a vortex as shown in Fig. 2.8 at point A. At the same time, the velocity in the radial direction near the cathode rapidly decreases, which is mainly caused by the rapid contraction of the high temperature region near the cathode in equilibrium state. For the two-temperature model, the jet flow is mainly concentrated near the cathode, thus forming a vortex as shown in Fig. 2.8 at point B. Under the action of the vortex, the high-temperature gas at the edge of the arc root on the cathode gradually diffuses outward. At the same time, the high-speed jet from the cathode begins to diverge near the anode, and the high-temperature gas on the axis diffuses towards the radial direction, forrming an anodic temperature distribution similar to LTE.

Figure 2.7a, b show the variation curves of SF_6 arc temperature with axial distance $r = 3.4$ mm and 0 mm respectively, at a certain time in the post-arc process. At the edge of the $r = 3.4$ mm area, there is an obvious thermal non-equilibrium phenomenon in the arc along the axial direction, that is electron temperature is significantly higher than the heavy particle temperature, especially in the area near the electrodes on both sides. Since the cooling effect of the electrode is strong, the non-equilibrium phenomenon is more apparent around the electrode, and the deviation degree of the anode is higher than that of the cathode. On the axis of symmetry of the arc centre, the electron temperature is basically the same as that of the heavy particle, and the thermal non-equilibrium phenomenon only occurs near the anode.

Figure 2.8 shows the curves of how T_e (solid line) and T_h (dotted line) change on the central axis of the arc with the axial position at different instants. Over the period, the joule heating source term of the electrons decreases rapidly, but electrons

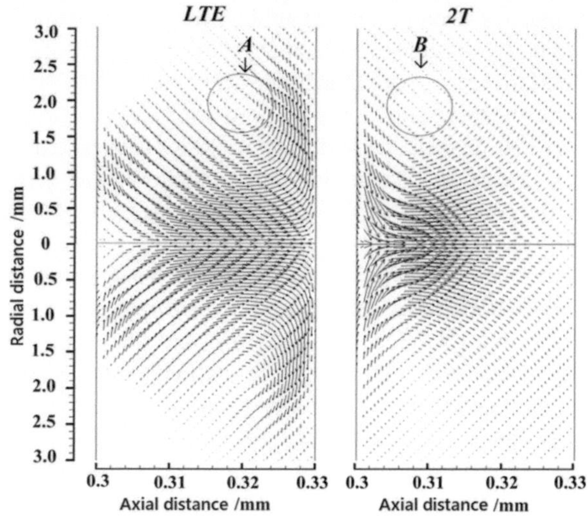

Fig. 2.6 Velocity distribution of flow field in the arc region of equilibrium and two-temperature models

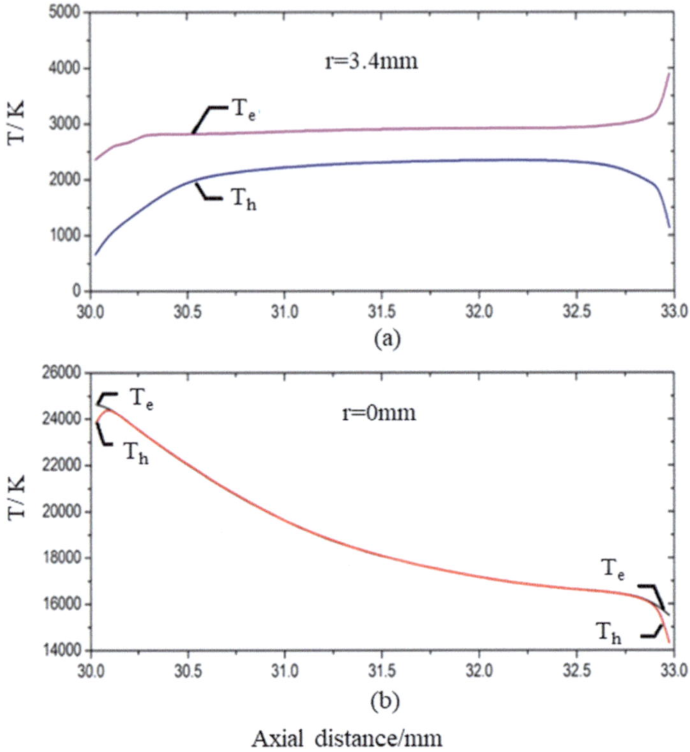

Fig. 2.7 Variation curves of SF6 arc temperature with axial distance at r = 3.4 mm and 0 mm at steady state

still have a high temperature. In this case, due to the small electrode spacing and the absence of strong external gas blast, radiation and elastic collisions between electrons and heavy particles dominate the electron energy dissipation, resulting in continuous electron temperature reduction. At the same time, due to the existence of the gas flow vortex near the anode, the electron temperature near the anode is significantly higher than that near the cathode within 40 μs of zero current, and the temperature drops at a slower rate. The temperature attenuation of heavy particles is mainly affected by the cooling effect of the electrodes on both sides, the temperature drop of heavy particles near the electrode is much faster than that of electrons. At the same time, the electron number density in the post-arc phase keeps decreasing, reaching a very low level at 40 μs and resulting in a decrease in the collision frequency between electrons and heavy particles. The electrons and heavy particles near the electrode cannot effectively exchange energy through elastic collisions. Therefore, the temperature deviation of electrons and heavy particles near the electrode increases.

It can be seen from the results obtained in this case that the thermal non-equilibrium effect of the arc mainly occurs in the regions with low arc temperature or large arc temperature gradient, that is, at the arc edge and in the near-electrode regions.

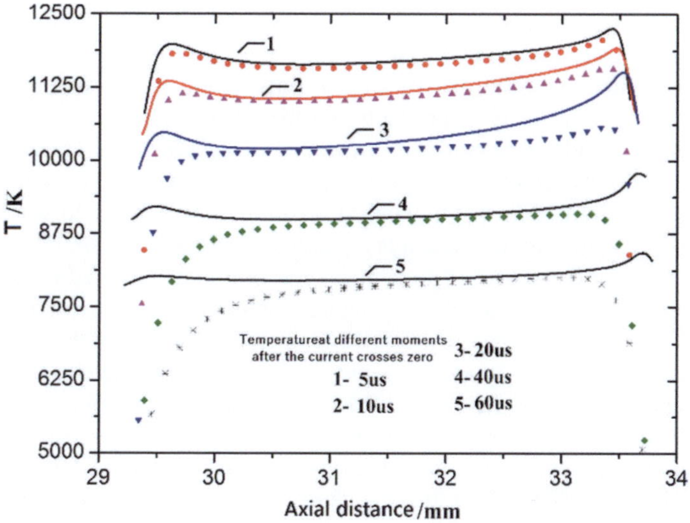

Fig. 2.8 Variation curves of T_e (solid line) and T_h (dotted line) on the central axis of the arc with axial position at different instants after the current zero

In addition, the electron number density and electron energy decrease sharply with the rapid decay of temperature in the post-arc phase, resulting in insufficient collisions frequency between electrons and heavy particles, and the thermal non-equilibrium effect will increase continuously with time in a period after zero.

2.3 Chemical Non-Equilibrium Behaviour

2.3.1 Causes of Chemical Non-Equilibrium

In the previous section, the thermal non-equilibrium effect caused by inadequate elastic collisions between electrons and heavy particles has been described. In addition to the elastic collisions that cause energy exchange, there are also inelastic collision processes inside the arc, such as excitation, ionisation, recombination, absorption and desorption of electrons, which are often called chemical reaction processes.

$$A + B \to C + D \tag{2.18}$$

$$\text{Forward reaction rate} = k_f n_A n_B \tag{2.19}$$

$$\text{Reverse reaction rate} = k_r n_C n_D \tag{2.20}$$

A typical reversible reaction process in arc is shown by Eqs. 2.18–2.20 above. For a reversible reaction, the forward reaction rate is determined by the forward reaction coefficients and the density of reactants, while the reverse reaction rate is determined by the density of products and the reverse reaction coefficients. The equilibrium state (and the forward and reverse reaction coefficients) are influenced by the global conditions such as temperature and pressure. An isotropic system reaches chemical equilibrium, after a sufficiently long time, in which the forward reaction rate of all the reactions in the system is equal to the reverse reaction rate.

For arc plasma, the speed of a chemical reaction is determined by the type of this chemical reaction, the local temperature (including the temperature of heavy species and electron temperature in the two-temperature case), and the concentration of reactants and products. In addition, the concentration of reactants and products will be affected by the gas flow and diffusion of species in the mixture. Therefore, the process of a chemical reaction in the arc is not only influenced by physical processes, including convection and diffusion of energy and mass, but also affected by chemical changes, including the change of chemical reaction time constant, the release or absorption of reaction heat, and the generation or loss of particles. So the effect of a chemical reaction deviating from equilibrium due to a physical or chemical change is generally referred to as a chemical non-equilibrium effect.

Under the following two conditions, chemical non-equilibrium effects will occur in arc plasma: firstly, on the time scale, the characteristic time of the slowest chemical reaction is comparable to that of the arc motion; secondly, on the spatial scale, the rate of convection and diffusion of a certain species due to its concentration gradient is comparable to the rate of generation or loss due to chemical reaction. In the process of dielectric recovery, the chemical reaction rate in the arc decreases rapidly after current zero and the joule heating rapidly attenuates. In terms of time and spatial scale, the plasma no longer meets the chemical equilibrium requirement. The specific aspect of the behaviour is as follows:

1. Time scale

The characteristic time of a chemical reaction describes the time required for a chemical reaction to tend to the equilibrium state from the non-equilibrium state when the original equilibrium state is disturbed. It should be noted that the characteristic time of chemical reactions change over time and location, and are different from each other since the chemical reaction rate is affected by local temperature and particle density. Numerous studies have summarised the characteristic time of various chemical reactions under different conditions. It turns out that the characteristic time of chemical reactions in the current zero period is $10^{-7} \sim 10^{-2}$ s for various gases used in the switchgear. For the current zero period, the characteristic time of arc motion ranges 10^{-6}–10^{-4} s. Therefore, it can be seen that the characteristic time of chemical reactions is greater than that of arc motion, which proves the necessity of introducing a chemical non-equilibrium effect in the arc model.

2. Spatial scale

Due to the spatial distribution of various quantities within a circuit breaker, convection and diffusion with the latter as a result of the spatial distribution of particle gradient lead to chemical reactions deviating from the equilibrium state. It can be found in the study that in the high temperature area (above 10,000 K), the rate of change of the mass of the particles as a result of convection and diffusion is about 10^0 ~ 10^2 kg/(m^3s), and the generation rate of the mass of particles due to chemical reaction is about 10^2 ~ 10^3kg/(m^3·s). Therefore, in the high temperature region, particles can be regarded as approximately satisfying the chemical equilibrium requirement on the spatial scale. However, in the low temperature region of 300 - 10,000 K, the magnitude of the mass change rate by convection and diffusion is still close to 10^0 ~ 10^2 kg/(m^3·s), while the magnitude of the mass change rate by chemical reaction is only 10^{-3} ~ 10^1 kg/(m^3·s) due to the rapid decay of chemical reaction rate at low temperatures. Therefore, due to the rapid attenuation of arc temperature in the current zero period arc, the chemical non-equilibrium effect should also be considered in the spatial scale.

It is worth noting that, from the initial steady-state arc to the current zero point, the arc can be approximately considered in equilibrium (in the time scale) due to the long arc duration, of about 10^{-2} ~ 10^{-1}s or longer. However, in the low temperature region at the arc boundary, the non-equilibrium effect due to particle convection and diffusion should also be considered.

2.3.2 Present Status of Research on Chemical Non-Equilibrium

For chemical non-equilibrium modelling, Murphy has had discussions in detail on the need to couple the MHD model with the model of chemical reaction [11]. It should be noted that Murphy's description is not just for arc plasma, but covers many aspects of hot plasma applications. According to his work, taking the N_2-Ar plasma torch as an example, there are ten species in the plasma, so the non-equilibrium model needs to solve the transport equations of all species, and each transport equation needs to include the effect of ten diffusion coefficients and all 55 chemical reactions.

Considering the difficulties in the calculation, the diffusion of particles was simplified by dividing the gas mixture into two categories based on N_2 and Ar. Tthe transport equations of particles were reduced from ten to two. Gonzalez et al. studied the thermal non-equilibrium and chemical non-equilibrium in the post-arc phase of SF_6 [12]. The important conclusion drawn from [12] is that compared with the LTE hypothesis the introduction of the non-equilibrium hypothesis reduces the recovery characteristics of the residue plasma after current zero. Teulet et al. established a 1-D coupled model of fluid mechanics and reaction kinetics [13, 14] and the object under study was a low voltage circuit breaker where air was mixed with PA66 nylon material and copper vapour as the arcing medium. Although only the radial distribution of each quantity is considered, the calculation methods for reaction kinetics, thermodynamic parameters, transport coefficients and radiation coefficient still provide references for the calculation of chemical non-equilibrium. Using a similar approach, Gleizes et al. described the modelling method for plasma where deviation from LTE and chemical equilibrium exists, including the arc in circuit breakers [15]. Tanaka et al. also studied chemical non-equilibrium modelling of air and SF_6 arcs, respectively [16, 17].

In other aspects of thermal plasma applications, the combustion model combined with fluid simulation is closest to the physical processes involved in the dynamic behaviour of the current zero phase of switching arc [18]. The combustion model focuses on the transport of different species due to chemical reactions, convection and diffusion at high temperatures. The mode of Tanaka et al. for plasma torches make references to such combustion models [19, 20]. All the conservation equations mentioned in this model is an important reference for the establishment of chemical non-equilibrium model.

In general, Work on the calculation of the chemical non-equilibrium behaviour of the current zero period is rather limited, and the main difficulty lies in the following aspects:

1. It is necessary to establish the transport equation for each species in the arc region, so the calculation effort will increase greatly compared to the case in chemical equilibrium.
2. The physical quantities involved in the species transport equation have a large span of magnitude. Therefore, the solution of the transport equation is easily prone to computational divergence.
3. A completely self-consistent chemical non-equilibrium model needs to consider the effect of composition on the material properties of the arcing gas. However, the calculation of material properties is a complicated process, which not only renders the modelling work extremely complex, but may also cause instability to the solution process of multiple equations.
4. Traditional arc measurement mainly focuses on arc voltage, current and other macroscopic quantities, which cannot provide direct experimental evidence for the results of chemical non-equilibrium model.

2.3.3 Chemical Non-Equilibrium Arc Modelling

1. Transport equations for species

The chemical non-equilibrium model couples chemical reactions with arc flow model. The key to this coupling is to establish the transport equations of species present in the arc. The transport equation of the jth species is shown in Eq. 2.21

$$\frac{\partial \rho Y_j}{\partial t} + \nabla \cdot \left(\mathbf{V} \rho Y_j\right) = \nabla \cdot \left(\rho D_j \nabla Y_j\right) + S_j \tag{2.21}$$

The physical meanings of the terms in the above equation, from left to right, are as follows: time-varying term of species mass in unit volume, change of mass caused by convection, change of mass caused by diffusion, and mass generated by chemical reactions. The calculation of the source term and diffusion coefficient is shown in Eqs. 2.22 and 2.23–2.25 respectively.

$$S_j = m_j \sum_{l}^{L} \left(\beta_{jl}^r - \beta_{jl}^f \right) \left(k_l^f \prod_{i=1}^{N} n_i^{\beta_{il}^f} - k_l^r \prod_{i=1}^{N} n_i^{\beta_{il}^r} \right) \tag{2.22}$$

$$D_j = \frac{1 - Y_j}{\sum_{k \neq j} \left(x_k / D_{jk} \right)} \tag{2.23}$$

$$D_{jk} = \frac{\kappa T}{p} \frac{1}{\Delta_{kj}^{(1)}} \tag{2.24}$$

$$\frac{1}{\Delta_{kj}^{(1)}} = \frac{3}{8} \left[\frac{\pi \kappa T \left(m_i + m_j \right)}{2 m_i m_j} \right]^{1/2} \frac{1}{\pi \Omega_{ij}^{(1,1)}} \tag{2.25}$$

The ideal gas equation is also included in the calculation, as in (2.26)

$$p = \sum_j n_j \kappa T \tag{2.26}$$

And the number density and of species j and total mass density are shown as (2.27) and (2.28), respectively

$$n_j = \frac{\rho Y_j}{m_j} \tag{2.27}$$

$$\rho_{\mathrm{GD}} = \sum_j n_j m_j \tag{2.28}$$

In all of the above equations from (2.21),ρ_{GD} is the gas density, Y_j is the mass fraction of species j, **V** is the velocity vector; m$_j$ stands for the mass of species j, β_{jl} is the stoichiometric number of species j in the l-th reaction; $k_l{}^f$ and $k_l{}^r$ are the forward and reverse reaction time constants in the l-th reaction respectively; x_k represents the mole fraction of species k, T is temperature; p for gas pressure, κ is Boltzmann constant; $\pi\Omega_{ij}{}^{(1},{}^{1)}$ is the collision integral for momentum transfer between species I and j; and n_j is for the number density of species j.

It should be noted that for the diffusion term of a species, the key problem is to solve for the diffusion coefficient D_j of all species. As can be seen from (2.23)–(2.25), for a system with j species, $j \times j$ diffusion coefficients need to be solved, which requires a large amount of calculation. In Murphy's work where a mixture of N_2 and Ar is taken as an example, in order to reduce the complexity of calculation the concept of combined diffusion coefficient is proposed. The particles in the system are classified into two categories, N_2 and Ar, and it only requires the solution of 2

× 2 diffusion coefficients [11]. However, in this book, in order to study the effect of changes in all species of the gas on the arcing behaviour, the combined diffusion coefficient proposed by Murphy is not used. The general diffusion coefficient as defined in Eq. 2.16 is used to solve the problem.

Equation 2.21 directly considers the species migration under the action of convection, diffusion and chemical reactions, and is the basis of chemical non-equilibrium calculation. In Chap. 4, the calculation method of material properties under chemical non-equilibrium is introduced. In addition to material properties, chemical reaction constant is another basic parameter in chemical non-equilibrium modelling. In order to ensure the accuracy of the calculation, values for this parameter obtained from experiment are preferred. Some chemical reaction constants can be accurately obtained theoretically in the absence of experimental results. At present, for arcs in SF_6, air, N_2, and CO_2, there has been a relatively complete database of chemical reaction paths.

The equation for calculating the forward reaction time constant is shown in (2.29) [21].

$$k_l^f = AT^B \exp\left(-\frac{C}{T}\right) \tag{2.29}$$

When considering the rate coefficient of electron-impact reactions, k_l^f can be calculated by the Boltzmann equation, which is given in Eq. 2.30 [22].

$$k_l^f = \left(\frac{2e}{m_e}\right)^{1/2} \int_0^\infty f_M(\varepsilon)\sigma_l(\varepsilon)\varepsilon d\varepsilon \tag{2.30}$$

where e is the electronic charge, $f_M(\varepsilon)$ is the Maxwellian distribution of the electron energy distribution function, $\sigma_l(\varepsilon)$ is the collision cross section of the l-th reaction, and ε is the electron energy. The early ELENDIF code or Bolsig + software can achieve accurate calculation of such chemical reaction constants [23] [74]. The specific calculation methods will be described in detail in the next section. Tables 2.1, 2.2, 2.3, 2.4, 2.5, 2.6, 2.7, 2.8 and 2.9 show the types of chemical reactions and chemical reaction constants which are to be considered in arc plasma burning in common gases.

After determining all of the forward reaction rate constants, the constants of the corresponding reverse reaction can be deduced from the principle of detailed balance [32]. Take the reaction AB + M → A + B + M as an example, when the reaction reaches equilibrium, it should satisfy Eq. 2.31

$$k^f n_{AB} = k^r n_A n_B \tag{2.31}$$

where n_{AB}, n_A and n_B are the number density of the species. Therefore, the forward and reverse reaction rate constants should satisfy the following relationship

Table 2.1 Relevant parameters of chemical reactions in SF_6 arc plasma

No	Chemical reaction	A/m^3/s,m^6/s	B	C/K	Reference
1	$F + F + F \rightarrow F_2 + F$	5.4×10^{-35}	0	0	[24]
2	$F + S + F \rightarrow SF + F$	2.8×10^{-34}	0	0	[24]
3	$S + S + F \rightarrow S_2 + F$	4.5×10^{-34}	0	0	[24]
4	$SF^+ + F \rightarrow F + S^+ + F$	3.92×10^{-12}	1/2	43750	[24]
5	$S + 2 + F \rightarrow S + S^+ + F$	3.81×10^{-12}	1/2	74735	[24]
6	$SF^- + F \rightarrow S + F^- + F$	3.92×10^{-12}	1/2	24950	[24]
7	$SF^- + F \rightarrow F + S^- + F$	3.92×10^{-12}	1/2	40849	[24]
8	$S^+ + S^- + F \rightarrow SF + F$	6.32×10^{-20}	−5/2	0	[24]
9	$F^- + S^+ + F \rightarrow SF + F$	7.32×10^{-20}	−5/2	0	[24]
10	$F^- + F \rightarrow F + e + F$	1.00×10^{-11}	0	40000	[24]
11	$S^- + F \rightarrow S + e + F$	4.23×10^{-12}	1/2	24105	[24]
12	$F^- + F \rightarrow F_2 + e$	1.4×10^{-10}	0	22000	[24]
13	$F^- + S \rightarrow SF + e$	1.4×10^{-10}	0	4642	[24]
14	$S + e \rightarrow S^+ + e + e$	5.02×10^{-1}	−3/2	120230	[24]
15	$S^+ + e \rightarrow S + h\upsilon$	3.46×10^{-6}	−1/2	0	[24]
16	$SF^+ + e \rightarrow S + F$	3.46×10^{-6}	−1/2	0	[24]
17	$S + 2 + e \rightarrow S + S$	3.46×10^{-6}	−1/2	0	[24]
18	$S^+ + S_2 \rightarrow S + 2 + S$	1.38×10^{-9}	0	0	[24]
19	$S^+ + SF \rightarrow SF^+ + S$	1.00×10^{-10}	0	0	[24]
20	$S^+ + SF \rightarrow S + 2 + F$	9.90×10^{-10}	0	0	[24]
21	$S^+ + F_2 \rightarrow SF^+ + F$	5.89×10^{-10}	0	0	[24]
22	$S^- + SF \rightarrow S + SF^-$	3.28×10^{-11}	1/2	0	[24]
23	$S + 2 + F^- + F \rightarrow S_2 + F + F$	6.61×10^{-10}	−5/2	0	[24]
24	$S + 2 + F^- + F \rightarrow SF + F + S$	6.61×10^{-10}	−5/2	0	[24]
25	$S + 2 + F^- + F \rightarrow S_2 + F_2$	6.61×10^{-10}	−5/2	0	[24]
26	$S + 2 + F^- + F \rightarrow SF + SF$	6.61×10^{-10}	−5/2	0	[24]

$$k^r = k^f \frac{n_{AB}}{n_A n_B} = k^f \left(\frac{h_p^2 m_{AB}}{2\pi m_A m_B k_B T} \right)^{3/2} \frac{Z_{AB}}{Z_A Z_B} \exp\left(\frac{\psi_{reac}}{k_B T} \right) \tag{2.32}$$

where m is the molecular mass of the corresponding species, Z is the partition function, ψ_{reac} is the reaction energy, and k_B is the Boltzmann constant.

2. Governing equations

Through the species transport equation, the composition under chemical non-equilibrium condition can be obtained, then the gas material properties necessary for the governing equations can be calculated. This section focuses on the basic governing equations.

Table 2.2 Parameters of chemical reactions in air or N_2 arc plasma

No	Chemical reaction	A/m^3/s,m^6/s$^-$	B	C/K	Reference
1	$O_2 + O_2 \rightarrow O + O + O_2$	3.33×10^{-9}	− 1.5	59,500	[21]
2	$O_2 + NO \rightarrow O + O + NO$	3.33×10^{-9}	− 1.5	59,500	[21]
3	$O_2 + N_2 \rightarrow O + O + N_2$	3.33×10^{-9}	− 1.5	59,500	[21]
4	$O_2 + O \rightarrow O + O + O$	1.67×10^{-8}	− 1.5	59,500	[21]
5	$O_2 + N \rightarrow O + O + N$	1.67×10^{-8}	− 1.5	59,500	[21]
6	$NO + O_2 \rightarrow N + O + O_2$	8.33×10^{-15}	0	75,500	[21]
7	$NO + NO \rightarrow N + O + NO$	1.83×10^{-13}	0	75,500	[21]
8	$NO + N_2 \rightarrow N + O + N_2$	8.33×10^{-15}	0	75,500	[21]
9	$NO + O \rightarrow N + O + O$	1.83×10^{-13}	0	75,500	[21]
10	$NO + N \rightarrow N + O + N$	1.83×10^{-13}	0	75,500	[21]
11	$N_2 + O_2 \rightarrow N + N + O_2$	1.17×10^{-8}	− 1.6	113,200	[21]
12	$N_2 + NO \rightarrow N + N + NO$	1.17×10^{-8}	− 1.6	113,200	[21]
13	$N_2 + N_2 \rightarrow N + N + N_2$	1.17×10^{-8}	− 1.6	113,260	[21]
14	$N_2 + O \rightarrow N + N + O$	5.0×10^{-8}	− 1.6	113,260	[21]
15	$N_2 + N \rightarrow N + N + N$	5.0×10^{-8}	− 1.6	113,260	[21]
16	$N_2 + e \rightarrow N + N + e$	5.0×10^{-6}	− 1.6	113,260	[21]
17	$N_2 + O \rightarrow NO + N$	1.07×10^{-12}	− 1.0	38,400	[21]
18	$NO + O \rightarrow N + O_2$	1.4×10^{-17}	0	19,400	[21]
19	$N + O \rightarrow NO^+ + e$	8.83×10^{-18}	0	31,900	[21]
20	$N + N \rightarrow {N_2}^+ + e$	3.33×10^{-17}	0	67,500	[21]
21	$O + O \rightarrow {O_2}^+ + e$	1.83×10^{-17}	0	80,600	[21]
22	$O + e \rightarrow O^+ + e + e$	6.5×10^3	− 3.78	158,500	[21]
23	$N + e \rightarrow N^+ + e + e$	4.17×10^3	0	168,200	[21]
24	$NO^+ + O \rightarrow N^+ + O_2$	1.67×10^{-18}	0.5	77,200	[21]
25	${O_2}^+ + N \rightarrow N^+ + O_2$ 2 + N → N + + O22 + N → N + + O2 2 + N → N + + O2	1.45×10^{-16}	0.14	28,600	[21]
26	$NO + O^+ \rightarrow N^+ + O_2$	2.33×10^{-25}	1.9	15,300	[21]
27	${O_2}^+ + N_2 \rightarrow {N_2}^+ + O_2$	1.65×10^{-17}	0	40,700	[21]
28	${O_2}^+ + O \rightarrow O^+ + O_2$	6.67×10^{-18}	− 0.09	18,000	[21]
29	$NO^+ + N \rightarrow O^+ + N_2$	5.66×10^{-17}	− 1.08	12,800	[21]
30	$NO^+ + O_2 \rightarrow {O_2}^+ + NO$	6.67×10^{-18}	− 0.09	18,000	[21]
31	$NO^+ + O \rightarrow {O_2}^+ + N$	1.2×10^{-17}	0.29	48,600	[21]
32	$O^+ + N_2 \rightarrow {N_2}^+ + O$	1.5×10^{-18}	0.36	22,800	[21]
33	$NO^+ + N \rightarrow {N_2}^+ + O$	1.2×10^{-16}	0	35,500	[21]
34	${N_2}^+ + N \rightarrow N_2 + N^+$	1.30×10^{-19}	0.5	0	[22]
35	${N_2}^+ + N_2 + e \rightarrow N_2 + N_2$	6.07×10^{-34}	− 2.5	0	[22]

(continued)

Table 2.2 (continued)

No	Chemical reaction	A/m^3/s,m^6/s$^-$	B	C/K	Reference
36	$N_2^+ + N + e \rightarrow N_2 + N$	1.66×10^{-35}	− 2.5	0	[22]
37	$N^+ + N_2 + e \rightarrow N + N_2$	6.07×10^{-34}	− 2.5	0	[22]
38	$N^+ + N + e \rightarrow N + N$	1.66×10^{-35}	− 2.5	0	[22]

Table 2.3 Parameters of chemical reactions involving C and O

No	Chemical reaction	A/m^3/s$^-$,m^6/s	B	C/K	Literature
1	$C + O_2 \rightarrow CO + O$	1.99×10^{-10}	0	2009.0	[25]
2	$O + CO_2 \rightarrow CO + O_2$	4.50×10^{-10}	0	33796.0	[25]
3	$CO_2 + e \rightarrow e + CO + O$	8.70×10^{-11}	0	0	[26]
4	$CO_2 + e \rightarrow O^- + CO$	3.60×10^{-13}	0	0	[26]
5	$CO_2 + e \rightarrow O^- + CO$	1.00×10^{-9}	0	0	[26]
6	$CO^+ + O \rightarrow O^+ + CO$	1.40×10^{-10}	0	0	[26]
7	$CO^+ + O_2 \rightarrow O + 2 + CO$	1.00×10^{-10}	0	0	[26]
8	$CO + CO \rightarrow C + O + CO$	7.14×10^{3}	−3.1	129000.0	[27]
9	$CO + e \rightarrow CO^+ + e + e$	6.72×10^{-10}	0.275	163076.6	[27]

Table 2.4 Parameters of chemical reactions involving C, O and N

No	Chemical reaction	A/m^3/s,m^6/s	B	C/K	Literature
1	$N + NCO \rightarrow CO + N_2$	3.32×10^{-11}	0	0	[25]
2	$N + NCO \rightarrow CN + NO$	1.66×10^{-12}	0	0	[25]
3	$CN + NO \rightarrow N_2 + CO$	4.17×10^{-14}	0	0	[25]
4	$NCO + NO \rightarrow CO_2 + N_2$	9.66×10^{-6}	−1.97	560	[25]
5	$O + CO + N_2 \rightarrow CO_2 + N_2$	6.60×10^{-33}	0	2173.0	[26]
6	$N + CO_2 \rightarrow NO + CO$	3.20×10^{-13}	0	1711.0	[26]
7	$CN + O_2 \rightarrow NCO + O$	1.20×10^{-11}	0	−210	[25]
8	$CN + O_2 \rightarrow CO + NC$	5.06×10^{-12}	0	0	[28]
9	$N^+ + O + CO_2 \rightarrow NO^+ + CO_2$	1.00×10^{-29}	0	0	[26]
10	$N^+ + CO \rightarrow CO^+ + N$	4.90×10^{-10}	0	0	[26]
11	$N^+ + CO_2 \rightarrow CO^+ + NO$	2.50×10^{-10}	0	0	[26]
12	$CO^+ + NO \rightarrow NO^+ + CO$	3.30×10^{-10}	0	0	[26]
13	$CO + N_2 \rightarrow C + O + N_2$	7.14×10^{-3}	−3.1	129000.0	[27]

The coupling of the flow governing equations bearing many features of the MHD approach with the electric and magnetic fields has been described in previous work. However, due to the unique features of the chemical non-equilibrium model, it is necessary to elaborate the details of the equations. The mass conservation equation,

Table 2.5 Parameters of chemical reactions involving H, O and N

No	Chemical reaction	A/m^3/s,m^6/s	B	C/K	Literature
1	$N + OH \rightarrow NO + H$	4.70×10^{-11}	0	0	[25]
2	$OH^+ + NO \rightarrow NO^+ + OH$	4.60×10^{-10}	0	0	[26]
3	$NH_2 + NO \rightarrow N_2 + H_2O$	2.08×10^{-11}	−1.61	150.0	[25]
4	$N^+ + OH \rightarrow NO^+ + H$	3.40×10^{-10}	0	0	[26]
5	$N^+ + OH \rightarrow OH^+ + N$	3.40×10^{-10}	0	0	[26]
6	$N^+ + H_2O \rightarrow NO^+ + H_2$	4.00×10^{-10}	0	0	[26]
7	$N^+ + O + H_2O \rightarrow NO^+ + H_2O$	1.00×10^{-29}	0	0	[26]

Table 2.6 Parameters of chemical reactions involving H and O

No	Chemical reaction	A/m^3/s,m^6/s	B	C/K	Literature
1	$H_2 + O_2 \rightarrow OH + OH$	3.16×10^{-10}	0	21890.0	[25]
2	$OH + H_2 \rightarrow H_2O + H$	4.65×10^{-19}	2.41	1240.0	[25]
3	$OH + H \rightarrow H_2 + O$	8.10×10^{-21}	2.8	1950.0	[25]
4	$OH + OH \rightarrow H_2O + O$	5.94×10^{-20}	2.4	−1060.0	[25]
5	$H + O_2 \rightarrow O + OH$	1.94×10^{-8}	−0.59	8152.0	[25]
6	$H_2O + e \rightarrow e + H + OH$	2.60×10^{-12}	0	0	[26]
7	$H_2O + e \rightarrow H^- + OH$	2.60×10^{-12}	0	0	[26]
8	$O^- + H_2 \rightarrow H_2O + e$	6.50×10^{-10}	0	0	[26]
9	$O^+ + H_2 \rightarrow OH^+ + H$	1.80×10^{-9}	0	0	[26]
10	$O^+ + OH \rightarrow OH^+ + O$	3.30×10^{-10}	0	0	[26]
11	$O^+ + OH \rightarrow O + 2 + H$	3.60×10^{-10}	0	0	[26]

momentum conservation equation and electri and magnetic field equations are similar to the traditional MHD equations, as shown in Eqs. 2.33–2.39. Mass conservation equation

$$\frac{\partial \rho}{\partial t} + \nabla \cdot \left(\vec{u} \rho\right) = 0 \tag{2.33}$$

Momentum conservation equation

$$\frac{\partial}{\partial t}\left(\rho \vec{u}\right) + \nabla \cdot \left(\rho \vec{u}\,\vec{u}\right) = -\nabla p + \nabla \cdot \left(\vec{\tau}\right) + \vec{j} \times \vec{B} \tag{2.34}$$

Electric and magnetic field equations

$$\vec{j} = \sigma \vec{E} = -\sigma \nabla \varphi = -\sigma \left(\frac{\partial \varphi}{\partial r}\vec{e}_r + \frac{\partial \varphi}{\partial z}\vec{e}_z\right) \tag{2.35}$$

Table 2.7 Parameters of chemical reactions involving H, O and C

No	Chemical reaction	A/m^3/s,m^6/s	B	C/K	Literature
1	$CH + O_2 \rightarrow CO + OH$	8.30×10^{-11}	0	0	[25]
2	$CH_2 + O \rightarrow CH + OH$	4.98×10^{-10}	0	6000.0	[25]
3	$CH_2 + O \rightarrow CO + H$	6.64×10^{-11}	0	0	[25]
4	$CH_2 + O \rightarrow HCO + H$	5.01×10^{-11}	0	0	[25]
5	$CH_2 + O_2 \rightarrow CO + H_2O$	4.00×10^{-13}	0	0	[25]
6	$CH_2 + O_2 \rightarrow H + H + CO$	8.30×10^{-14}	0	0	[25]
7	$CH_2 + O_2 \rightarrow CH_2O + O$	6.64×10^{-14}	0	0	[25]
8	$CH_2 + OH \rightarrow CH_2O + H$	3.01×10^{-11}	0	0	[25]
9	$CH_2 + H_2O \rightarrow CH_3 + OH$	1.60×10^{-16}	0	0	[25]
10	$CH_2 + CO_2 \rightarrow CH_2O + CO$	1.40×10^{-14}	0	0	[25]
11	$HCO + H \rightarrow CO + H_2$	1.83×10^{-10}	0	0	[25]
12	$HCO + HCO \rightarrow CH_2O + CO$	4.48×10^{-11}	0	0	[25]
13	$HCO + OH \rightarrow H_2O + CO$	1.69×10^{-10}	0	0	[25]
14	$HCO + CH2 \rightarrow CH_3 + CO$	3.01×10^{-11}	0	0	[25]
15	$HCO + CH_3 \rightarrow CH_4 + CO$	2.01×10^{-10}	0	0	[25]
16	$CH_3 + O_2 \rightarrow CH_2O + OH$	3.11×10^{-13}	0	4953.0	[25]
17	$CH_3 + OH \rightarrow CH_2O + H_2$	5.30×10^{-12}	−0.53	5440.0	[25]
18	$O + CH \rightarrow H + CO$	1.02×10^{-10}	0	914.0	[25]
19	$O + CH \rightarrow C + OH$	2.52×10^{-11}	0	2380.0	[25]
20	$CH_4 + O \rightarrow CH_3 + OH$	1.15×10^{-15}	1.56	4270.0	[25]
21	$CH_4 + OH \rightarrow CH_3 + H_2O$	1.68×10^{-18}	2.18	1232.0	[25]
22	$HCO + H2O \rightarrow CH_2O + OH$	3.90×10^{-16}	1.35	13110.0	[25]
23	$HCO + H2 \rightarrow CH_2O + H$	3.00×10^{-18}	2.0	8970.0	[25]
24	$HCO + CH_4 \rightarrow CH_2O + CH_3$	1.21×10^{-20}	2.85	11300.0	[25]
25	$HCO + H \rightarrow O + CH_2$	6.61×10^{-11}	0	51597.0	[25]
26	$CO_2 + H \rightarrow CO + OH$	3.32×10^{-10}	0	13470.0	[25]
27	$CO^+ + OH \rightarrow OH^+ + CO$	2.50×10^{-10}	0	0	[26]

Table 2.8 Parameters related to Cu-related chemical reactions

No	Chemical reaction	A/m^3/s,m^6/s	B	C/K	Literature
1	$Cu_2 + N_2 \rightarrow Cu + Cu + N_2$	8.18×10^{-13}	0.31	24270.0	[29]
2	$CuH + N_2 \rightarrow Cu + H + N_2$	9.19×10^{-13}	0.34	31793.0	[29]
3	$CuO + N_2 \rightarrow Cu + O + N_2$	8.04×10^{-13}	0.34	32939.0	[29]

Table 2.9 Parameters of chemical reactions involving H and C

No	Chemical reaction	A/m^3/s,m^6/s	B	C/K	Literature
1	$C + H_2 \rightarrow CH + H$	6.64×10^{-10}	0	11700.0	[25]
2	$CH_2 + CH_2 \rightarrow CH_3 + CH$	3.99×10^{-10}	0	5000.0	[25]
3	$CH_2 + H_2 \rightarrow CH_3 + H$	6.910×10^{-15}	0	0	[25]
4	$CH_2 + CH_4 \rightarrow CH_3 + CH_3$	3.01×10^{-19}	0	0	[25]
5	$H + CH_4 \rightarrow H_2 + CH_3$	7.49×10^{-19}	2.59	5057.0	[25]
6	$CH_3 + CH \rightarrow C_2H_2 + H_2$	2.69×10^{-24}	2.2	5800.0	[30]
7	$CH_3 + CH \rightarrow C_2H_2 + H_2$	2.99×10^{-11}	0	0	[30]
8	$CH_3 + C \rightarrow C_2H_2 + H$	1.33×10^{-11}	0	0	[30]
9	$CH_3 + C \rightarrow CH_2 + CH$	6.64×10^{-10}	0	20146.0	[30]
10	$CH_3 + C \rightarrow C_2H + H_2$	3.65×10^{-11}	0	0	[30]
11	$CH_2 + CH \rightarrow C_2H_2 + H$	1.16×10^{-10}	0	0	[30]
12	$CH_3 + H \rightarrow CH_2 + H_2$	3.32×10^{-10}	0	7605.0	[30]
13	$CH_2 + C \rightarrow CH + CH$	2.69×10^{-12}	0	23573.0	[25]
14	$CH_2 + C \rightarrow C_2H + H$	3.32×10^{-10}	0	0	[30]
15	$CH_2 + C \rightarrow C_2 + H_2$	3.32×10^{-10}	0	0	[30]
16	$CH_2 + H \rightarrow CH + H_2$	1.00×10^{-11}	0	−900.0	[25]
17	$CH + CH \rightarrow C_2H + H$	2.66×10^{-10}	0	0	[30]
18	$CH + CH \rightarrow C_2 + H_2$	9.96×10^{-12}	0	0	[30]
19	$CH + C \rightarrow C_2 + H$	4.15×10^{-10}	0	0	[30]
20	$CH + H \rightarrow H_2 + C$	1.31×10^{-10}	0	81.0	[25]
21	$CH_4 + CH_3 \rightarrow CH_3 + H + CH_3$	5.00×10^{-8}	0	40500.0	[31]
22	$CH_4 + H_2 \rightarrow CH_3 + H + H_2$	5.00×10^{-8}	0	40500.0	[31]
23	$CH_4 + CH_4 \rightarrow CH_3 + H + CH_4$	5.00×10^{-8}	0	40500.0	[31]
24	$CH_4 + H \rightarrow CH_3 + H_2$	1.10×10^{-15}	1.62	5460.0	[31]

$$\nabla \cdot \vec{j} = 0 \tag{2.36}$$

$$\nabla^2 A_r = \mu_0 j_r - \frac{A_r}{r^2} \tag{2.37}$$

$$\nabla^2 A_z = \mu_0 j_z \tag{2.38}$$

$$\overrightarrow{B} = \nabla \times \overrightarrow{A} \tag{2.39}$$

where τ stands for a stress tensor, $\overrightarrow{B}$ is used for the magnetic flux density vector, $\overrightarrow{j}$ for the current density vector, $\overrightarrow{E}$ for the electric field vector, φ for potential, σ for electrical conductivity, and μ_0 for permeability of vacuum, which has a value of

1. A_r and A_z are the components of the magnetic potential vector in the radial and axial directions, respectively.

The energy equation in the chemical non-equilibrium model needs to be modified, as shown in Eqs. 2.40–2.43.

$$\frac{\partial}{\partial t}(\rho U) + \nabla \cdot (\vec{v}(\rho U + p)) = \nabla \cdot \left(k \nabla T + (\vec{\tau} \cdot \vec{v}) + \sum_{j}^{N} \rho D_j h_j \nabla Y_j \right) - P_{rad} + \sigma U^2 \quad (2.40)$$

$$U = h - \frac{p}{\rho} + \frac{u^2}{2} \quad (2.41)$$

$$h_j = \frac{1}{m_j} \left(\frac{5}{2} \kappa T + \kappa T^2 \frac{\partial \ln Z_j}{\partial T} + \Delta H_{fj} \right) \quad (2.42)$$

$$h = \sum_j Y_j h_j \quad (2.43)$$

where k represents the translational thermal conductivity, h_j for enthalpy of species j, P_{rad} for the radiative energy source. ΔH_{fj} stands for the standard formation enthalpy of species j.

3. Input of material properties in chemical non-equilibrium modelling

Figure 2.9 shows the flow chart of the chemical non-equilibrium modelling. The specific process is as follows:

(1) As the first step of the calculation, the arc region is to be initialised. The initialised data include flow field, temperature field, particle number density, and relevant material properties. For the analysis of the decaying process of the current zero phase, the initial conditions are the calculation results of the preceding arcing period.
(2) The solution process follows the initialisation of relevant quantities. In each iteration of a time step, all conservation equations relating to the flow are first solved, including the above equations from (2.23)–(2.33).
(3) After obtaining the temperature, velocity and pressure fields, the species transport Eqs. 2.14–2.18 are solved to obtain the concentrations or densities of the species with chemical non-equilibrium, convection and diffusion taken into account.
(4) Based on the updated non-equilibrium composition of the plasma, the thermodynamic properties and transport coefficients in the plasma region are updated in preparation for the next iteration in the solution process.
(5) Repeat step (2) until solution in a time steps converges before moving to the next step.

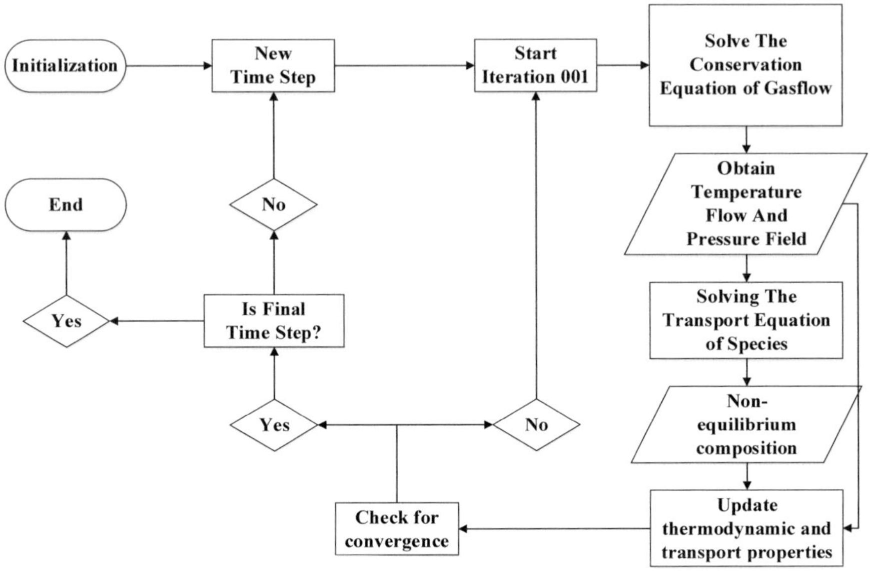

Fig. 2.9 Calculation flow chart of chemical non-equilibrium model

The steps shown inside the red dotted box of Fig. 2.9 are the fundamental differences between the chemical non-equilibrium model and the equilibrium model: in thermodynamic equilibrium models [33–42] and two-temperature models [2, 3], the updating of material properties in each iteration loop is obtained by interpolating the pre-calculated data table with the most recent temperature, velocity and pressure fields; In the chemical non-equilibrium model, the transport equation of each species is required to be solved. The calculation work is multiple times that of the equilibrium model and two-temperature model. Secondly, the material properties need to be calculated and updated in each iteration, requiring a huge amount of calculation, which renders the solution of the chemical non-equilibrium model a prohibitively costly and difficult talk for switching arcs.

2.3.4 *Exemplary Calculation*

The calculation model, discharge conditions and boundary conditions in this example are shown in Fig. 2.10, where the arcing medium is air. Free recovery is assumed in the post-arc phase.

During the actual breaking of a circuit breaker, the recovery voltage is immediately applied to the contact gap after the current reaches its final zero point. The specific structure, a simple nozzle model, is shown in Fig. 2.10. The model is completely axisymmetric and a cylindrical polar coordinate system is used in the calculation.

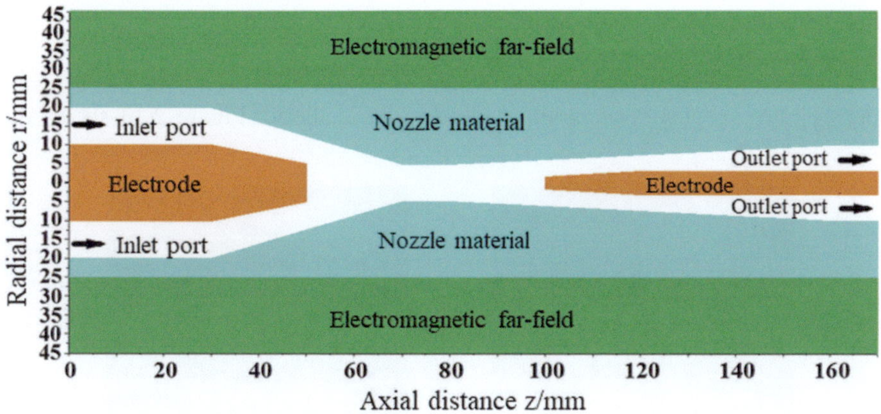

Fig. 2.10 Model diagram of calculation region

The solid region in the model includes two electrodes and the nozzle, where the electrode material is copper-tungsten alloy and the nozzle material is PTFE, both are also used in actual high voltage circuit breakers. The left side of the fluid area is the gas inlet and the right side is the gas outlet. In addition, the area marked in green is the far-field domain required for electromagnetic field calculation. In the arcing process, an initial DC current of 100 A is applied, and the current then decreases to zero during the current zero period.

In the simulation, it is necessary to define a current conducting path in the gas to initialise the arc. In existing work, the conducting column is defined by a hot gas column with high temperature, such as 15,000 K or above. This approach removes the difficulty in the solution process to breakdown the cold gas to form an arc. For chemical non-equilibrium models, there is additional issues. When the concentrations of species are given distinct values in in the conducting column and the surrounding insulating region, the species concentration gradient across the two regions will be too large to maintain solution stability. Experience shows that initial conditions can be defined according to the following equation at the beginning of the solution process:

$$T = 15000 \times \exp(-790r) \tag{2.44}$$

where r is the radial distance from the axis of symmetry.

Figure 2.11 shows the temperature distribution at different times in the current zero phase for two different calculation models. It can be seen from the figure that at the times $t < 50$ μs after current zero, the conducting zone above 8,000 K disappears rapidly. This is mainly due to the radiative dissipation of arc energy in the area where temperature is above 8,000 K. During this period, especially with the temperature field at $t = 20$ μs, it can be seen that the decay of the high temperature zone in the chemical non-equilibrium model lags behind the equilibrium model. For the period of $t > 50$μs, the rate of temperature decay decreases. By comparing the two models, it is shown that the arc column shrinks faster in the region of $z = 50$–70 mm in the chemical non-equilibrium model (Fig. 2.12).

In order to better explain this difference in temperature decay, the mechanisms rooting from species concentrations need to be analysed. Figure 2.13 shows the

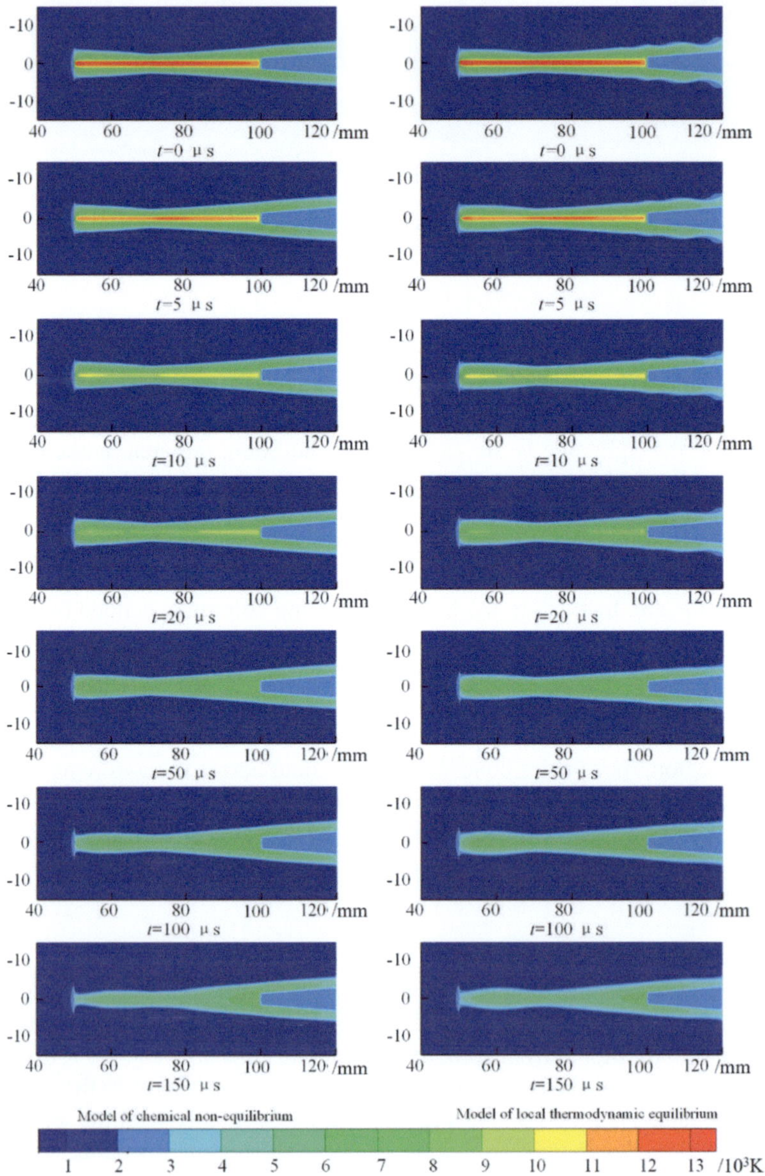

Fig. 2.11 Temperature distribution of attenuation process in the zero current phase of arc in the two models

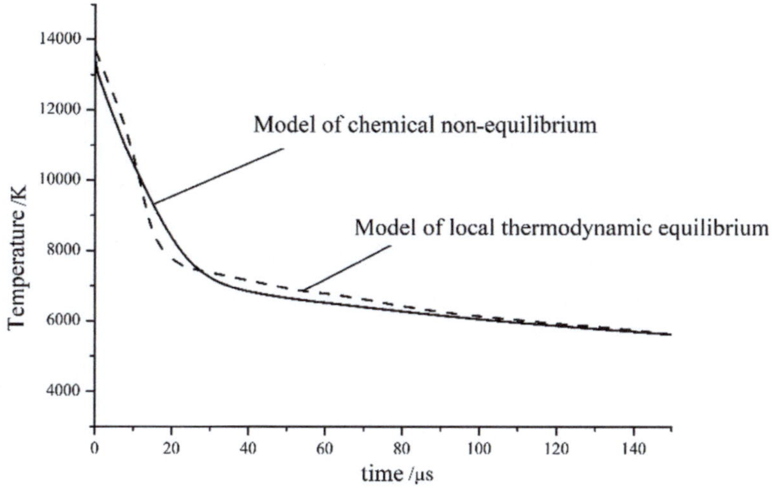

Fig. 2.12 Attenuation of temperature with time at $r = 0$ mm and $z = 80$ mm

trend of species number density change with time at $r = 0$ mm and $z = 80$ mm. In the first 15 μs after arc current, the variation of charged species is obviously different from the neutral species. The reduction in the number density of the charged species due to chemical non-equilibrium lags that in the LTE model. Meanwhile the combination reaction rate associated with the loss of charged species is still higher in the chemical non-equilibrium case. The combination reaction is an exothermic reaction, so the temperature decay rate in this period is lower than that in the LTE model. In the period of $t > 15$ μs, the exothermic chemical reaction of atoms starts to dominate, causing the temperature attenuation of the chemical non-equilibrium model to accelerate in this period.

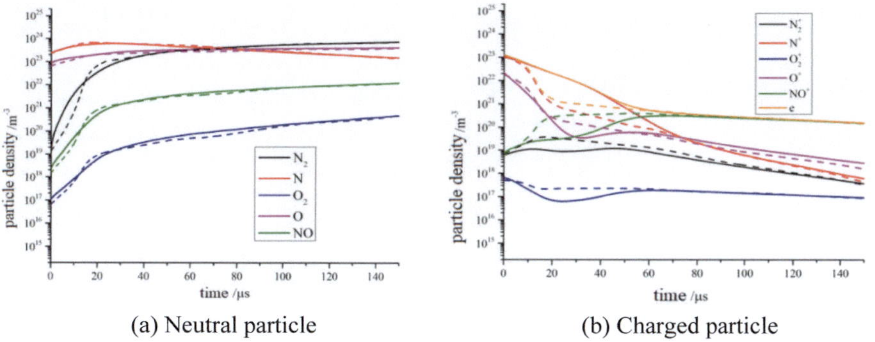

Fig. 2.13 Variation trend of particle number density with time at $r = 0$ mm and $z = 80$ mm (solid line: chemical non-equilibrium model; dotted line: LTE model)

It is evident that the change in the heat released from the decaying exothermic reactions in the chemical non-equilibrium model is the decisive factor in producing the difference in the arc temperature after current zero.

For the post-arc phase of air arc, the key chemical reactions that determine the plasma composition are summarized in Table 2.10. It is worth noting that in the air plasma, the formation or consumption of nearly all species are affected by reactions involving N and O.

Table 2.10 Key chemical reactions in N_2 and O_2 arc plamas

Types of particles	Generation	Consumption
N_2	$NO + N \rightarrow N_2 + O$	$N_2 + N^+ \rightarrow N_2^+ + N$
		$N_2 + e \rightarrow N + N + e$
N	$N^+ + e + e \rightarrow N + e$	$N + NO \rightarrow N_2 + O$
	$N_2 + e \rightarrow N + N + e$	$N + O + O \rightarrow NO + O$
	$e + N_2^+ \rightarrow N + N$	$N + O + N \rightarrow NO + N$
O_2	$O + O + N \rightarrow O_2 + N$	$N + O_2 \rightarrow NO + O$
	$O + O + O \rightarrow O_2 + O$	$N^+ + O_2 \rightarrow O_2^+ + N$
	$O + O + N_2 \rightarrow O_2 + N_2$	
O	$NO + N \rightarrow N + O + N$	$N + O + N \rightarrow NO + N$
	$NO + O \rightarrow N + O + O$	$N + O + O \rightarrow NO + O$
	$NO + N \rightarrow N_2 + O$	$N_2 + O \rightarrow NO + N$
	$O^+ + e + e \rightarrow O + e$	$O + e \rightarrow O^+ + e + e$
NO	$N + O + N \rightarrow NO + N$	$NO + N \rightarrow N + O + N$
	$N + O + O \rightarrow NO + O$	$NO + O \rightarrow N + O + O$
	$N_2 + O \rightarrow NO + N$	$NO + N \rightarrow N_2 + O$
N^+		$N^+ + e + e \rightarrow N + e$
		$N_2 + N^+ \rightarrow N_2^+ + N$
O^+		$O^+ + e + e \rightarrow O + e$
		$N_2 + O^+ \rightarrow N_2^+ + O$
NO^+	$NO^+ + N \rightarrow N_2^+ + O$	$NO^+ + e \rightarrow N + O$
	$N + O \rightarrow NO^+ + e$	
e	$N + O \rightarrow NO^+ + e$	$N^+ + e + e \rightarrow N + e$
		$O^+ + e + e \rightarrow O + e$
		$NO^+ + e \rightarrow N + O$
		$e + N_2^+ \rightarrow N + N$

2.4 Non-Maxwellian Distribution of Electron Energy

2.4.1 Causes of Electron Energy Deviation from Maxwellian Distribution

In an arc system without external electric field, it is generally believed that the electron energy distribution in plasma follows the Maxwellian distribution. However, in the post-arc phase, the recovery voltage accelerates the free electrons, causing a deviation of their energy distribution function from the Maxwellian distribution. The change of the electron energy distribution function will further affect the generation and loss of electrons in the plasma, for example the ionisation and attachment frequencies.

1. Calculation of electron energy distribution function

(1) Boltzmann analysis

The electron energy distribution function is obtained by the Boltzmann equation of 0-D space through the second-order spherical harmonic approximation. The algorithm based on this method is solved by ELENDIF code or Bolsig + software.

The specific equation in the Boltzmann analysis is:

$$\frac{\partial n(\varepsilon)}{\partial t}=-\frac{\partial J_f}{\partial \varepsilon}-\frac{\partial J_{el}}{\partial \varepsilon}+\frac{\partial J_{inel}}{\partial \varepsilon}+\frac{\partial J_{ion}}{\partial \varepsilon}+I_{att}+\left(\frac{\partial n}{\partial t}\right)_{e-e} \tag{2.45}$$

The above equation represents Boltzmann's transport equation. The left side represents the rate of change with time of the number density of electrons corresponding to a particular energy value. The right side represents the effect of processes relating to electron-impact. The specific meanings are as follows:

$$J_f=\frac{2Ne^2(E/N)^2\varepsilon}{3m_e(2\varepsilon/m_e)^{1/2}\sum_s x_s\sigma_s^m(\varepsilon)}\left(\frac{n}{2\varepsilon}-\frac{\partial n}{\partial \varepsilon}\right) \tag{2.46}$$

where J_f is the rate of change of electron number density corresponding to the particular energy due to the acceleration of the electrons by the externally applied electric field. N denotes the total number density of all local particles, and E/N is the reduced electric field strength, in units of Td (1Td = 1 × 10^{-21}V·m$^{2)}$. x_s and $\sigma_s{}^m$ are, respective, the mole fraction and momentum transfer collision cross section with electrons of the s-th species.

$$J_{el}=-N\left(\frac{2\varepsilon}{m_e}\right)^{1/2}\sum_s x_s\frac{2m_e}{M_s}\sigma_s^m(\varepsilon)\left[n\left(\frac{\kappa T_h}{2}-\varepsilon\right)-\kappa T_h\varepsilon\frac{\partial n}{\partial \varepsilon}\right] \tag{2.47}$$

where J_{el} is the effect of change due to the elastic collisions between electrons and heavy particles. M_s is the molar mass of heavy particle species s, and T_h is the local temperature of heavy particles.

$$J_{inel} = \left(\frac{2}{m_e}\right)^{1/2} \sum_s \sum_j N_s^0 \int_{\varepsilon}^{\varepsilon+\varepsilon_{s,j}^{ex}} \varepsilon^{1/2} \sigma_{s,j}^{ex}(\varepsilon) n(\varepsilon) d\varepsilon \tag{2.48}$$

where J_{inel} is the consequence of the inelastic collisions between electrons and heavy particles. In the calculations shown in this book, only excitation is considered for the inelastic collisions between electrons and atoms. Three processes are included in the inelastic collisions between electrons and diatomic molecules, i.e. rotational excitation, vibrational excitation and electronic excitation; $\sigma_{s,j}{}^{ex}$ is the collision cross section of the *j*-th excited state of the *s*-th species, and $\varepsilon_{s,j}{}^{ex}$ is the threshold energy of the *j*-th excited state of the *s*-th species.

$$J_{ion} = \left(\frac{2}{m_e}\right)^{1/2} \sum_s N_s^0 \int_{\varepsilon}^{\varepsilon+\varepsilon_s^i} \varepsilon^{1/2} \sigma_s^i(\varepsilon) n(\varepsilon) d\varepsilon \tag{2.49}$$

where J_{ion} is the contribution of electron impact ionisation with heavy particle towards the change of the electron energy distribution function., $\sigma_s{}^i$ is the ionisation collision cross section of the *s*-th species, and $\varepsilon_s{}^i$ is the ionisation threshold energy of the *s*-th species.

$$I_{att} = -\left(\frac{2}{m_e}\right)^{1/2} \sum_s N_s^0 \varepsilon^{1/2} \sigma_s^a(\varepsilon) n(\varepsilon) \tag{2.50}$$

where I_{att} is the rate of change due to electron attachment collisions with heavy particles. $\sigma_s{}^a$ is the attachment collision cross section of the *s*-th species.

$$\left(\frac{\partial n}{\partial t}\right)_{e-e} = \frac{2\pi e^4}{3m_e} \ln\left(\frac{3\kappa T_e}{2e^2}\sqrt{\frac{\kappa T_e}{4\pi n_e e^2}}\right)\left[\frac{3}{\varepsilon^{1/2}} n(\varepsilon)^2 - 2\varepsilon^{3/2}\frac{\partial \varphi}{\partial \varepsilon}\frac{\partial \beta}{\partial \varepsilon} - \frac{\varphi}{\varepsilon^{1/2}}\beta\right] \tag{2.51}$$

$$\varphi = 3\int_0^{\varepsilon} n(\varepsilon')d\varepsilon' - \frac{1}{\varepsilon}\int_0^{\varepsilon} \varepsilon' n(\varepsilon')d\varepsilon' + 2\varepsilon^{1/2}\int_{\varepsilon}^{\infty} \frac{n(\varepsilon')}{\varepsilon'^{1/2}} d\varepsilon' \tag{2.52}$$

$$\beta = \frac{n(\varepsilon)}{2\varepsilon} - \frac{\partial n(\varepsilon)}{\partial \varepsilon} \tag{2.53}$$

$$T_e = \frac{2}{3\kappa}\int_0^{\infty} f_0(\varepsilon)\varepsilon^{3/2} d\varepsilon \tag{2.54}$$

The four Eqs. 2.51–2.54 above represent the influence of the interaction between electron themselves on the electron energy distribution function.

Finally through the transformation process of (2.55) and (2.56), the obtained $f_0(\varepsilon)$ is the expression of the electron energy distribution function.

$$n(\varepsilon) = n_e f_0(\varepsilon)\varepsilon^{1/2} \tag{2.55}$$

$$\int_0^\infty f_0(\varepsilon)\varepsilon^{1/2} d\varepsilon = 1 \tag{2.56}$$

(2) Collision cross sections

The basic data to characterise the interaction between electrons and heavy particles are the collision sections, which are also essential to calculate the electron energy distribution function. The collision cross-sections mainly include the following categories: momentum transfer collision cross-section, rotation, vibration and electronic excitation collision cross-sections, ionisation collision cross-section and attachment/recombination collision cross-sections. There are collision cross section data from different sources and they differ considerably. Large discrepancies will lead to inaccurate calculation of the electron energy distribution function, so it is necessary to verify the suitability of the data before performing the calculation. Pitchford proposed the "Plasma Data Exchange Project" [75], which aims at the determination of the optimal collision cross section data through comparison of data from different sources and comparison of calculated data with experimentally obtained results. A library of cross section data is compiled in this project, also referred as the Laplace laboratory database [76].

(3) Calculation of ionisation and attachment coefficients

In the calculation of the critical breakdown field strength, it is common to assume that ionisation controls the generation of electrons and attachment dominates the loss of electrons. Therefore, obtaining the reaction constants of the two processes is an important step in the analysis of the critical breakdown field strength, requiring the use of the relationship:

$$\frac{\alpha}{N} = \frac{1}{v_d}\sum_s \left(\frac{2}{m_e}\right)^2 \int_0^\infty x_s \sigma_s^i(\varepsilon) f_0(\varepsilon)\varepsilon d\varepsilon \tag{2.57}$$

$$\frac{\eta}{N} = \frac{1}{v_d}\sum_s \left(\frac{2}{m_e}\right)^2 \int_0^\infty x_s \sigma_s^a(\varepsilon) f_0(\varepsilon)\varepsilon d\varepsilon \tag{2.58}$$

$$v_d = -\left(\frac{2}{m_e}\right)^{1/2}\left(\frac{eE}{3N}\right)\int_0^\infty \frac{1}{\sum_s x_s \sigma_s^m(\varepsilon)} \frac{df_0(\varepsilon)}{d\varepsilon}\varepsilon d\varepsilon \tag{2.59}$$

2.5 Summary

The existing studies on the current zero phase of the switching process are reviewed in this chapter. Attention has been paid to the effects of thermal and chemical non-equilibria and the calculation of critical breakdown field strength of gases at elevated temperature. A significant amount of simulation and experimental work has been conducted to understand the behaviour of the discharge in the current zero phase and to establish relevant theoretical models in a rigorous manner. The work provides important reference knowledge for switching arc research and also theoretical support to the design of circuit breakers with enhanced breaking capacity. For example, the close agreement of measured electron number density with prediction based on the non-equilibrium hypothesis demonstrates the significance of the latter as the theoretical basis for more accurate prediction of the electron density decay. Further improvement towards the practical application of the non-equilibrium hypothesis is still needed in the following aspects:

1. More knowledge is needed on the mechanisms important to the non-equilibrium plasma in the post arc phase, which enables the derivation of more accurate governing equations. A typical example is the calculation of the radiation emitted by electrons and heavy particles.
2. The calculation is computationally costly in comparison with the LTE model. For example, only 4 partial differential equations need to be solved in an axisymmetric case (2D) for the plasma flow in equilibrium. While for the non-equilibrium case tens of chemical species equations need to be solved. This demands much more efficient numerical algorithms as well as powerful parallel computing facilities and strategies, such as the use of Graphics Processing Unit (GPU) accelerated computation.
3. More reliable experimental results are needed to verify the theoretical hypothesis. Issues to be addressed include the accuracy of the measurement that is related to both equipment and theory, as well as the control of the discharge conditions to keep the target plasma stable. Therefore more experimental effort is still needed.

References

1. Ma Z Y. Zero current phase phenomena of AC arc and its analysis [J]. Journal of Xi'an Jiaotong University, 1983, 17(3): 71-78.
2. Girard R, Gonzalez J J, Gleizes A. Modelling of a two-temperature SF6 arc plasma during extinction [J]. Journal of Physics D: Applied Physics, 1999, 32(11): 1229.
3. Hsu K C, Pfender E. Two-temperature modeling of the free-burning, high-intensity arc [J]. Journal of Applied Physics, 1983, 54(8): 4359-4366.
4. Gonzalez J, Girard R, Gleizes A. Decay and post-arc phases of a SF6 arc plasma: a thermal and chemical non-equilibrium model [J]. Journal of Physics D: Applied Physics, 2000, 33(21): 2759-2768.

5. Wang W, Yan J D, Rong M, et al. Theoretical investigation of the decay of a SF6 gas-blast arc using a two-temperature hydrodynamic model [J]. Journal of Physics D: Applied Physics, 2013, 46(6): 065203.
6. Freton P, Gonzalez J, Ranarijaona Z, et al. Energy equation formulations for two-temperature modelling of "thermal" plasmas [J]. Journal of Physics D: Applied Physics, 2012, 45(46): 465206.
7. Colombo V, Ghedini E, Boselli M, et al. 3D static and time-dependent modelling of a DC transferred arc twin torch system [J]. Journal of Physics D: Applied Physics, 2011, 44(19): 194005.
8. Wang W Z, Rong M Z, Wu Y, et al. Thermodynamic and transport properties of two-temperature SF6 plasmas [J]. Physics of Plasmas, 2012, 19(8): 083506.
9. Trelles J P, Chazelas C, Vardelle A, et al. Arc plasma torch modeling [J]. Journal of Thermal Spray Technology, 2009, 18(5-6): 728-752.
10. Ghorui S, Heberlein J V, Pfender E. Thermodynamic and transport properties of two-temperature oxygen plasmas [J]. Plasma Chemistry and Plasma Processing, 2007, 27(3): 267-291.
11. Murphy A B Thermal plasmas in gas mixtures [J]. Journal of Physics D: Applied Physics, 2001, 34(20): R151–R173.
12. Gonzalez J J, Girard R, Gleizes A. Decay and post-arc phases of a SF6 arc plasma: a thermal and chemical non-equilibrium model[J]. Journal of Physics D: Applied Physics, 2000. 33(21): 2759-2768.
13. Teulet P, Gonzalez J J, Mercado-Cabrera A, et al. One-dimensional hydro-kinetic modelling of the decaying arc in air–PA66–copper mixtures: I. Chemical kinetics, thermodynamics, transport and radiative properties[J]. Journal of Physics D: Applied Physics, 2009. 42(17): 175201.
14. Teulet P, Gonzalez J J, Mercado-Cabrera A, et al. One-dimensional hydro-kinetic modelling of the decaying arc in air–PA66–copper mixtures: II. Study of the interruption ability[J]. Journal of Physics D: Applied Physics, 2009. 42(18): 185207.
15. Gleizes A, Gonzalez J J, Freton P. Thermal plasma modelling[J]. Journal of Physics D: Applied Physics, 2005. 38(9): R153–R183.
16. Tanaka Y, Michishita T, Uesugi Y. Hydrodynamic chemical non-equilibrium model of a pulsed arc discharge in dry air at atmospheric pressure[J]. Plasma Sources Science and Technology, 2005. 14(1): 134–151.
17. Tanaka Y, Suzuki K. Development of a chemically nonequilibrium model on decaying SF6 arc plasmas[J]. IEEE Transactions on Power Delivery, 2013. 28(4): 2622-2629.
18. Norton D G, Vlachos D G. Combustion characteristics and flame stability at the microscale: a CFD study of premixed methane/air mixtures[J]. Chemical Engineering Science, 2003. 58(21): 4871-4882.
19. Tanaka Y. Time-dependent two-temperature chemically non-equilibrium modelling of high-power Ar–N2 pulse-modulated inductively coupled plasmas at atmospheric pressure[J]. Journal of Physics D: Applied Physics, 2006. 39(2): 307–319.
20. Tanaka Y. Two-temperature chemically non-equilibrium modelling of high-power Ar–N2 inductively coupled plasmas at atmospheric pressure[J]. Journal of Physics D: Applied Physics, 2004. 37(8): 1190–1205.
21. Park C. Nonequilibrium Hypersonic Aerothermodynamics[M]. New York: Wiley, 1990.
22. Phelps A V. Electron impact cross section[DB]. JILA Los Alamos: ftp://jila.colorado.edu/. 1994.
23. Morgan W L, Penetrante B M. ELENDIF: A time-dependent Boltzmann solver for partially ionized plasmas[J]. Computer Physics Communications, 1990. 58(1-2): 127-152.
24. Brand K P, Kopainsky J. Particle densities in a decaying SF6 plasma[J]. Applied Physics, 1978. 16(4): 425-432.
25. NIST Chemical Kinetics Database version 6.0, NIST Reference Database 17 version 7.0[DB]. 2001. http://kinetics.nist.gov/kinetics/index.jsp
26. Eichwald O, Yousfi M, Hennad A, et al. Coupling of chemical kinetics, gas dynamics, and charged particle kinetics models for the analysis of NO reduction from flue gases[J]. Journal of Applied Physics, 1997. 82(10): 4781-4794.

27. Mick H J, Burmeister M, Roth P. Atomic resonance absorption spectroscopy measurements on high-temperature CO dissociation kinetics[J]. AIAA Journal, 1993. 31(4): 671-676.
28. Rim K T, Hershberger J F. Temperature Dependence of the Product Branching Ratio of the CN+ O2 Reaction[J]. The Journal of Physical Chemistry A, 1999. 103(19): 3721-3725.
29. Park C. Assessment of a two-temperature kinetic model for dissociating and weakly ionizing nitrogen[J]. Journal of Thermophysics and Heat Transfer, 1988. 2(1): 8-16.
30. Markus M W, Roth P. On the reaction of CH with CO studied in high temperature CH4/Ar+ CO mixtures[J]. International Journal of Chemical Kinetics, 1992. 24(5): 432-445.
31. Koike T, Kudo M, Maeda I, et al. Rate constants of CH4+ M CH3+ H+ M and CH3OH+ M CH3+ OH+ M over 1400–2500 K[J]. International Journal of Chemical Kinetics, 2000, 32(1): 1-6.
32. Capellos C, Bielski B H J. Kinetic systems: mathematical description of chemical kinetics in solution [M]. Wiley-Interscience, 1972.
33. Fang M T C, Lin W Y. Current zero behaviour of a gas-blast arc. Part 1: Nitrogen [J]. IEE Proceedings, 1990, 137(4): 175–183.
34. Fang M T C, Zhuang Q. Current zero behaviour of an SF6 gas-blast arc. part I: laminar flow [J]. Journal of Physics D: Applied Physics, 1992, 25(8): 1197–1204.
35. Fang M T C, Zhuang Q, Guo X J. Current-zero behaviour of an SF6 gas-blast arc. part II: turbulent flow [J]. Journal of Physics D: Applied Physics, 1994, 27(1): 74–83.
36. Franck C M, Seeger M. Application of high current and current zero simulations of high-voltage circuit breakers [J]. Contributions to Plasma Physics, 2006, 46(10): 787-797.
37. Gleizes A, Mbolidi F, Razafinimanan M, et al. Cooling and chemical kinetics in a wall-stabilized SF6 arc in extinction [J]. Journal of Physics D: Applied Physics, 1991, 24(8): 1333.
38. Schade E, Ragaller K. Dielectric recovery of an axially blown SF6-arc after current zero: part I-experimental investigations [J]. IEEE Transactions on Plasma Science, 1982, 10(3): 141-153.
39. Ragaller K, Egli W, Brand K P. Dielectric recovery of an axially blown SF6-arc after current zero: Part Il-Theoretical investigations [J]. IEEE Transactions on Plasma Science, 1982, 10(3): 154-162.
40. Brand K P, Egli W, Niemeyer L, et al. Dielectric recovery of an axially blown SF6-arc after current zero: Part III-Comparison of experiment and theory [J]. IEEE Transactions on Plasma Science, 1982, 10(3): 162-172.
41. Kline L E, Davies D K, Chen C L, et al. Dielectric properties for SF6 and SF6 mixtures predicted from basic data [J]. Journal of Applied Physics, 1979, 50(11): 6789-6796.
42. Yan J D, Fang M T C, Liu Q S. Dielectric breakdown of a residual SF6 plasma at 3000 K under diatomic equilibrium [J]. IEEE Transactions on Dielectrics and Electrical Insulation, 1997, 4(1): 114-119.

Chapter 3
Calculation of Material Properties of Arcing Gas

3.1 Introduction

Switching arc is a neutral aggregate composed of charged and neutral particles and under certain conditions can behave like a magnetic fluid with multiple physical fields coupled to each other at the macroscopic level. The research on arc plasma focuses on both the microscopic and macroscopic aspects, and an increasing amount of research has been devoted to the collision mechanisms [1]. At the macroscopic level, the description of the arc behaviour has evolved from the 1-D Elenbaas-Heller equation [2, 3] to 2-D and 3-D equations, or more precisely partial differential equations, representing the conservation laws for mass, momentum and energy [4–6]. These equations are called governing equations.

With the development of computer simulation technology, great achievements have been made in the study and prediction of arc characteristics. Obtaining the relevant material properties of the arcing gas is a prerequisite to the solution of the governing equations for the arc flow. The material properties include thermophysical properties, transport coefficients (viscosity, thermal conductivity and diffusion coefficient) and radiation characteristics. In arc modelling and simulation, the plasma state is commonly assumed to be in local thermodynamic equilibrium (LTE) state. This assumption is safe in the main arc column at sufficiently high pressure and current while deviation exists at the arc edge of high current arcs or during the current zero period of a switching process. Therefore, it is important to study and quantify the effect of non-equilibrium state. In the second chapter of this book, the non-equilibrium effect of arc has been described in detail. Arc plasma is divided into LTE plasma, non-thermodynamic equilibrium plasma and non-chemical equilibrium plasma.

M. Rong et al., *Computing Methodology for Switchgear*,
https://doi.org/10.1007/978-981-95-2840-0_3

3.2 Calculation of Material Properties for Arcing Gas in Local Thermodynamic Equilibrium

The characteristics of arc plasma are closely related to its thermodynamic state. When the temperature is very high and all particles have a uniform thermodynamic temperature, the plasma is in a thermodynamic equilibrium state, the velocity of particles follows the Maxwell–Boltzmann distribution, and the internal energy level of particles obeys the Boltzmann distribution. However, plasmas in full thermodynamic equilibrium rarely exist in natural and laboratory environments (only really occurring in the interior of cosmic stars) [1]. For arc plasma, it is generally assumed to be in LTE state. The electron temperature is very close to the temperature of heavy particles (including molecules, atoms and ions), and the energy level distribution of the particles approximately follows the Boltzmann distribution. The partition functions and the composition and thermodynamic properties of the plasma can be obtained based on the Boltzmann distribution.

3.2.1 *Partition Functions*

Partition function is an important concept in statistical physics, serving as the bridge between the microscopic physical state of the particles and the macroscopic physical properties. The partition function Q_t consists of a translational partition function and an internal partition function.

$$Q_t = Q_{tr} \cdot Q_{int} = Q_{tr} \cdot Q_{rot} \cdot Q_{vib} \cdot Q_{el} \cdot Q_{nucl} \tag{3.1}$$

where Q_{tr} is the translational partition function and Q_{int} the internal partition function. Q_{rot} is the rotation partition function, Q_{vib} the vibrational partition function Q_{el} the electronic partition function, and Q_{nucl} the kernel partition function which is usually a constant.

Unless otherwise stated, the partition functions in the calculation of the plasma composition and thermodynamic properties refer only to the internal partition functions.

1. Monoatomic species

Monoatomic species have no rotational and vibrational energy levels and their internal partition function depends only on the electronic energy levels.

$$Q_{int} = Q_{el} = \sum_{i}^{\varepsilon_{ei}<E_{ieff}} g_{ei} e^{-\frac{\varepsilon_{ei}}{kT}} \tag{3.2}$$

where ε_{ei} is the electron energy level (J), E_{ieff} the effective ionisation energy (J) of the atomic species, g_{ei} the degeneracy of level i and T the temperature (K).

In the plasma, due to the Debye shielding effect, the effective ionisation energy E_{ieff} of atoms is lower than its value, E_i, in a pure neutral gas environment. The Debye radius is a typical length scale describing the range of influence of charged particles in plasma, i.e. the plasma–charge shielding effect. When the system dimension under study is much larger than the Debye radius, the plasma as a whole can be considered as electrically neutral, otherwise, the plasma is considered charged.

$$E_{ieff} = E_i - \frac{q_e^2}{4\pi \varepsilon_0} \frac{Z+1}{\lambda_D} \tag{3.3}$$

where q_e is the electronic charge (1.6×10^{-19}C), ε_0 the permitivity in vacuum (/8.854×10^{-12} F·m^{-1}), Z the charge of species for the partition function calculation, and λ_D the Debye radius (m).

2. Diatomic species

For diatomic species, the partition function consists of those of the electronic, vibrational and rotational energy levels. It can be calculated by the molar potential minimization method [7]. For polyatomic molecules, the calculation method is introduced in next section.

$$\begin{aligned} Q_{int} &= Q_{el} \cdot Q_{vib} \cdot Q_{rot} \\ &= \frac{1}{\sigma_c} \sum_j g_{e,j} \exp(-\varepsilon_j/kT) \cdot \sum_v^{v_{max}} \exp(-\varepsilon_{vib}(n, v)/kT) \cdot \\ &\quad \sum_J^{J_{max}} (2J+1) \exp(-\varepsilon_{rot}(n, v)/kT) \end{aligned} \tag{3.4}$$

$$\frac{\varepsilon_{vib}(n, v)}{hc} = \omega_e(v + \frac{1}{2}) - \omega_e x_e (v + \frac{1}{2})^2 + \omega_e y_e (v + \frac{1}{2})^3 + \omega_e z_e (v + \frac{1}{2})^4 \tag{3.5}$$

$$\frac{\varepsilon_{rot}(n, v, J)}{hc} = B_v J(J+1) - D_v (J(J+1))^2 \tag{3.6}$$

$$B_v = B_e - \alpha_e (v + 1/2) \tag{3.7}$$

$$D_v = D_e - \beta_e (v + 1/2) \tag{3.8}$$

where k is Boltzmann constant (J·K^{-1}), $\varepsilon_{vib}(n, v)$ the energy level (J) of the *v-th* order vibration and the *n-th* order electronic state, $\varepsilon_{rot}(n, v, J)$ the energy level (J) of the *J*-th order rotation, the *v-th* order vibration, and the *n-th* order electronic state, σ_c the symmetric constant which takes the value of 2 and 1 for homonuclear and heteronuclear molecules, respectively, *v* the vibrational quantum number, and *J* the rotational quantum number. The parameters B_v, B_e, α_e, D_v, D_e, β_e, ω_e, $\omega_e x_e$, $\omega_e y_e$, and $\omega_e z_e$ are all constants and the method of obtaining these data can be found in [7].

The maximum vibrational quantum number $J_{\max}$ used for the calculation in this book was calculated by the method of Drellish et al. [7]. When the equilibrium radius r_e (the radius where the potential energy reaches the minimum value), the potential well depth $D_{e'}$, the spectral constant B_e, and the *n-th* order electronic state energy level ω_e are known, the potential energy of the unrotating diatomic molecule ($J = 0$) is written as a function of the distance between the nuclei, *r*, by the molar potential approach.

$$U_0(r) = D_{e'}(1 - e^{-2\beta\xi}) \tag{3.9}$$

$$\xi = (r - r_e)/r_e \tag{3.10}$$

$$\beta = \frac{\omega_e}{4\sqrt{B_e D_{e'}}} \tag{3.11}$$

When $J \neq 0$, the potential energy can be calculated from the molar potential and centrifugal potential.

$$U_J(r) = D_e'(1 - e^{-2\beta\xi}) + \frac{h}{8\pi^2 c\mu r^2} J(J + 1) \tag{3.12}$$

where $U_J(r)$ is the potential function of species (J), h the Planck constant (J·s), and μ the reduced mass of the diatomic molecules (kg), c is the speed of light.

Given *n* and *v*, different potential energy curves can be obtained by changing the value of *J*. When $0 < J < J_{\max}$, these potential energy curves have a maximum value in the region $r > r_e$. When $J = 0$, the molar potential curve doesn't have a maximum value. When $J > J_{\max}$, the potential energy curves have no minimum value for stable states. For the given *n* and *v*, $J_{\max}$ is the minimum value when the vibrational energy is greater than the maximum value of the potential energy. The radius r_m at the maximum potential energy is calculated by taking the derivative of Eq. 3.12 with respect to the radius *r* and set it to zero. Vibrational rotational energy can be obtained by summing up all vibrational and rotational energy levels.

3. Polyatomic species

For polyatomic species, due to the lack of spectral data, the partition function is usually approximated using the theory of rigid rotational harmonic oscillators [7].

$$Q_{int} = \sum_j g_{e,j} \exp(-\varepsilon_j/kT) \prod_i [\frac{\exp(-\omega_i/2kT)}{1 - \exp(-\omega_i/kT)}]^{g_i} (T/\theta_{rot})^{3/2} \tag{3.13}$$

where g_i is the degeneracy of vibration, i.e. degeneracy, ω_i the basic vibration frequency (Hz), and θ_{rot} the rotational constant.

The rotational constant of polyatomic species is usually calculated by three moments of inertia I_A, I_B and I_C, with units of kg.m^2.

$$\left(\frac{T}{\theta_{rot}}\right)^{3/2} = \frac{\pi^{1/2}}{\sigma_c h^3}(I_A I_B I_C)^{1/2}(8\pi kT)^{3/2} \tag{3.14}$$

3.2.2 Species Compositions

The calculation of composition of arc plasma is based on the assumption of LTE, namely all electrons, ions or neutral species in the arc have the same kinetic energy and temperature, and the whole system is in chemical equilibrium where the material properties (including chemical composition, thermodynamic properties and transport coefficients of the arc plasma) are only the function of temperature and pressure.

The calculation of plasma composition is not only the first step in calculating the thermophysical properties and transport coefficients of the plasma, but also an important step to understand the microscopic processes occurring within the plasma. The minimum Gibbs free energy method is commonly used [8] which is a standard method for calculation of chemical equilibrium [9, 10] and applicable to closed isothermal and constant pressure systems. The total Gibbs free energy of a system, g (/J/kg), is determined by

$$g = \sum_{i=1}^{w} n_i^{kg} \mu_i \tag{3.15}$$

where w is the number of species in the system, n_i^{kg} the number of moles of species i per kilogram of the plasma gas (mol/kg) and μ_i the chemical potential of species i (/J/mol).

The chemical equilibrium composition can be determined by minimizing the total Gibbs free energy of the system under the constraint of mass conservation:

$$\sum_{j=1}^{w} a_{ij} n_j^{kg} = b_i^0 \quad i = 1, 2, ..., l \tag{3.16}$$

where a_{ij} is the stoichiometric coefficient of element i in chemical species j, b_i^0 the number of moles of element i in all the chemical reactants, which is determined by the initial conditions, and l is the number of types of elements in the system.

By defining the expression $G = g + \sum_{i=1}^{l} \lambda_i (\sum_{j=1}^{w} a_{ij} n_j^{kg} - b_i^0)$, the chemical equilibrium condition is equivalent to

$$\delta G = \sum_{j=1}^{w} (\mu_j + \sum_{i=1}^{l} \lambda_i a_{ij}) \delta n_j^{kg} + \sum_{i=1}^{l} (\sum_{j=1}^{w} a_{ij} n_j^{kg} - b_i^0) \delta \lambda_i = 0 \tag{3.17}$$

where λ_i is the Lagrange multiplier.

Taking δn_j^{kg} and $\delta \lambda_i$ as the independent variables, the following constraint relation can be obtained:

$$\mu_j + \sum_{i=1}^{l} \lambda_i a_{ij} = 0 \tag{3.18}$$

At the same time, the constraint conditions of stoichiometric conservation (mass conservation) can also be obtained.

The chemical potential of chemical species *j* is calculated by the following expression, with Eq. 3.18 for gas-phase species and Eq. 3.19 for non-gas-phase species, respectively. The standard state for the gas-phase species is the one where particles follow the ideal gas law. For solid species, the standard state is the one where all elements have the most stable solid state at the standard atmospheric pressure:

$$\mu_j = \mu_j^0 + RT \ln(P/P_0) + RT \ln(n_j^{kg} / \sum_{\xi=1}^{w_{max}^g} n_\xi^{kg}) \tag{3.19}$$

$$\mu_j = \mu_j^0 \tag{3.20}$$

$$\mu_j^0 = H_{T,j}^0 - T \cdot S_{T,j}^0 \tag{3.21}$$

where R is the gas constant (8.314 J/(mol·K)), μ_j^0 the chemical potential of the standard state of the species (J/mol), P_0 the standard reference pressure (Pa), $S_{T,j}^0$ the standard entropy of species *j* (J/(mol·K)), $H_{T,j}^0$ the standard enthalpy of species *j* (J/mol) and $w_{\max}^g$ is the number of gas phase species in the reaction system.

Under the condition of chemical equilibrium, in addition to the constraints of system minimum Gibbs free energy and stoichiometric conservation (mass conservation), the composition of arc plasma must also satisfy Dalton's law of partial pressure (gas state equation) and charge quasi-neutral conditions [11]. The interaction between charged particles in arc plasma results in Debye–Hückel correction of the above equilibrium constraint conditions [12]. The modified chemical potential expression of gas-phase particles and the expression of Dalton's law of partial pressure are shown below.

$$\mu_j = \mu_j^0 + RT \ln(P/P_0) + RT \ln(n_j^{kg} / \sum_{\xi=1}^{w_{max}^g} n_\xi^{kg}) - \mathrm{R}T(eZ_j)^2/(8\pi \varepsilon_0 kT \lambda_D) \tag{3.22}$$

$$P + kT/(24\pi\lambda_D^3) = \sum_{j=1}^{w_{max}^g} n_j kT \tag{3.23}$$

where n_j is the number density of species j (/m^{-3}), Z_j the charge number of species j and λ_D the Debye radius (m).

The constraints (minimum Gibbs free energy, stoichiometric conservation, Dalton's law of partial pressure and charge quasi-neutral conditions) describing the plasma composition in chemical equilibrium form a set of nonlinear equations and are difficult to solve directly. It is normally solved by numerical iteration with a given convergence accuracy. Gordon and McBride [8] used the Lagrange multipliers and Newton–Raphson iteration method with steepest descent to obtain the numerical solution of the non-linear equations, which is often used to solve the composition of arc plasma in LTE state.

In addition to the minimum Gibbs free energy method, the composition of arc plasma in LTE state can also be calculated by the mass action law method. With this method, the Saha equation and Guldberg-Waage equation are used to describe the proportional relationship between different species in chemical equilibrium. Combined with stoichiometric conservation, Dalton's law of partial pressure and charge quasi-neutralily conditions, the composition of the arc plasma under the assumption of LTE can be solved. The method based on the law of mass action is described in detail in Sect. 3.2.

3.2.3 Thermophysical Properties

After the composition of the arc plasma is obtained, the thermodynamic properties can be obtained directly by using standard statistical thermodynamic relations.

1. Mass density

$$\rho = \sum_{i=1}^{w_{\max}^g} m_i n_i \tag{3.24}$$

ρ is the mass density of arc plasma (kg·m^{-3}), $w_{\max}^g$ the number of gas-phase species in the reaction system, m_i the mass of species i (kg) and n_i the number density of species i (m^{-3}).

2. Specific enthalpy

$$h = \sum_{i=1}^{w_{\max}^g} x_i H_{T,i}^0 / M \tag{3.25}$$

$$x_i = n_i / \sum_{i=1}^{w_{\max}^g} n_i \tag{3.26}$$

$$M = \sum_{i=1}^{w_{\max}^g} x_i M_i \tag{3.27}$$

h is specific enthalpy of arc plasma (J·kg^{-1}), x_i the molar fraction of species I, $H_{T,i}^0$ the enthalpy of species i in standard state (J·mol^{-1}), and M_i the molar mass of species i (kg·mol^{-1}).

3. Specific heat at constant pressure

$$C_p = (\frac{\partial h}{\partial T})_p = \sum_{i=1}^{w_{\max}^g} n_i c_{p,i}^0 + \sum_{i=1}^{w_{\max}^g} n_i H_{T,i}^0 (\frac{\partial \ln n_i}{\partial \ln T})_P / T \tag{3.28}$$

$c_{p,i}^0$ is specific heat of species i at constant pressure in the standard state (/J·K^{-1}·mol^{-1}).

The first term on the right hand side of Eq. 3.28 is the specific heat of the mixture when no chemical reactions occur. The second term is the reactive contribution to the specific heat. The specific heat in the standard state at constant pressure (the species index i is dropped0 can be obtained by the partition function.

$$\frac{C_P^0}{\mathrm{R}} = T^2 \frac{d^2(\ln Q_{\mathrm{int}})}{dT^2} + 2T \frac{d(\ln Q_{\mathrm{int}})}{dT} + 2.5 \tag{3.29}$$

3.2.4 Collision Integrals and Transport Coefficients

Transport coefficients, which include viscosity, thermal conductivity and electrical conductivity, are related to the transport of momentum, energy and charge through the random motion and collision between particles, as described by the Boltzmann equation [7]. This equation is a complex integral differential equation and difficult to solve directly. Based on the assumption of LTE state, the Chapman-Enskog expansion approximation is used to obtain the transport coefficient [13–15]. Based on the first order perturbation approximation of the Maxwellian distribution, the distribution function is expanded into a finite series of Sonine polynomials and the Boltzmann equation is linearised. By using the Sonine polynomial, the finite term of the first order perturbation of the distribution function is expanded and the expressions of the transport coefficients are obtained. The Chapman-Enskog method establishes a connection between the classical Boltzmann equation and the Navier–Stokes hydrodynamic equation.

1. Collision integral

Obtaining the collision integral of particle interaction is the basis and premise of using the Chapman-Enskog method. The collision integral is obtained by the potential energy integral of particle interaction, reflecting the influence of potential energy function on the distribution function. Its accuracy has a significant impact on the reliability of the calculated transport properties. The collision integral is computed by the following expression

$$\Omega_{ij}^{(l,s)} = \sqrt{\frac{\mathrm{k}T}{2\pi\mu_{ij}}} \int_0^\infty \exp(-\gamma_{ij}^2)\gamma_{ij}^{2s+3} Q_{ij}^l(g) d\gamma_{ij} \tag{3.30}$$

where $\Omega_{ij}^{(l,s)}$ is the collision integral of order (l, s) of the interaction between particle i and j (/m^2), $Q_{ij}^l(g)$ is the collision cross section of order (l) of interaction between particle i and j (/m^2), γ_{ij} is the reduced initial velocity between particle i and j (/m·s^{-1}), and μ_{ij} is the reduced mass between particles i and j (/kg).

The reduced initial velocity and the reduced mass are calculated by the following formula

$$\frac{1}{\mu_{ij}} = \frac{1}{m_i} + \frac{1}{m_j} \tag{3.31}$$

$$\gamma_{ij} = \sqrt{\frac{\mu_{ij}}{2\mathrm{k}T}} g_{ij} \tag{3.32}$$

The collision cross section $Q_{ij}^l(g)$ is calculated by the following expression

$$Q_{ij}^l(g) = 2\pi \int_0^\infty (1 - \cos^l \chi) b db \tag{3.33}$$

$$\chi = \pi - 2b \int_{r_m}^\infty \frac{\mathrm{d}r/r^2}{\sqrt{1 - \left[\varphi_{ij}(r)/0.5\mu_{ij}g_{ij}^2\right] - (b^2/r^2)}} \tag{3.34}$$

where, χ is the deflection angle of the colliding particle relative to the center of gravity coordinate system (rad); $\varphi_{ij}(r)$ is the interaction potential between colliding particles (J); b is the collision parameter, and r_m is the smallest positive root of the following equation

$$1 - \frac{\varphi_{ij}(r)}{\frac{1}{2}\mu_{ij}g_{ij}^2} - \frac{b^2}{r_m^2} = 0 \tag{3.35}$$

Collisions between particles can be divided into four typical forms: collision between neutral particles, collision between neutral particles and ions, collision between neutral particles and electrons, and collision between charged particles.

(1) Collisions between neutral particles

There are various types of classical interaction potentials in the literature that describe the interactions between neutral particles. For example, Aubreton [16–18] et al. used the known bound and unbound neutral particle interaction potential curves to fit the Hulbert-Hirschfelder potential and exponential repulsive potential, and obtained numerous collision integrals for neutral particle interaction. The accuracy of this method depends on the accuracy of the original potential curve and the accuracy of the non-linear fitting. A good fit must be able to recover the original potential function. However, whether the potential energy function can accurately estimate the interaction outside the fitting interval is always debatable. André [19] et al. calculated the collision integral of neutral partical interactions in N_2 and H_2 arc plasma using the classical Lennard–Jones potential function based on simple combination rules. The method is extended to the calculation of the arc plasma transport coefficient for the atmospheric composition of Mars (95% CO_2, 2% N_2, 2% Ar) and Saturn (99% N_2, 1% CF_4), and the results are consistent with of other complex interaction potential functions.

Capitelli [20, 21] and his colleagues built a new phenomenological potential model whose interaction potential function is and improvement and refinement of the Lennard–Jones potential. The result shows that the phenomenological potential energy can accurately reproduce the potential energy results calculated ab initio in quantum chemistry [22], which can provide accurate estimation for action potentials which have not previously been reliably measured in experiments or theoretically calculated. The expression of the phenomenological potential function is as follows:

$$\varphi(x) = \varepsilon_b \left[\frac{m}{n(x) - m} \left(\frac{1}{x} \right)^{n(x)} - \frac{n(x)}{n(x) - m} \left(\frac{1}{x} \right)^{m} \right] \tag{3.36}$$

$$n(x) = \beta + 4x^2 \tag{3.37}$$

$$x = r/r_e \tag{3.38}$$

where, r_e is the equilibrium distance (/Å); m is the potential energy parameter, with a value of 6 for the interaction between neutral particles.

The parameter β is obtained from the empirical formula using the polarisation ratio α as the input parameter:

$$\beta = 6 + \frac{5}{s_i + s_j} \tag{3.39}$$

The parameter s_i is defined as the cube root of the polarisability α_i of particle i. For open-shell atoms and ions, the ground state spin multiplicity should be considered simultaneously.

(2) Collisions between neutral particles and ions

There are two different types of interactions between ions and neutral particles, one is an elastic collision, the other is an inelastic collision with charge transfer. For odd l ($l = 1$ or 3), inelastic collision of charge transfer plays a large part in the calculation of the collision integral $\Omega_{ij}^{(l,s)}$. According to the recommendation of Murphy [29], elastic collision and inelastic collision processes are combined together through the following expression:

$$\Omega^{(l,s)} = \sqrt{\left(\Omega_{el}^{(l,s)}\right)^2 + \left(\Omega_{in}^{(l,s)}\right)^2} \tag{3.40}$$

where, the subscripts *in* and *el* respectively represent inelastic collision and elastic collision.

The elastic interaction between neutral particles and ions can be described by the phenomenological potential model given by Eq. 3.36, where the potential energy parameter is equal to 4.

For ions with unknown polarisability, the elastic interaction potential with neutral particles can be represented by the Polaris ability potential model:

$$\varphi_{in}(r) = (\frac{1}{4\pi\varepsilon_0})^2 \frac{(Z_i e)^2 \alpha_n}{2r^4} \tag{3.41}$$

where, α_n is the polarisability of neutral particles. Compared to the phenomenological potential, the polarisation potential model is not related the polarisation rate of ions, and its collision integral is calculated of a closed form [23]:

$$\begin{array}{lll} \Omega_{i,n}^{(1,1)} = 424.443 Z_i \frac{\sqrt{\alpha_n}}{T} & \Omega_{i,n}^{(1,2)} = 0.833\Omega_{i,n}^{(1,1)} & \Omega_{i,n}^{(2,2)} = 0.870\Omega_{i,n}^{(1,1)} \\ \Omega_{i,n}^{(1,3)} = 0.729\Omega_{i,n}^{(1,1)} & \Omega_{i,n}^{(2,3)} = 0.761\Omega_{i,n}^{(1,1)} & \Omega_{i,n}^{(1,4)} = 0.656\Omega_{i,n}^{(1,1)} \\ \Omega_{i,n}^{(2,4)} = 0.685\Omega_{i,n}^{(1,1)} & \Omega_{i,n}^{(1,5)} = 0.602\Omega_{i,n}^{(1,1)} & \Omega_{i,n}^{(3,3)} = 0.842\Omega_{i,n}^{(1,1)} \end{array} \tag{3.42}$$

In Eq. 3.40, the inelastic collision integral between neutral particles and ions is obtained by the charge transfer collision section integral. The collision cross-section between the neutral particle X and the first-order ion of its cokernel X^- or X^+ is calculated from the following expression:

$$Q^l(g) = 2Q_{ex} \tag{3.43}$$

In the formula, the collision cross-section of charge transfer Q_{ex} is approximately expressed by the following relation

$$Q_{ex} = (\mathrm{A} - \mathrm{B}\ln E)^2 \tag{3.44}$$

where, E is collision energy (/eV); and A and B are constants determined either by experiment or calculation.

Inelastic collisions of charge transfer between neutral particles and heteronuclear first order ions $X - Y^{+} (X \neq Y)$ and between neutral particles and their higher order ions $X - X^{n+}$ $(n \geq 2)$ have very little impact on the collision integral and can be ignored. When the value of l is even, the collision integral is only dependent on elastic collisions.

(3) Collisions between electrons and neutral particles

The interaction between electrons and neutral particles can be assumed to conform to the Boltzmann distribution, and can be obtained by direct integration of the energy of the collision cross-sections of various orders $Q_1(E)$, $Q_2(E)$ and $Q_3(E)$. Different techniques have been developed in the literature for estimating $Q_2(E)$ and $Q_3(E)$ when they are unknown.

The collision cross-section can be obtained according to the technique recommended by Laricchiuta et al. [24]. The ratios $Q_2(E)/Q_1(E)$ and $Q_3(E)/Q_1(E)$ are obtained from the known calculation $Q_1(E)/Q_0(E)$ by assuming that the differential collision cross-section is a function of the collision angle.

$$\frac{1}{(1 - h(E)\cos\theta)^2} \tag{3.45}$$

The adjustable parameter, h, is a function of electron energy, determined by the ratio $Q_1(E)/Q_0(E)$. If the collision cross section $Q_0(E)$ is unknown, its value is obtained by the Bohr collision of the Coulomb shielded potential.

$$\frac{1}{(1 + 2\eta)^2}\eta = 10.9Z^{2/3}/E \tag{3.46}$$

For interaction between other electrons and neutral particles, there are no relevant experimental measurements and theoretical derivations, which are approximated by the polarisation effect assuming that the collision cross-section is independent of energy.

(4) Collisions between charged particles

The collision integral between charged particles is calculated by the shielded Coulomb potential;

$$\varphi_{ij}(r) = \frac{Z_i Z_j e^2}{4\pi\varepsilon_0 r}\exp(-r/\lambda_D) \tag{3.47}$$

where, Z_i is the charge of charged particle i and Z_j for particle j.

The shielding effect of ions is considered in the calculation of Debye radius:

$$\lambda_{d}^{-2} = \frac{e^2}{\varepsilon_0 k}\left[\frac{n_e}{T_e} + \sum_{t=1,t\neq e}^{w} \frac{z_t^2 n_t}{T_h}\right] \tag{3.48}$$

If only considering electron shielding, the second term in the above expression can be omitted.

2. Transport coefficient
(1) Electrical conductivity

The lectrical conductivity is a parameter that describes the migration of charged particles under potential gradients. In practical calculation, the effect of ions is often ignored because electrons have much smaller mass and drift faster than ions. The conductivity in the third order approximation is given by [25–28]

$$\sigma = \frac{3}{2} q_e^2 n_e^2 \sqrt{\frac{2\pi}{m_e kT}} \frac{\begin{vmatrix} q^{11} & q^{12} \\ q^{12} & q^{22} \end{vmatrix}}{\begin{vmatrix} q^{00} & q^{01} & q^{02} \\ q^{01} & q^{11} & q^{12} \\ q^{02} & q^{12} & q^{22} \end{vmatrix}} \tag{3.49}$$

where, n_e is electron density (m^{-3}); m_e is electron mass (kg); q^{ij} is the collision coefficient, a function of species composition and collision integral, and the specific form of which is given by Devoto [29].

(2) Viscosity

The viscosity is a parameter that describes the viscous effect of the fluid. Since the electron mass is much smaller than that of heavy particles, the contribution of electrons to the viscosity is generally not included in the actual calculation, and so it is calculated by using the following formulae [25–28]

$$\eta = -\frac{\begin{vmatrix} H_{11} & \cdots & H_{1v} & x_1 \\ \vdots & & \vdots & \vdots \\ H_{v1} & \cdots & H_{vv} & x_v \\ x_1 & \cdots & x_v & 0 \end{vmatrix}}{\begin{vmatrix} H_{11} & \cdots & H_{1v} \\ \vdots & & \vdots \\ H_{v1} & \cdots & H_{vv} \end{vmatrix}} \tag{3.50}$$

$$H_{ii} = \frac{x_i^2}{\eta_i} + \sum_{k=1,k\neq i}^{v} \frac{2x_i x_k}{\eta_{ik}} \frac{M_i M_k}{(M_i + M_k)^2}\left(\frac{5}{3A_{ik}^*} + \frac{M_k}{M_i}\right) \tag{3.51}$$

$$H_{ij} = -\frac{2x_i x_j}{\eta_{ij}} \frac{M_i M_j}{\left(M_i + M_j\right)^2} \left(\frac{5}{3A_{ij}^*} - 1 \right) \quad (i \neq j) \tag{3.52}$$

$$\eta_i = \frac{5}{16} \frac{1}{\overline{\Omega}_{ii}^{(2,2)}} \sqrt{\frac{\mathrm{k}T}{\pi N_a}} \sqrt{M_i} \tag{3.53}$$

$$\eta_{ij} = \frac{5}{16} \frac{1}{\overline{\Omega}_{ij}^{(2,2)}} \sqrt{\frac{\mathrm{k}T}{\pi N_a}} \sqrt{\frac{2M_i M_j}{M_i + M_j}} \tag{3.54}$$

where, v is the number of heavy particles; x_i is the mole fraction of species i; M_i is the molar mass of species i (kg·mol^{-1}); N_a is Avogadro constant (mol^{-1}); A_{ij}^* is the collision parameter, the function of the collision integral, $A_{ij}^* = \overline{\Omega}_{ij}^{(2,2)} \Big/ \overline{\Omega}_{ij}^{(1,1)}$.

(3) Thermal conductivity

Thermal conductivity is a parameter that describes the conductivity of thermal energy. The calculation of thermal conductivity of arc plasma is rather complicated. The thermal conductivity of plasma consists of three parts, namely: translational thermal conductivity κ_{tr}, internal thermal conductivity κ_{int} and thermal conductivity due to chemical reaction κ_{reac} [25, 26]:

$$\kappa = \kappa_{tr} + \kappa_{int} + \kappa_{reac} \tag{3.55}$$

① Translational thermal conductivity

Due to the existence of different types of moving species, translational thermal conductivity κ_{tr} can be divided into translational thermal conductivity of electrons (κ_{tr}^e) and translational thermal conductivity of heavy particles (κ_{tr}^h) [25–28].

$$\kappa_{tr} = \kappa_{tr}^e + \kappa_{tr}^h \tag{3.56}$$

$$\kappa_{tr}^e = \frac{75}{8} n_e^2 \mathrm{k} \sqrt{\frac{2\pi \mathrm{k}T}{m_e}} \frac{q^{22}}{q^{11} q^{22} - \left(q^{12}\right)^2} \tag{3.57}$$

$$\kappa_{tr}^h = 4 \frac{\begin{vmatrix} L_{11} & \cdots & L_{1v} & x_1 \\ \vdots & & \vdots & \vdots \\ L_{v1} & \cdots & L_{vv} & x_v \\ x_1 & \cdots & x_v & 0 \end{vmatrix}}{\begin{vmatrix} L_{11} & \cdots & L_{1v} \\ \vdots & & \vdots \\ L_{v1} & \cdots & L_{vv} \end{vmatrix}} \tag{3.58}$$

$$L_{ii} = -4\frac{x_i^2}{k_{ii}} - \sum_{k=1,k\neq i}^{v} \frac{2x_i x_k\left(\frac{15}{2}M_i^2 + \frac{25}{4}M_k^2 - 3B_{ik}^* M_k^2 + 4A_{ik}^* M_i M_k\right)}{k_{ik} A_{ik}^* (M_i + M_k)^2} \tag{3.59}$$

$$L_{ij} = \frac{2x_i x_j M_i M_j}{k_{ij} A_{ij}^* \left(M_i + M_j\right)^2}\left(\frac{55}{4} - 3B_{ij}^* - 4A_{ij}^*\right) \ (i \neq j) \tag{3.60}$$

$$k_{ij} = \frac{75}{64}\frac{\mathrm{k}}{\overline{\Omega}_{ij}^{(2,2)}}\sqrt{\frac{N_a \mathrm{k}T}{\pi}}\sqrt{\frac{M_i + M_j}{2M_i M_j}} \tag{3.61}$$

where, x_i is the molar fraction of species i; M_i is the molar mass of species i (kg·mol^{-1}); B_{ij}^* is the collision parameter, the function of the collision integral, $B_{ij}^* = \left(5\overline{\Omega}_{ij}^{(1,2)} - 4\overline{\Omega}_{ij}^{(1,3)}\right)\Big/\overline{\Omega}_{ij}^{(1,1)}$.

② Internal thermal conductivity

The internal thermal conductivity κ_{int} is a representation of the internal energy with multiple degree of freedom and can be approximated by the following formula [25–28]

$$\kappa_{int} = \sum_{i=1}^{N} \frac{\kappa_{int}^i}{1 + \sum\limits_{j=1,j\neq i}^{N} \frac{x_j D_{ii}}{x_i D_{ij}}} \tag{3.62}$$

$$\kappa_{int}^i = \frac{PD_{ii}}{T}\left(\frac{C_{pi}}{R} - \frac{5}{2}\right) \tag{3.63}$$

$$D_{ij} = \frac{3}{8}\frac{\mathrm{k}T}{P\overline{\Omega}_{ij}^{(1,1)}}\sqrt{\frac{N_a \mathrm{k}T}{\pi}}\sqrt{\frac{M_i + M_j}{2M_i M_j}} \tag{3.64}$$

where, P is pressure (Pa); C_{pi} is the specific heat at constant pressure of species i (J·K^{-1}·mol^{-1}). The specific heat at constant pressure of the species can be obtained by the partition function.

③ Thermal conductivity of chemical reaction

The thermal conductivity of chemical reaction κ_{reac} is a characterisation of chemical reaction heat in an arc plasma. Considering an arc plasma that is composed of N species, in which μ independent (linearly independent) chemical reactions take place, according to Butler and Brokaw [62], the thermal conductivity of the reaction κ_{reac} can be expressed as [25–28]

$$\kappa_{reac} = -\frac{1}{RT^2}\frac{\begin{vmatrix} A_{11} & \cdots & A_{1\mu} & \Delta H_1 \\ \vdots & & \vdots & \vdots \\ A_{\mu 1} & \cdots & A_{\mu\mu} & \Delta H_\mu \\ \Delta H_1 & \cdots & \Delta H_\mu & 0 \end{vmatrix}}{\begin{vmatrix} A_{11} & \cdots & A_{1\mu} \\ \vdots & & \vdots \\ A_{\mu 1} & \cdots & A_{\mu\mu} \end{vmatrix}} \tag{3.65}$$

$$A_{ij} = \sum_{k=1}^{N-1}\sum_{l=k+1}^{N} \frac{RT}{PD_{kl}} x_l x_k \left(\frac{a_{ik}}{x_k} - \frac{a_{il}}{x_l}\right)\left(\frac{a_{jk}}{x_k} - \frac{a_{jl}}{x_l}\right) \tag{3.66}$$

where, H_i is the enthalpy change of chemical reaction i (J·mol^{-1}) and a_{ij} is the chemical reaction measurement coefficient.

3.3 Calculation of Material Properties of Arcing Gas in Non-thermodynamic Equilibrium

In the previous section, the calculation method of material properties of LTE arc plasma was introduced. However, in the actual switching process, the arc plasma cannot always meet the LTE hypothesis, and the electron temperature in the arc plasma can be higher than the heavy particle temperature. Under such conditions, the material properties and numerical models based on LTE hypothesis cannot accurately simulate the arc plasma [30].

In order to study the micro characteristics of arc plasma deviating from LTE and to provide necessary input parameters for the establishment and solution of a model for the non-LTE arc plasma, a two temperature plasma model was proposed to account for the thermodynamic non-equilibrium [31–39]. In this model, the energy of electrons and heavy particles are described by the electron temperature T_e and heavy particle temperature T_h, respectively, with $T_{e \geq} T_h$. Other temperatures in the system (such as electron excitation temperature, vibrational excitation temperature, rotational excitation temperature, reaction excitation temperature, etc.) are represented by T_e or T_h or their functions. This section mainly describes the calculation method of material properties based on this model using examples.

3.3.1 *Basic Assumptions of the Two-Temperature Model*

The two-temperature model can be used to calculate the plasma composition, thermophysical properties and transport coefficients for a given pressure *P*, electron temperature T_e, and non-equilibrium degree $\theta = T_e/T_h$ (P, T_e and T_h can also be used in equation of state). The model is based on the following basic assumptions:

(1) Electrons and heavy particles in the system satisfy their respective Maxwell velocity distributions, and have a temperature of T_e and T_h, respectively, with $T_e \geq T_h$.
(2) The electronic, vibrational and rotational energy levels of the particles all satisfy the Boltzmann distribution, but their temperatures may be different. This assumption is a necessary condition for calculating the partition function.
(3) The system satisfies the local chemical equilibrium (LCE) assumption, that is, the time scale involved in the model is much larger than that of the chemical reactions. This assumption is a necessary condition for establishing the equation of mass action or the minimum Gibbs free energy equation.
(4) The system satisfies (or approximately satisfies) Dalton's law of partial pressure, and when the system pressure is high, the classical Dalton's law of partial pressure is no longer applicable. Dalton's law of partial pressure thereby needs to be modified to reflect the actual situation.
(5) The system is quasi-neutral, that is, the spatial scale involved in the model is much larger than the Debye radius. When the system of plasma contains a variety of elements, the conservation of stoichiometric number should be satisfied.

3.3.2 Calculation of Composition of Two-Temperature Arc Plasma

Plasma composition in thermodynamic non-equilibrium can be obtained through the two temperature mass action law. Using simultaneous equations concerning Dalton's law of partial pressures, charge quasi-neutrality, stoichiometric number conservation and chemical reaction balance, the plasma composition at given P, T_e and θ (or P, T_e and T_h) can be obtained. Among them, the charge quasi-neutral equation and stoichiometric conservation equation in the two-temperature model are the same as for the LTE model (refer to Sect. 1.2 for details). Here, the Dalton's law of partial pressures with two temperatures and the equilibrium equation of chemical reaction will be introduced.

1. Dalton's law of partial pressures for a two-temperature system

Dalton's law of partial pressure in the two-temperature model is similar to the LTE model. The difference is that T_e and T_h are different in a non-LTE system. Therefore, the contribution of electrons and heavy particles to the system pressure needs to be calculated separately.

$$P + \Delta P = \sum_{i \neq e} n_i \mathrm{k} T_h + n_e \mathrm{k} T_e \tag{3.67}$$

where, n_i ($i \neq e$) and n_e are the number density of heavy species i and electrons respectively. ΔP is the Debye pressure correction, representing the effect of interactions between charged particles. In the two-temperature system, the expression of ΔP is

$$\Delta P = \frac{1}{24\varepsilon_0 \pi \lambda_d^3} \sum_i Z_i^2 n_i \tag{3.68}$$

where, Z_i represents the charge of species i, λ_d is the Debye radius of the system, which can be obtained by the following formula under the assumption of a two–temperature system

$$\lambda_d^{-2} = \frac{e^2}{\varepsilon_0 k}[\frac{n_e}{T_e} + \sum_{i \neq e} \frac{Z_i^2 n_i}{T_h}] \tag{3.69}$$

where the two terms within the square brackets are the contributions of electrons and ions to the Debye radius. It is worth mentioning that Murphy's research [40] indicates that theoretical calculation results are more consistent with experimental results when only the contribution of electrons to the Debye radius is considered.

2. Two-temperature chemical equilibrium equation

Over the years, there has been much debate on how to describe the chemical equilibrium in two-temperature plasma, and different methods have been developed based on different hypothesis, which can be roughly divided into the following two categories:

(a) The two-temperature modified mass action law (Saha equation and Guldberg-Waage equation) is used to describe the proportional relationship between species in chemical equilibrium. This mass action law equation is combined with Dalton's law of partial pressure equation, charge quasi-neutral equation and stoichiometric conservation equation to obtain the plasma composition.
(b) Describing the chemical equilibrium by using the two-temperature Gibbs free energy minimisation method. The Gibbs free energy expression of the system is derived, and Dalton's law of partial pressure, charge quasi-neutral condition and stoichiometric number conservation are taken as the constraint conditions, and the results of the composition with the least Gibbs free energy of the system are obtained by using the Lagrange multiplier and the fastest descent algorithm.

In the mid-twentieth century, Prigogine [41] and Potapov [42] developed the two-temperature mass action law equation by finding the zero of the first derivative of the Helmholtz free energy and solving for the minimum value of Helmholtz free energy, respectively. This makes it possible to solve the non-LTE plasma composition. Van de Sanden et al. [43] then modified Potapov's mass action law equation by assuming that the electronic excitation process of the particle was completely dominated by the electron temperature. The Gibbs free energy minimisation method proposed by White et al. [44], which is suitable for the calculation of LTE plasma composition, was extended to the two-temperature plasma by André et al. [45, 46].

Although there has been a long debate about the description of chemical equilibrium in a two-temperature system, and many researchers have put forward different theories, the debate continues due to the lack of reliable experimental data to verify the theoretical calculation. In this book, three of the most commonly used two-temperature chemical equilibrium equations are listed, and a brief introduction to other methods of describing chemical equilibrium is given.

(1) Potapov method

Potapov used the Helmholtz free energy minimisation method to derive the mass action law equation for the chemical equilibrium of a two-temperature plasma. The particle number density of reactants and products for dissociation reactions (A + B ↔ C) satisfies the Guldberg-Waage equation

$$\frac{n_A n_B}{n_C} = \frac{Z_A^* Z_B^*}{Z_C^*}\left(\frac{2\pi \mathrm{k} T_h}{\mathrm{h}^2}\right)^{3/2}\left(\frac{m_A m_B}{m_C}\right)^{3/2}\exp\left(-\frac{E_D}{\mathrm{k}T_h}\right) \tag{3.70}$$

The number density of reactants and products of ionisation reaction ($A^{z+} \leftrightarrow A^{(z+1)+} + e$) satisfies the Saha equation

$$n_e\left(\frac{n_{A^{(z+1)+}}}{n_{A^{z+}}}\right)^{1/\theta} = 2\left(\frac{Z^*_{A^{(z+1)+}}}{Z^*_{A^{z+}}}\right)^{1/\theta}\left(\frac{2\pi m_e \mathrm{k} T_e}{\mathrm{h}^2}\right)^{3/2}\exp\left(-\frac{E_I - \Delta E_{I,z+1}}{\mathrm{k}T_{ex}}\right) \tag{3.71}$$

E_D and E_I are the reaction energy of the dissociative reaction and of the ionisation reaction respectively, $\Delta E_{I,z+1}$ is the lowering of the ionisation energy by charged particle interaction when z + 1 order ionisation occurs.

$$\Delta E_{I,z+1} = (z+1)\frac{\mathrm{e}^2}{4\pi\varepsilon_0}\frac{1}{\lambda_d} \tag{3.72}$$

with Z_i^* as the effective partition function of species *i*. The effective partition functions of single atom and multi-atom species are expressed as

$$Z_i^* = \left[Z_i^e(T_e)\right]^\theta \tag{3.73}$$

and

$$Z_i^* = \left[Z_i^e(T_e)\right]^\theta Z_i^v(T_v) Z_i^r(T_r) \tag{3.74}$$

with T_{ex}, T_v and T_r being the reaction excitation temperature, vibration excitation temperature and rotation temperature, respectively. In the Potapov method, it is assumed that $T_v = T_r = T_h$. The selection of excitation temperature in the calculation of the two-temperature composition has a great influence on the results. The determination of excitation temperature in the two-temperature composition calculation is described below.

(2) Van de Sanden method

Different from the Potapov method, Van de Sanden et al. [43] believed that the electronic excitation energy level of the heavy particles in the plasma is completely dominated by electrons. Therefore, the plasma was divided into two subsystems: the electron translation and electronic excitation energy levels of heavy particles are dominated by T_e, and the translational, vibrational and rotational excitation energy levels of heavy particles are controlled by T_h. Based on this assumption, Van de Sanden gave different mass action law expressions, such as Eqs. 3.75 and 3.76, in which the reaction excitation temperature of the Guldberg-Waage equation is changed from T_h in the Potapov method to $T_{ex,\mathrm{C}}$ for species C

$$\frac{n_A n_B}{n_C} = \frac{Z_A Z_B}{Z_C}\left(\frac{2\pi \mathrm{k} T_h}{\mathrm{h}^2}\right)^{3/2}\left(\frac{m_A m_B}{m_C}\right)^{3/2} \exp\left(-\frac{E_D}{\mathrm{k} T_{ex,C}}\right) \tag{3.75}$$

$$\frac{n_e n_{A^{(z+1)+}}}{n_{A^{z+}}} = 2\left(\frac{Z_{A^{(z+1)+}}}{Z_{A^{z+}}}\right)\left(\frac{2\pi m_e \mathrm{k} T_e}{\mathrm{h}^2}\right)^{3/2} \exp\left(-\frac{E_I - \Delta E_{I,z+1}}{\mathrm{k} T_{ex}}\right) \tag{3.76}$$

It should be noted that, when $T_e = T_h$, the two-temperature mass action law method will recover into the single-temperature mass action law, and the calculated result is the composition of the plasma under the assumption of thermodynamic equilibrium state.

(3) The two-temperature Gibbs free energy minimisation method

Unlike in the law of mass action, the Gibbs free energy minimum method does not need knowledge of the specific chemical reaction paths, which leads to a greater simplicity. André et al. [45, 46] derived the Gibbs free energy expression for a two-temperature system

$$G = \sum_i n_i \left[\mu_i^0 + RT_i \ln\left(\frac{p}{\mathrm{p}_0}\right) + RT_i \ln\left(\frac{n_i T_i}{n_e T_e + \sum_{k\neq e} n_k T_h}\right)\right] \tag{3.77}$$

where, p and p_0 represent the actual pressure and reference pressure of the plasma (10^5 Pa is generally used). $\mathrm{n_i}$, $\mathrm{T_i}$ and μ_i^0 represent the number density, translational temperature and the chemical potential of species i at pressure p_0 respectively. When species i is electron, $T_i = T_e$. When the species is heavy particle, $T_i = T_h$. The chemical potential of electrons and heavy particles at reference pressure can then be expressed as

$$\mu_e^0 = -\mathrm{R}T_e \ln\left(\frac{Z_i^{tr}(T_e)}{N}\right) - \mathrm{R}T_e \ln 2 \tag{3.78}$$

$$\mu_i^0 = -\mathrm{R}T_h \ln\left(\frac{Z_i^{tr}(T_h)}{N}\right) - RT_{ex} \ln Z_i^e(T_{ex}) - \mathrm{R}T_v \ln Z_i^v(T_v)$$

$$- \mathrm{R}T_r \ln Z_i^r(T_r) + e_i^0 \tag{3.79}$$

where, Z_{itr}, Z_{ie}, Z_{iv} and Z_{ir} represent, respectively, translational, electron, vibrational and rotational partition functions. e_i^0 is the enthalpy of formation of species i in the standard state (298.15 K, 10^5 Pa).

3. Determination of temperature in two-temperature plasma composition calculation

In the composition calculation of two-temperature arc plasma, the excitation temperature, vibrational temperature and rotational temperature need to be determined to complete the particle partition function calculation and the construction of mass action law equation (or Gibbs free energy). Different temperature choices will directly affect the partition function and the reaction equilibrium constant, thus significantly affecting the result of the composition.

(1) Determination of excitation temperature

A common assumption, including that in [xx,Van de Sanden] is that ionization reaction is dominated by the collisions between electrons and heavy particles, while the dissociation reaction is mainly controlled by the collisions between heavy particles. Therefore, in most studies, it is assumed that $T_{ex} = T_e$ in the Saha equation and $T_{ex} = T_h$ in the Guldberg-Waage equation.

According to Wang et al. [47], ionisation of molecules usually occurs at a low temperature stage with low electron density, so the process of molecular ionisation is mainly enabled by collisions between heavy particles. Therefore, T_e is proposed as the excitation temperature of a single atomic ionisation reaction, and T_h is proposed as the excitation temperature of the dissociation and molecular ionisation reactions.

Tanaka and Gleizes argue that the excitation temperature of the ionisation and decomposition reactions should be equivalent to the excitation temperature of the electron energy level of the heavy particles, which is determined by the combination of the heavy particle—heavy particle and electron-heavy particle collisions. Tanaka et al. [48] gave an expression for the excitation temperature by considering the ratio of kinetic energy absorption of heavy particles in the process of electron—heavy particle elastic collision

$$T_{ex,i} = \frac{\upsilon_{e-i} T_e + \sum_{j \neq e} \upsilon_{j-i} T_h}{\upsilon_{e-i} + \sum_{j \neq e} \upsilon_{j-i}} \tag{3.80}$$

where, $T_{ex,\,i}$ is the excitation temperature of the heavy species i, and υ_{j-i} is the collision frequency of species j and species i.

However, the definition of excitation temperature by Eq. 3.80 in [xx, Tanaka et al.] is challenged in the work of [49]. Gleizes et al. argues that the excitation process of heavy species cannot be described by an elastic collision process. On the basis of combining reaction dynamics, Gleizes et al. gave a new definition of excitation temperature by calculating the reaction rates of excitation and de-excitation reactions caused by electron collisions and heavy particle collisions, as shown in Eq. 3.81.

$$\exp\left(-\frac{\Delta E}{kT_{ex}}\right) \cong \frac{10^4 n_e \exp\left[\frac{\Delta E(T_e - T_h)}{kT_e T_h}\right] + n_A}{10^4 n_e \exp\left[\frac{\Delta E(T_e - T_h)}{kT_e T_h}\right] \exp\left(-\frac{\Delta E}{kT_e}\right) + n_A \exp\left(-\frac{\Delta E}{kT_h}\right)} \tag{3.81}$$

In addition, André et al. [50] introduced the concept of equivalent temperature related to particle flow, which was calculated by T_e, T_h and n_e. The equivalent temperature is between T_e and T_h. When the plasma is fully ionised (n_e is sufficiently large), $T_{ex} = T_e$, and when the system is pure gas ($n_e = 0$), $T_{ex} = T_h$.

$$T_{ex} = T_e - (T_e - T_h) \times \exp\left[-\left(\frac{T_e}{T_h}\right)^{1/2} \frac{n_e/m_e^{1/2}}{\sum_{i \neq e} n_i/m_i^{1/2}}\right] \tag{3.82}$$

(2) Determination of vibrational and rotational temperatures

The population density of the electronic energy level, vibrational energy level and rotational energy level of two-temperature plasma all satisfy the Boltzmann distribution, but the electron, vibrational and rotational temperatures are not equal. Therefore, it is necessary to determine the temperatures according to the type of the partition function that is to be calculated for the two-temperature plasma [51]. It is generally assumed that the rotational temperature of the molecules in the plasma is $T_r = T_h$. Compared with the rotational energy level, the vibrational energy level of the molecule is affected by both heavy particles and electrons, so the determination of the vibrational temperature is more complicated. This is because molecules generally appear in the lower temperature range where the influence of electrons on the vibrational energy level is not obvious. Therefore, it is commonly assumed that the vibrational temperature is equal to the temperature of heavy particles, $T_v = T_h$. André et al. [46] and Aubreton et al. [52] proposed a method to calculate the vibrational temperature taking into account the influence of electrons and heavy particles

$$T_v = T_h \times \gamma_v = T_h \times \sqrt{\theta} \tag{3.83}$$

Aubreton et al. [52] found that changing the value of T_v has a great impact on the result of the plasma composition and material properties. There is no firm conclusion on the selection or definition of the vibrational temperature at present.

3.3.3 *Calculation of Thermophysical Properties of Two-Temperature Arc Plasma*

Similar to the calculation of the material properties of LTE arc plasma, the thermophysical properties of the two temperature plasma can be calculated by standard thermodynamic expressions after the species composition is obtained. The calculation methods for mass density and molar mass are the same as those in the LTE model and will not be repeated. This section introduces the calculation methods of specific enthalpy and specific heat at constant pressure in the two-temperature arc plasma.

1. Calculation of specific enthalpy of two-temperature plasma

Trelles et al. [53] proposed a specific enthalpy definition considering only the contribution of translational energy and chemical reactions. Colombo et al. [34] proposed a more complete definition of specific enthalpy that takes into account translational energy, chemical reactions, and the contribution of internal energy levels. However, Wang et al. [47] questioned Colombo et al.'s definition, arguing that it is inappropriate to directly take the partial derivative of the internal partition function with respect to the temperature of heavy particles in the internal energy level term. The contribution of electronic excitation energy level, vibrational energy level and rotational energy level should be considered separately. According to the definition of Wang et al. [47], the specific enthalpy expressions of electrons, monoatomic species and polyatomic species are

$$h_e = \frac{5}{2}\frac{\mathrm{k}}{\rho}n_e T_e \tag{3.84}$$

$$h_{i,mon} = \frac{5}{2}\frac{\mathrm{k}}{\rho}n_i T_h + \frac{1}{\rho}n_i e_i + \frac{\mathrm{k}}{\rho}n_i\left(T_e^2\frac{\partial \ln Z_i^e}{\partial T_e}\right) \tag{3.85}$$

$$h_{i,poly} = \frac{5}{2}\frac{\mathrm{k}}{\rho}n_i T_h + \frac{1}{\rho}n_i e_i + \frac{\mathrm{k}}{\rho}n_i\left(T_e^2\frac{\partial \ln Z_i^e}{\partial T_e} + T_h^2\frac{\partial \ln\left(Z_i^r Z_i^v\right)}{\partial T_h}\right) \tag{3.86}$$

where, e_i is the enthalpy of formation of species i. It should be pointed out that the present description of the enthalpy contribution of the internal energy levels of heavy particles in the two-temperature plasma is still controversial. Considering that the contribution of the internal energy levels to enthalpy is relatively small compared with the translational term and the chemical reaction term, sometimes the contribution of the internal energy levels when calculating the specific enthalpy is ignored. When the specific enthalpy of a species is obtained, the total specific enthalpy of a two-temperature plasma can be calculated by summation.

$$h = \sum_i h_i \tag{3.87}$$

When a two-temperature arc plasma is modelled, the specific enthalpy of heavy particles and electrons need to be distinguished in order to establish the energy conservation equation of electrons and heavy particles, respectively. The specific enthalpy of heavy particles can be expressed as in Eq. 3.88,

$$h_h = \sum_{i \neq e} h_i \tag{3.88}$$

2. Calculation of specific heat of two-temperature plasma at constant pressure

Colombo et al. [34] generalised the definition of specific heat at constant pressure for LTE plasma to obtain the total increment of specific enthalpy and the definition of total specific heat at constant pressure in a two-temperature plasma is

$$dh = \left.\frac{\partial h}{\partial P}\right|_{T_e,\theta} dP + \left.\frac{\partial h}{\partial T_e}\right|_{P,\theta} dT_e + \left.\frac{\partial h}{\partial \theta}\right|_{P,T_e} d\theta \tag{3.89}$$

$$C_{P,total} = \left.\frac{\partial h}{\partial T_e}\right|_{P,\theta} \tag{3.90}$$

Similar to the calculation of the specific enthalpy of a two-temperature plasma, in order to solve the energy conservation equations of electrons and heavy particles, the specific heat of heavy particles and the specific heat of electrons need to be separately defined. Trelles et al. [53] used P, T_e and T_h as state parameters to express the total derivatives of electron specific enthalpy and heavy particle specific enthalpy

$$dh_e = \left.\frac{\partial h_e}{\partial P}\right|_{T_e,T_h} dP + \left.\frac{\partial h_e}{\partial T_e}\right|_{P,T_h} dT_e + \left.\frac{\partial h_e}{\partial T_h}\right|_{P,T_e} dT_h \tag{3.91}$$

$$dh_h = \left.\frac{\partial h_h}{\partial P}\right|_{T_e,T_h} dP + \left.\frac{\partial h_h}{\partial T_e}\right|_{P,T_h} dT_e + \left.\frac{\partial h_h}{\partial T_h}\right|_{P,T_e} dT_h \tag{3.92}$$

3.3.4 Calculation of Transport Coefficients of Two-Temperature Arc Plasma

The methods to calculate the transport coefficients of two-temperature arc plasma (diffusion coefficient, viscosity, thermal conductivity and electrical conductivity) are the same as those of LTE arc plasma, and the classical Chapman-Enskog method [13–15] is adopted where it is assumed that the particle velocity distribution function is a Maxwell distribution with first-order perturbation. To date, there are two main approaches for calculating the transport coefficient of two-temperature arc plasma, namely the simplified approach of Devoto [29, 54–56] and Bonnefoi [57] which ignores the interaction between electrons and heavy particles, and that by Rat's [58] adopting a fully coupled approach. The latter considers the interaction between electrons and heavy particles. Colombo et al. [59] found that apart from the normal diffusion coefficient, the two approaches leads to negligible difference in the modelling results. In this section, the transport coefficient calculation method derived by Rat, simplified calculation method derived by Devoto, and other commonly used calculation methods are introduced.

1. Calculation of collision integral

The collision integral calculation of the two-temperature plasma is basically the same as that of the LTE plasma with the only difference being that the reduced temperature $T_{ij}{}^{*}$ of particle collision is used to replace the gas temperature T.

$$T_{ij}^{*} = \frac{\left(m_i T_j + m_j T_i\right)}{m_i + m_j} \tag{3.93}$$

where, T_i is the temperature of species i. When i represents the electrons, $T_i = T_e$; when i represents the heavy particles, $T_i = T_h$.

2. Diffusion coefficient

The normal diffusion coefficient and thermal diffusion coefficient are important parameters in the calculation of thermal conductivity. The normal diffusion coefficient can be obtained using Hirschfelder's theory [15]. The normal diffusion coefficient between species i and j is expressed as

$$D_{ij} = \frac{F^{ji} - F^{ii}}{m_j |F|} \tag{3.94}$$

where, F_{ij} is an element of the matrix F which is defined as

$$F_{ij} = \frac{1}{\rho}\left[\frac{n_i}{D_{ij}^{b}} + \sum_{l \neq i} \frac{n_l m_j}{m_i D_{il}^{b}}\right]\left(1 - \delta_{ij}\right) \tag{3.95}$$

The binary diffusion coefficient D_{ij}^{b} of a two-temperature plasma considering the interaction between electrons and heavy particles is derived by Rat [60]

$$D_{ij}^{b} = \left(\frac{2\pi \mathrm{k}}{\mu_{ij} T_{ij}^{*}}\right)^{1/2} \frac{3 T_i g_{ij}}{16 n Q_{ij}^{(1,1)}} \tag{3.96}$$

$$g_{ij} = \frac{\left(m_j + m_i (T_i/T_j)\right)^2 \left(m_i + m_j (T_j/T_i)\right)^3}{\left(T_j/T_i\right)^{5/2} \left(m_i + m_j\right)^5} \tag{3.97}$$

The thermal diffusion coefficient of a two-temperature plasma particle is expressed as

$$D_i^{T} = n_i m_i \left(\frac{\mathrm{k} T_i}{2 m_i}\right)^{1/2} a_{i0} \tag{3.98}$$

where, the coefficient a_{i0} can be obtained by solving the linear equations with a definite integral. Ignoring the interaction between electrons and heavy particles, the solution of the Boltzmann equation is greatly simplified such that an analytical expression of the coefficient a_{i0} can be obtained. Devoto [29] presents the fourth-order approximation of the thermal diffusion coefficients of electrons and heavy particles under this simplified condition as follows.

$$D_e^T = \frac{15n_e^2\sqrt{2\pi \mathrm{k}T_e}}{4|q|}\left(\begin{vmatrix} q^{01} & q^{02} & q^{03} \\ q^{21} & q^{22} & q^{23} \\ q^{31} & q^{31} & q^{32} \end{vmatrix}\right) \tag{3.99}$$

$$D_i^T = \frac{15n_i\sqrt{2\pi m_i \mathrm{k}T_h}}{4|q|}\left(\begin{vmatrix} q_{ij}^{00} & q_{ij}^{01} & q_{ij}^{02} & q_{ij}^{03} & 0 \\ q_{ij}^{10} & q_{ij}^{11} & q_{ij}^{12} & q_{ij}^{13} & n_j \\ q_{ij}^{20} & q_{ij}^{21} & q_{ij}^{22} & q_{ij}^{23} & 0 \\ q_{ij}^{30} & q_{ij}^{31} & q_{ij}^{32} & q_{ij}^{33} & 0 \\ \delta_{ji} & 0 & 0 & 0 & 0 \end{vmatrix}\right) \tag{3.100}$$

Under the condition that the collision integral is known, the element q^{mp} and q_{ij}^{mp} can be obtained by solving the linear equations consisting of definite integrals. Its calculation method and the specific expression of the fourth-order approximation expansion are given by Devoto [29, 54]. For the thermal diffusion coefficient considering the interaction between electrons and heavy particles, the analytical expression cannot be given directly due to the complicated calculation process. Rat [58] provides the solution method of square bracket integral linear equations under this condition. Interested readers are directed to the reference [58] for fuller details.

3. Viscosity

Following the study of Hirschfelder et al. [15], the viscosity of the two-temperature plasma can be expressed as

$$\mu = \sum_i \mu_i = \frac{1}{2}\mathrm{k}\sum_i n_i T_i b_{i0} \tag{3.101}$$

Coefficient b_{i0} can be obtained by solving definite linear equations. The contribution of electrons to viscosity is usually negligible due to their very small mass. When only the contribution of heavy particles is considered, the formula of viscosity can be obtained directly by following Devoto's derivation [29]. The expression based on the second-order approximation is

$$\mu = -\frac{5\sqrt{2\pi \mathrm{k}T_h}}{2|q|}\left(\begin{vmatrix} q_{ij}^{00} & q_{ij}^{01} & n_i m_i^{1/2} \\ q_{ij}^{10} & q_{ij}^{11} & 0 \\ n_j & 0 & 0 \end{vmatrix}\right) \tag{3.102}$$

Rat et al. [58] gave an expression for the viscosity considering the contribution of electrons, but found that the results including the electron contribution are very close to those obtained by using the Devoto simplified method. This means the simplified method of Devoto is acceptable in the calculation of viscosity of the two-temperature plasma.

4. Electrical conductivity

A complete definition of the electrical conductivity of a two-temperature plasma can be derived directly from the particle flux [29], and its expression is

$$\sigma = \sum_i \sigma_i = -\sum_i \frac{\mathrm{e}^2 n Z_i}{\rho \mathrm{k}} \sum_j \frac{n_j m_j Z_j D_{ij}}{T_{ij}^*} \tag{3.103}$$

where, n is the total number density of the particles in plasma, Z_i is the charge of species i, and the charge of an electron is $Z_e = -1$. Since the ionic mass is much larger than the electron's, their ability to carry charge for drift is much less than that of the electrons. So in the actual calculation, the contribution of ions is often ignored, and the electrical conductivity is thus calculated with only the contribution of electrons. Devoto only considers the electron contribution in deriving a simplifies expression for electrical conductivity and that based on the third order approximation is

$$\sigma = \frac{\mathrm{e}^2 n n_e m_e}{\rho \mathrm{k} T_e} D_{11} = \frac{3e^2 n_e}{2\mathrm{k}T_e}\sqrt{\frac{2\pi \mathrm{k} T_e}{m_e}} \begin{vmatrix} q^{11} & q^{12} \\ q^{21} & q^{22} \end{vmatrix} \left(\begin{vmatrix} q^{00} & q^{01} & q^{02} \\ q^{10} & q^{11} & q^{12} \\ q^{20} & q^{21} & q^{22} \end{vmatrix} \right)^{-1} \tag{3.104}$$

5. Thermal conductivity

The thermal conductivity of a two-temperature plasma is divided into electron thermal conductivity κ_e and heavy particle thermal conductivity κ_h, which are defined as

$$\kappa_e = \kappa_{tr,e} + \kappa_{reac,e} \tag{3.105}$$

$$\kappa_h = \kappa_{tr,h} + \kappa_{reac,h} + \kappa_{\mathrm{int}} \tag{3.106}$$

where, $\kappa_{tr,e}$, $\kappa_{reac,e}$, $\kappa_{tr,h}$, $\kappa_{reac,h}$ and κ_{int} are, respectively, the translational thermal conductivity of electrons, the reactional thermal conductivity of electron, the translational thermal conductivity of heavy particles, the reactional thermal conductivity of heavy particles and the internal thermal conductivity of heavy particles.

(1) Translational thermal conductivity

The translational thermal conductivity describes the energy transport process caused by particle diffusion due to elastic collisions, which is defined as

$$\kappa_{tr} = -\frac{5\mathrm{k}}{4}\sum_i n_i \left(\frac{2\mathrm{k}T_i}{m_i}\right)^{1/2} a_{i1} \tag{3.107}$$

where, the coefficient a_{i1} can be obtained by solving the linear equations with definite integral. Devoto [29, 54] developed the expression of translational thermal conductivity of electrons and heavy particles ignoring the interaction between electrons and heavy particles

$$\kappa_{tr,e} = \frac{75n_e^2\mathrm{k}}{8}\sqrt{\frac{2\pi \mathrm{k}T_e}{m_e}} \begin{vmatrix} q^{22} & q^{23} \\ q^{32} & q^{33} \end{vmatrix} \left(\begin{vmatrix} q^{11} & q^{12} & q^{13} \\ q^{21} & q^{22} & q^{23} \\ q^{31} & q^{32} & q^{33} \end{vmatrix}\right)^{-1} \tag{3.108}$$

$$\kappa_{tr,h} = -\frac{75k\sqrt{2\pi kT_h}}{8|q|} \left(\begin{vmatrix} q_{ij}^{00} & q_{ij}^{01} & q_{ij}^{02} & q_{ij}^{03} & 0 \\ q_{ij}^{10} & q_{ij}^{11} & q_{ij}^{12} & q_{ij}^{13} & n_i \\ q_{ij}^{20} & q_{ij}^{21} & q_{ij}^{22} & q_{ij}^{23} & 0 \\ q_{ij}^{30} & q_{ij}^{31} & q_{ij}^{32} & q_{ij}^{33} & 0 \\ 0 & \frac{n_j}{m_j^{1/2}} & 0 & 0 & 0 \end{vmatrix}\right) \tag{3.109}$$

(2) Internal thermal conductivity

The internal thermal conductivity describes the energy transport process caused by energy level changes in heavy particles, following the definition given by Hirschfelder et al. [15]

$$\lambda_{\text{int}} = \sum_{i\neq e} x_i D_{ii}^b k\left(C_{P,i}/R - 2.5\right)\left(\sum_{j\neq e}\frac{x_j D_{ii}^b}{x_i D_{ij}^b}\right)^{-1} \tag{3.110}$$

where, x_i and $C_{P,i}$ are the molar ratio and specific heat at constant pressure of species i respectively.

(3) Thermal conductivity of reaction

The reactive thermal conductivity of the two-temperature plasma can be calculated by establishing the heat flux expression ignoring the external force, pressure gradient and thermal diffusion term, and establishing the relationship between the diffusion velocity and the temperature gradient. In this method, the particle system needs to be divided into independent species and non-independent species. The chemical formula of a non-independent species can be expressed as a linear addition of the chemical formula of an independent species. In the actual calculation, the reference state species with zero enthalpy of formation is generally selected as the independent species. Suppose there are N types of species in the system, among which μ types are independent species and the rest $M = N - \mu$ kinds are non-independent species, then the chemical formula of non-independent species X_m (the value range of m is 1 – M) can be expressed as

$$X_m = \sum_{k=1}^{\mu} \alpha_{mk} X_k \tag{3.111}$$

where, α_{mk} is the coefficient of the chemical formula linear relationship between independent and non-independent species. Rat et al. [60] and Chen and Li [61] gave definitions of the thermal conductivity of electrons, the thermal conductivity of heavy particles and the thermal conductivity of the total reaction in a two-temperature plasma.

$$\kappa_{reac,e} = \sum_{i=1}^{M} \Delta h_i \frac{n}{\rho \mathrm{k} T_h} \sum_{j=1}^{N} \frac{T_h}{T_j} m_j D_{ij}^a \frac{\partial p_j}{\partial T_e} \tag{3.112}$$

$$\kappa_{reac,h} = \sum_{i=1}^{M} \Delta h_i \frac{n}{\rho \mathrm{k} T_h} \sum_{j=1}^{N} \frac{T_h}{T_j} m_j D_{ij}^a \frac{\partial p_j}{\partial T_h} \tag{3.113}$$

$$\lambda_{reac} = \theta \lambda_{reac,e} + \lambda_{reac,h} \tag{3.114}$$

where, p_i and Δh_i are, respectively, the partial pressure and enthalpy of formation of species i in the system. It is important to note that the definition of species partial pressure (component) is given as a function of T_e and T_h, rather than the usual T_e and non-equilibrium degree θ. When calculating the partial derivative of partial pressure, the original component equations need to be rewritten, the number density of species in the mass action law equation (or the Gibbs free energy equation) is transformed into the partial pressure of species, and the partial derivatives with regard to T_e and T_h are calculated, respectively, to establish new non-linear equations. Rat et al. [58] provided a detailed introduction to this calculation process.

In addition to the above methods, the LTE plasma reactive thermal conductivity calculation method derived by Brokaw and Butler [62] was also extended to the calculation of the reactive thermal conductivity of the two-temperature plasma. The method is based on species concentration, diffusion velocity and the Van't Hoff equation. Chen and Li [61] extended this method to the calculation of the reactive thermal conductivity of two-temperature monatomic plasma. Wang et al. [63] further extended this method to the calculation of the reactive thermal conductivity of two-temperature polyatomic plasma, with the following expression

$$\lambda_{reac} = \sum_{i} \Delta h_i \overrightarrow{\Psi_i} / \nabla T_h = \sum_{i} \Delta h_i \frac{|H_i|}{|H|} \tag{3.115}$$

where, $\overrightarrow{\Psi_i}$ is the diffusion flux of species, $|H_i|$ and $|H|$ are the determinants related to the diffusion and heat transfer of arc plasma species and the specific expression is given by Wang [63]. The results show that this method is in good agreement with the method proposed by Rat et al. [58], but the limitation of this method is that it is unable to distinguish between the contributions of electrons and heavy particles.

So the calculated values cannot be used to provide input material properties for the modelling of two-temperature arc plasma.

3.4 Calculation of Radiation Emission Coefficient for Arcing Gas

Radiation transfer is one of the key processes of energy transport in an arc plasma. For example, in the process of breaking a short circuit current by a circuit breaker, the energy loss from the arc through radiation contributes to the rate of arc cooling, and it also causes ablation of the electrodes and the nozzle [64]. Gleizes [65], Cressault [66, 67], Aubrecht [68], and Billoux [69] et al. found that the metal vapor from electrode erosion, due to its low ionisation energy and high spectral radiation intensity, significantly affects the radiation process of the arc plasma.

Radiation is a collection of photons, and the equation describing the transfer of radiation (the radiation transfer equation) is a form of particle transport equation. Compared with ordinary particles, the behaviour of photons is not only a function of space, time and velocity, but also a function of frequency or wavelength, and their transfer equation is much more complicated than that of ordinary particles. In arc plasma, the wavelength range of photons is usually from far ultraviolet to infrared [68], which makes it necessary to calculate the radiation process over a wide wavelength range, and the computational complexity is accordingly extremely high.

In order to reduce the computational complexity of radiation transfer in arc plasma, some simplified models, such as the Net Emission model [66, 70], Partial Characteristics model [71, 72] and P-1 model [73], have been proposed and applied with different degrees of success. The Net Emission model and P-1 model are widely used in numerical simulation of switching [5, 74–76] and welding arcs [77, 78]. In the NEC model, the estimation of radiation loss is ultimately reduced to the calculation of net emission coefficient (NEC). In the P-1 model, the key model parameter is the mean absorption coefficient (MAC).

In this book, two radiation transfer models (Net Emission model and P-1 model) are considered for SF_6-Cu arcs. On the basis of theoretical studies of spectral radiation such as atomic line spectrum, atomic continuum spectrum and molecular spectrum, the NEC of the Net Emission model and the mean absorption coefficient of the P-1 model are calculated respectively. The influence of arc radius, copper concentration and arc pressure on the NEC was analysed, and the influence of copper concentration on the Planck average absorption coefficient was studied.

3.4.1 Net Emission and Net Emission Coefficient

1. Overview of the model

Under the assumption of LTE state, if photon diffusion can be ignored, the radiation transfer equation can be simplified to [79]

$$\vec{n} \cdot \vec{\nabla} I_\lambda(\vec{r}, \vec{n}) = \kappa'_\lambda (B_\lambda - I_\lambda) \tag{3.116}$$

where, $\vec{n}$ is the unit vector representing the radiation direction; I_λ is the spectral intensity (/W·m^{-3}·sr^{-1}), which is a function of wavelength λ, position $\vec{r}$ and direction $\vec{n}$; κ'_λ is the absorption coefficient (m^{-1}); B_λ is the blackbody radiation intensity (W·m^{-3}·ster^{-1}), which is given by Planck law:

$$B_\lambda = \frac{2\mathrm{hc}^2}{\lambda^5}\left(\exp\left(\frac{\mathrm{hc}}{\mathrm{k}T\lambda}\right) - 1\right)^{-1} \tag{3.117}$$

where, c is the speed of light (m·s^{-1}).

If the radiation is isotropic, Eq. 3.116 is integrated in the solid angle of Ω to obtain the divergence of the radiation flux $\vec{F_R}$ [79],

$$\nabla \cdot \vec{F_R} = 4\pi \left(\varepsilon_\lambda - \kappa'_\lambda J_\lambda\right) \tag{3.118}$$

where, J_λ is the average spectral intensity (W·m^{-2}·ster^{-1}), defined as follows:

$$J_\lambda = \frac{1}{4\pi} \int_{4\pi} I_\lambda \, d\Omega \tag{3.119}$$

Without additional assumptions, the radiation equation described in Eq. 3.119 is almost impossible to solve directly. In Lowke's [66] model based on net emission, the effect of radiation transfer is described by the following term for an isothermal sphere with a radius of *R*:

$$\nabla \cdot \vec{F_R} = 4\pi \varepsilon_N \tag{3.120}$$

$$\varepsilon_N = \int_0^\infty B_\lambda \kappa_\lambda \exp(-\kappa_\lambda \mathrm{R}) d\lambda \tag{3.121}$$

where, ε_N is the net emission coefficient (W·m^{-3}·ster^{-1}); κ_λ is the absorption coefficient taking into accounting the induced emission (m^{-2}).

The physical definition of the net emission coefficient, ε_N, is the radiative power in a unit solid angle from a unit volume of space at the arc centre that leaves the isothermal sphere. It can be seen from the above equation that the calculation of ε_N ultimately boils down to the acquisition of the absorption coefficient κ_λ. In order to simplify the calculation, we assume that the spectral lines are separated from each other, and κ_λ can be obtained from the following equation [79].

$$\kappa_\lambda = \left[\kappa_c' + \kappa_0' P(v)\right]\left[1 - \exp\left(-\frac{\mathrm{hc}}{\mathrm{k}\lambda T}\right)\right] \tag{3.122}$$

where, κ_c' is the continuum spectral absorption coefficient (m^{-1}); κ_0' is the line spectral absorption coefficient (m^{-1}); $P(v)$ is the function describing the geometric contour of the line spectrum. The term in the second pair of square brackets is a multiplier to take into account of the induced emission, which is described by Kirchhoff's law [79–81].

Let us consider the net emission coefficient, $\varepsilon_{N\lambda_0}$, associated with the central wavelength of the line spectrum, λ_0. Compared with the whole line width, the spectral bandwidth, $\Delta\lambda$, associated with λ_0 is very narrow, so the blackbody radiation intensity B_λ and exponential factor exp(-$hc/k\lambda T$) can be considered to remain the same within the bandwidth, so that the continuum contribution towards the net emission coefficient can be calculated separately. Based on this consideration, $\varepsilon_{N\lambda_0}$ can be obtained from the following formula [64, 79].

$$\varepsilon_{N\lambda_0} = B_{\lambda_0}\kappa_0'\left[1 - \exp\left(-\frac{\mathrm{hc}}{\mathrm{k}\lambda T}\right)\right]\int_0^\infty P(\lambda)\exp(-\kappa_\lambda \mathrm{R})d\lambda \tag{3.123}$$

By combining Eq. 3.123 with Eq. 3.121 and assuming that the continuum spectral absorption coefficient κ_c' remains constant within the spectral line bandwidth, Eq. 3.124 can be obtained [79]

$$\varepsilon_{N\lambda_0} = B_{\lambda_0}\kappa_0'\left[1 - \exp\left(-\frac{\mathrm{hc}}{\mathrm{k}\lambda T}\right)\right]\exp\left(-\kappa_c' R\right)\Lambda_{ul} \tag{3.124}$$

where, Λ_{ul} is the escape factor, as defined below by

$$\Lambda_{ul} = \int_0^\infty P(\lambda)\exp\left(-\kappa_0' P(\lambda)\mathrm{R}\right)d\lambda \tag{3.125}$$

To sum up, the calculation of the emission coefficient (NEC) mainly includes two parts: line spectral absorption coefficient and continuum spectral absorption coefficient. To calculate the escape factor, a line spectral profile function $P(\lambda)$ is required. The Gaussian, Lorentzian, and Voigt functions are commonly used.

2. Atomic and molecular radiation calculation

(1) Atomic line spectrum

Atomic line spectra are associated with atomic energy level transition. Spontaneous emission occurs when an atom transitions from a high energy level u to a low energy level l, and its emission coefficient ε_{ul} can be expressed as in Eq. 3.126 [64]

$$\varepsilon_{ul}(\lambda) = \frac{\mathrm{hc}}{4\pi\lambda_{ul}} A_{ul} N_u P(\lambda) \tag{3.126}$$

where, λ_{ul} is the wavelength (m) of the spontaneously emitted line that transitions from high energy level u to low energy level l; A_{ul} is the probability (s^{-1}) of spontaneous transition and N_u is the population number density of the atom at energy level u (m^{-3}).

The spectral lines corresponding to atomic level transitions are not strictly monochromatic, but have certain geometric profile. The function $P(\lambda)$ describes the spectral profile of a line spectrum. This function is the key to calculating the escape factor Λ_{ul}. Introducing optical thickness τ_{ul}, the definition of Λ_{ul} can be rewritten as follows in Eq. 3.127 [64]

$$\Lambda_{ul} = \int_0^{\infty} P(\lambda)\exp\left(-\tau_{ul}\frac{P(\lambda)}{P(\lambda = \lambda_{ul})}\right)d\lambda \tag{3.127}$$

The optical thickness of the line spectrum λ_{ul} can be calculated by the following

$$\tau_{ul} = \frac{\pi \mathrm{e}^2}{m_e \mathrm{c}} f_{lu} N_l \mathrm{R} P(\lambda = \lambda_{ul}) \tag{3.128}$$

where, f_{lu} is the absorption oscillator strength; N_l is the particle density of atoms in energy level l (m^{-3}); and m_e is the mass of an electron (kg).

Each spectral line has a definitive profile. The profile of the atomic spectral line can be characterised by the maximum intensity of the spectral line I_0 and its full-width at half maximum (FWHM), $\Delta\lambda$, as shown in Fig. 3.1. In the case of a pure line spectral, the FWHM of the spectral line, $\Delta\lambda$, is the corresponding wavelength range when the intensity of the spectral line drops to half of the maximum value. If the radiation spectrum is made up of a continuous spectrum and a line spectrum, the continuum spectrum as the background needs to be removed from the line, and then the FWHM of the line can be calculated.

The profile of the spectral line is related to the broadening of the spectral line. In plasma, the origin of line emission disturbance and the resultant line broadening are the result of the interaction between the radiating particle and its surrounding particles. The more frequently the radiating particles are disturbed, the shorter the emission period is and the wider the spectral line profile is. Conversely, with less disturbance, the emission period is longer, and the spectral line profile is narrower. As the emission period approaches infinity, the line becomes a sharp line (the Dirac function) [82].

The mechanisms of spectral line broadening include natural broadening, Doppler broadening, pressure broadening and Stark broadening. These are discussed separately below.

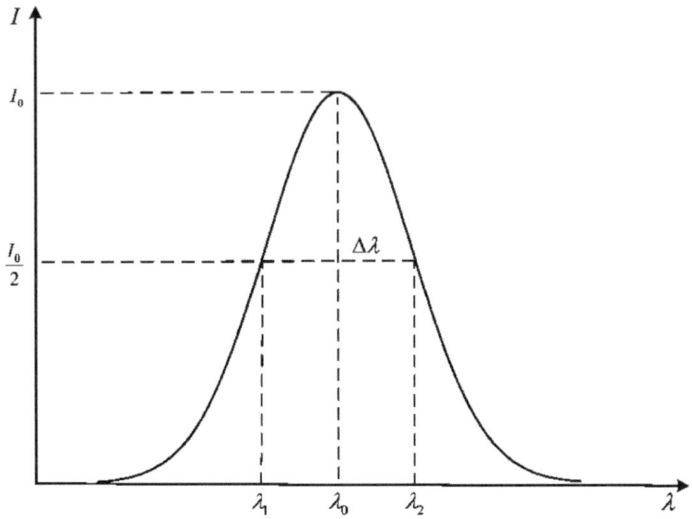

Fig. 3.1 Outline of spectral lines

① Natural broadening

Natural broadening is the simplest broadening mechanism of spectral lines, which is caused by spontaneous emission. The average time, Δt, that electrons stay at a certain energy level, E, follows the Heisenburg uncertainty principle [83].

$$\Delta E \cdot \Delta t \propto \frac{\mathrm{h}}{2\pi} \tag{3.129}$$

where ΔE is the uncertainty of energy (J). This indicates that the photon energy corresponding to the energy level transition is not a specific value, but a range of energy, resulting in a natural broadening of the spectral line width. The natural broadening profile is described by the Lorentzian function [79].

$$P_N(\lambda) = \frac{\delta_N/2}{\pi\left[(\lambda - \lambda_0)^2 + (\delta_N/2)^2\right]} \tag{3.130}$$

where, δ_N is the FWHM. The natural broadening width is much smaller than other widths and is not easily measured, so it is often overlooked.

② Doppler broadening

Doppler broadening is caused by the random thermal motion of particles. According to the Doppler theorem, when a particle moves at a certain speed relative to the receiver, the wavelength of the photon emitted by the particle changes, and this is called Doppler broadening.

Assuming that the velocity of particles conforms to Maxwell–Boltzmann distribution, the profile function of Doppler broadening can be obtained as [79].

$$P_D(\lambda) = \frac{2\sqrt{\ln 2}}{\delta_D\sqrt{\pi}} \exp\left[-\ln 2\left(\frac{\lambda - \lambda_0}{\delta_D/2}\right)^2\right] \tag{3.131}$$

where δ_D is the FWHM, given by the following expression [64]

$$\delta_D = \lambda_0\sqrt{\frac{8\ln 2 \cdot kT}{mc^2}} \tag{3.132}$$

where m is the mass of the emitting atom (kg). Doppler broadening as given by Eq. 3.132 is a Gaussian function. Doppler broadening is generally larger than natural broadening by two orders of magnitude [83].

③ Pressure broadening

Pressure broadening is caused by the interaction between emitting particles and surrounding particles, also known as collisional broadening. It depends mainly on the plasma pressure and the number density of particles [83]. It is very difficult to describe the profile of the pressure broadening which is affected by the action potential of the atoms or ions colliding with each other [84]. In the simplified model, the pressure broadening is usually divided into van der Waals broadening and resonance broadening according to the types of colliding particles. Stark broadening is also listed under the category of pressure broadening in some literature, but it is placed separately in this book and discussed below.

When the emitting particle and the interfering particle are neutral heteronuclear molecules, or both are co-nuclear molecules, but the emitting particle is still in an excited state after emitting a photon, the interaction between the emitting particle and the interfering particle is described by van der Waals force, and the spectral line broadening is represented by van der Waals broadening [64]. In practical calculation, the van der Waals force is characterised by the sixth order Lennard–Jones potential function.

The van der Waals broadening profile is described by the Lorentzian function. Based on the work of Walkup et al. [85], the FWHM, δ_V, of van der Waals broadening can be estimated by the following expression [81].

$$\delta_V = 8.16 \cdot C_6^{2/5}\overline{V}^{3/5}N^* \frac{\lambda_0^2}{2\pi c} \tag{3.133}$$

where, C_6 is the Van der Waals constant (m^6); $\overline{V}$ is the average velocity of the colliding particles ($m{\cdot}s^{-1}$); and N^* is the number density of the interfering particle (m^{-3}).

When the emitting particle and the interfering particle are neutral homonuclear atoms or molecules, and the emitting particle is in steady state after emitting a photon, the broadening of the spectral line is called resonance broadening [64]. In practical calculation, third order Lennard–Jones action potentials are used to describe the resonant impact forces.

The spectral profile of resonance broadening is also described by the Lorentzian function, whose FWHM, δ_R, is given by the following expression [64, 81].

$$\delta_R = 5.48 \cdot \sqrt{\frac{g_0}{g_r}} N_0 \frac{e^2}{4\varepsilon_0} \frac{f_r}{m_e} \frac{\lambda_0^3}{(2\pi c)^2} \tag{3.134}$$

where, g_0 is the degeneracy of the ground state energy level of the particle; g_r is the degeneracy of the resonant state of the particle; N_0 is the number density of the ground state particles (m^{-3}); f_r is the strength of the resonant state oscillator; and m_e is electron mass (kg).

④ Stark broadening

When the electric field of a charged particle interferes with the energy level of the emitting particle, the spectral line will also be affected. In this case, the spectral line broadening is called Stark broadening. When the emitting particle interacts with the interfering particle for longer than the particle collision period, Stark broadening needs to be considered [82]. For example, when the interfering particle is an ion, its velocity is much lower than that of an electron, and the emitting particle is subjected to an ionic electric field before colliding with it. In this case Stark broadening should be considered. For plasma spectra, compared to van der Waals broadening and resonance broadening, Stark broadening plays a dominant role when the ionisation degree exceeds 1% [83].

In practical calculation, fourth order Lennard–Jones potentials are used to describe the forces involved in Stark broadening. For neutral atoms, the FWHM of Stark broadening can be expressed as [64, 81]

$$\delta_S^{atom} = 11.37 \cdot C_4^{2/3} V_e^{1/3} N_e \frac{\lambda_0^2}{2\pi c} \tag{3.135}$$

For ions, the FWHM of Stark broadening is obtained by the following expression [64, 81]

$$\delta_S^{ion} = \frac{200.8}{631} \cdot C_4^{2/5} Z^{4/5} \frac{N_e}{\sqrt{T}} \frac{\lambda_0^2}{2\pi c} \tag{3.136}$$

where, C_4 is the Stark constant describing the force (m^4); V_e is the average velocity of the electron ($m \cdot s^{-1}$); N_e is electron number density (m^{-3}); and Z is the charge of the emitted particle.

(2) Atomic continuum spectrum

In addition to the line spectrum, atomic radiation in the plasma also produces a continuum spectrum. Examples of work on this topic are [80, 81, 86–88]. The constituting mechanisms for atomic continuum spectrum mainly includes recombination radiation and braking radiation. In the arc, these two processes take place at the same time to form the continuum.

① Recombination radiation

When a free electron is captured by a heavy particle, the recombination energy is emitted as a photon, called the recombination radiation. The energy conservation equation of this process is [83]

$$\frac{1}{2}m_e v_e^2 + \Delta E = \mathrm{h}\frac{\mathrm{c}}{\lambda} \tag{3.137}$$

where, m_e is electron mass (kg); v_e is the electron velocity (m s^{-1}); and ΔE is the recombination energy (eV).

The recombination energy is generally fixed for a particular energy level, while the kinetic energy of the free electron is continuous, so the wavelength of the emitted photon is also continuous.

Depending on the type of particles that capture the electrons, recombination radiation is divided into two categories:

In the first type of recombination radiation, a neutral atom captures a free electron and becomes an ion, releasing a photon [87]: $\mathrm{A} + \mathrm{e}^- \leftrightarrow \mathrm{A}^- + hv$. The emission coefficient under this mechanism, ε_{att}, can be calculated as [80].

$$\varepsilon_{att} = \frac{2\mathrm{hc}^2}{\lambda^5}\exp\left(-\frac{\mathrm{hc}}{\mathrm{k}\lambda T}\right)N_{A^-}\sigma_{\det}(\lambda) \tag{3.138}$$

where, N_{A-} is the number density (m^{-3}) of A^-; and $\sigma_{\det}$ is the photo-induced detachment cross-section (m^2) of A^-.

In the second type of recombination radiation, an ion charged at Z+ captures a free electron and becomes an ion charged at (Z − 1)+, releasing a photon [87]: $\mathrm{A}^{Z+} + \mathrm{e}^- \leftrightarrow \mathrm{A}^{(Z-1)+} + hv$. In this type of radiation, the emission coefficient ε_{rec} is calculated according to the work in [80].

$$\varepsilon_{rec} = \frac{16\pi\left(e^2/4\pi\varepsilon_0\right)^3}{3\mathrm{c}^3\sqrt{6\pi k m_e^3}}\frac{\mathrm{c}}{\lambda^2}\frac{N_e N_{Z+}}{Q_{Z+}^{\mathrm{int}}}\frac{Z^2}{\sqrt{T}}\left[1-\exp\left(-\frac{\mathrm{hc}}{\mathrm{k}\lambda T}\right)\right]g_0^{Z+}\xi^{(Z-1)+}(\lambda, T) \tag{3.139}$$

where, N_e is the number density (m^{-3}) of electrons; N_{Z+} is the number density (m^{-3}) of charged particles A^{Z+}; Q^{int}_{Z+} is the internal partition function of charged particle A^{Z+}; g_0^{Z+} is the degeneracy of the ground state energy level of charged particle A^{Z+}; $\xi^{(Z-1)+}$ is the Bibermann-Schlüter factor of charged particle $A^{(Z-1)+}$, which is a compensation for the non-hydrogen-like structure in the atomic structure and can be obtained by summation of the photo-ionisation cross-sections of each energy level of the atom [86].

② Braking radiation

In plasma, free-moving electrons are affected by an electric field. When the electron decelerates in the electric field and changes from a high-energy free state to a low-energy free state, the energy difference is emitted as photons. This radiation is called bremsstrahlung [86]. Since the electron energy is continuous, the spectral produced by this mechanism is also continuous.

In bremsstrahlung, according to different types of interfering particles, the calculation of the emission coefficient should be divided into two categories: neutral atom, ε_{ea}, and ion, ε_{ei}. Their calculation is given by in [80, 86] and [87] et al., and is listed below

$$\varepsilon_{ea} = \frac{32}{3c^3}\frac{e^2}{4\pi\varepsilon_0}\left(\frac{k}{2\pi m_e}\right)^{3/2}\frac{c}{\lambda^2}N_A N_e T^{3/2}\exp\left(-\frac{hc}{k\lambda T}\right)G_{ea}(\lambda, T) \tag{3.140}$$

$$\varepsilon_{ei} = \frac{16\pi\left(e^2/4\pi\varepsilon_0\right)^3}{3c^3\sqrt{6\pi k m^3}}\frac{c}{\lambda^2}\frac{Z^2 N_{A+}N_e}{\sqrt{T}}\exp\left(-\frac{hc}{k\lambda T}\right)G_{ei}(\lambda, T) \tag{3.141}$$

where, N_A is the number density (m^{-3}) of the neutral atom; N_{A+} is the number density (m^{-3}) of the ions; G_{ea} is the collision cross section for the interaction between electrons and neutral atoms (m^2); and G_{ei} is the Gaunt factor describing the "free-free" transition of atoms.

(3) Molecular spectrum

When the temperature of plasma is low, molecules are the dominant components in the composition and molecular spectra dominate the radiation transfer. However, the effects of molecular radiation are often ignored in studies of arc radiation emission in the literature. For example, when arc plasma radiation in SF_6-N_2 [64, 86] is studied, N_2 and SF_6 molecules and their decomposition products were not included in the calculation. Only N, S, F and their ions were considered. This is mainly because molecular structure is much more complex in comparison with the atomic structure. The internal energy of a molecule includes not only electronic energy, but also vibrational and rotational energies. Each electronic energy level of the molecule contains a group of vibrational energy levels, and each vibrational energy level contains a group of rotational energy levels. The molecular spectrum is therefore much more complex than the atomic spectrum. In addition, due to the lack of fundamental spectral data, the emission coefficients of many molecules, especially polyatomic molecules, are difficult to calculate.

For a diatomic molecule, a transition between two rotational energy levels (J' and J'') emits a photon, resulting in a spectrum of molecular bands, whose emission coefficients are shown in the following expression [79, 89].

$$\varepsilon_{band} = \frac{\mathrm{hc}}{4\pi\lambda} A_{n'',v'',K'',J''}^{n',v',K',J'} N\left(n', v', K', J'\right) \quad (3.142)$$

where, $N\left(n', v', K', J'\right)$ is the number density of particles in energy level $\left(n', v', K', J'\right)$ (m^{-3}); $A_{n'',v'',K'',J''}^{n',v',K',J'}$ is the transition probability related to the energy level transition (s^{-1}).

For polyatomic molecules, the calculation of banded spectra is very difficult due to the lack of spectral data, so only the contribution of continuum spectra is usually considered in the actual calculation. The molecular continuum spectra are generally produced by two mechanisms: photo-decomposition ($hv + \mathrm{AB} \leftrightarrow \mathrm{A} + \mathrm{B}$) and photo-ionisation ($hv + \mathrm{AB} \leftrightarrow \mathrm{AB}^{+} + \mathrm{e}^{-}$). With these two mechanisms, the absorption coefficient corresponding to the molecular continuum spectra can be obtained from the photo-decomposition and photo-ionisation cross-sections of each particle [90].

$$\kappa_{molecule} = \sum_{i} N_i \sigma_i \quad (3.143)$$

where, N_i is the number density of particle i (m^{-3}); σ_i is photo-decomposition cross-section or photo-ionization cross-section (m^2).

3.4.2 P-1 Radiation Model and Average Absorption Coefficient

1. Overview of the model

The radiation transfer equation is a complicated non-linear differential-integral equation. The net emission coefficient (NEC) model is one of the most widely used simplified models. In the NEC model, the radiative energy is calculated by the NEC without solving the radiation transfer equation. However, the NEC model assumes that the arc plasma is axisymmetric and that the NEC is related to the radius of the cylindrical arc. This limits its use to certain extent. For example, free-burning arc and low-voltage switching arc do not maintain axisymmetry. In addition, the NEC model can hardly handle radiation re-absorption, especially at the arc edge. Therefore, a more elaborate model is needed to solve the radiation energy transfer problem of arcs with complex shapes.

The spherical harmonic method (P-N method) is an approximate method of the radiation transfer equation. The P-N method is to expand the radiation intensity according to the solid angle into a spherical harmonic function and transforms the radiation transfer equation into a set of differential equations with the expansion coefficient. N is the approximate order of the system. When N approaches infinity, the approximate solution from the P-N method is the exact solution of the radiation transfer equation [82]. In the P-N method, the spectral line intensity, I_λ, is obtained by the following expression.

$$I_\lambda(\vec{r}, \vec{n}) = \sum_{l=0}^{\infty} \sum_{m=-l}^{l} I_l^m(\vec{r}) Y_l^m(\vec{n}) \tag{3.144}$$

where, $\vec{n}$ is the unit vector representing the radiation direction; I_l^m is the Fourier series related to spatial position; Y_l^m is a spherical harmonic function, which usually takes the form of Legendre polynomials.

When the value of N is 1, the first order approximation of the P-N method is obtained, namely the P-1 radiation model. The expression of its spectral intensity is [73]:

$$I_\lambda(\vec{r}, \vec{n}) = \frac{1}{4\pi}\left(G_\lambda + 3\vec{F_R} \cdot \vec{n}\right) \tag{3.145}$$

$$G_\lambda = \int_{4\pi} I_\lambda(\vec{r}, \vec{n})\, d\Omega \tag{3.146}$$

$$\vec{F_R} = \int_{4\pi} I_\lambda(\vec{r}, \vec{n})\, \vec{n}\, d\Omega \tag{3.147}$$

where, G_λ is the incident radiation (W·m^{-3}); and $\vec{F_R}$ is the radiative flux (W·m^{-3}).

Thus, the conservation equation of the P-1 model can be obtained as [73]

$$\nabla \cdot \left(\frac{1}{\kappa_\lambda} \nabla G_\lambda\right) = 3\kappa_\lambda (G_\lambda - 4\pi B_\lambda) \tag{3.148}$$

where, κ_λ is the absorption coefficient (m^{-1}) and B_λ is the spectral black-body intensity (W·m^{-3}·ster^{-1}).

The divergence of the radiative flux $\vec{F_R}$ is obtained by the incident radiation G_λ [73]

$$\nabla \cdot \vec{F_R} = \kappa_\lambda (4\pi B_\lambda - G_\lambda) \tag{3.149}$$

The complete P-1 model should also include the boundary conditions of the conservation equation. Common boundaries include: Dirichlet condition and Marshak condition. This has come under the category of computational MHD and will not be discussed in this chapter.

Although the P-1 model is only the first-order approximation of the P-N method, it can obtain high approximation accuracy in many cases [82]. Compared with the

NEC model, the P-1 model needs to solve additional conservation equations, which increases the cost of numerical calculation.

2. Calculation of mean absorption coefficient

In order to solve the conservation Eq. 3.149, it is necessary to integrate over the whole wavelength range, which undoubtedly makes the calculation extremely burdensome. In practical calculation, the multi-band method is generally adopted: the wavelength is divided into several wavelength bands. In each band, all photons are treated exactly the same, that is, it is assumed that these photons have average properties, such as average absorption coefficient [82]. In the study of SF_6 arc radiation transport [73, 91], the wavelength was divided into five bands [73]. In the numerical simulation of air arc [5], the wavelength was divided into six bands.

Assuming that the radiation wavelength is divided into N_{bands} bands, and the wavelength range of the i-th band is $[\lambda_a, \lambda_b]$, then the average incident radiation of the band, G_i, and average radiative flux of the band, $\overrightarrow{F_i}$, are respectively [73]

$$G_i = \int_{\lambda_b}^{\lambda_a} G_\lambda d\lambda \tag{3.150}$$

$$\overrightarrow{F_i} = \int_{\lambda_b}^{\lambda_a} \overrightarrow{F_\lambda} d\lambda \tag{3.151}$$

The total incident radiation, G, and radiative flux, $\overrightarrow{F}$, are respectively [73]

$$G = \sum_{i=1}^{N_{bands}} G_i \tag{3.152}$$

$$\overrightarrow{F} = \sum_{i=1}^{N_{bands}} \overrightarrow{F_i} \tag{3.153}$$

In order to solve the conservation Eq. 3.149 under the multi-band method, it is necessary to calculate the mean absorption coefficient, $\overline{\kappa_i}$. The radiation absorption coefficient is a function of wavelength, and different mean absorption coefficients can be obtained when different weighting functions are used. The Planck average and Rosseland average are commonly used for the absorption coefficient.

The Planck average is generally suitable for optically thin systems. The absorption coefficient based on the Planck average is defined as [92]:

$$\overline{\kappa_i} = \frac{\int_{\lambda_a}^{\lambda_b} \kappa_\lambda B_\lambda d\lambda}{\int_{\lambda_a}^{\lambda_b} B_\lambda d\lambda} \tag{3.154}$$

Rosseland average is generally applicable to re-absorption dominated systems. The absorption coefficient based on Rosseland average is defined as [92]:

$$\overline{\kappa_i^{-1}} = \frac{\int_{\lambda_a}^{\lambda_b} \kappa_\lambda^{-1} \frac{dB_\lambda}{dT} d\lambda}{\int_{\lambda_a}^{\lambda_b} \frac{dB_\lambda}{dT} d\lambda} \tag{3.155}$$

There is no definitive answer for the suitability of the two methods for radiation calculation of arc plasma. Randrianandraina et al. [93] compared the effects of the Rosseland average and Planck average on the radiation transfer in an SF_6 arc. It is found that the P-1 model based on the mean absorption coefficient of Rosseland performed poorly in the high temperature region of the arc, and the calculated flux divergence was much smaller than the result of accurate calculation, whereas the Planck average produces result that was closer to the accurate value. For this reason, the Planck average method is used in this book to calculate the mean absorption coefficient.

3. Classification and division of radiative spectrum

The absorption coefficient of arc plasma is composed of three parts: absorption coefficient of continuum spectrum (including molecular and atomic continuum spectra), absorption coefficient of atomic line spectrum and absorption coefficient of molecular band spectrum. The radiation absorption coefficient is a function of wavelength, temperature, pressure and composition. Figure 3.2 shows the variation of the absorption coefficient of an SF_6-Cu arc with wavelength at different temperatures. The volume concentration of copper in the composition is 10%. Due to the combined effect of various spectral lines (especially atomic lines), the change of radiation absorption coefficient for an SF_6-Cu arc with wavelength is not a smooth function, but shows a drastic fluctuation in some wavebands, even a step change. It is therefore not straightforward to average such a quantity, and it's difficult to be precise. In order to reduce the uncertainty, the wavelength should be divided into as many bands as possible in the presence of abrupt changes.

According to the work in [94], the radiation spectral of the SF_6-Cu arc plasma is divided into six bands, as shown in Table 3.1.

3.5 Summary

This chapter is devoted to the calculation methods of material properties of arc plasma, which include composition, thermophysical properties, transport coefficients (including the combined diffusion coefficient), and radiation emission coefficients. Local thermodynamic Equilibrium (LTE) may not hold at the edge of the hot core of a high current arc column where temperature is fast decreasing or in the arc column during the current zero period. It is very important to consider the non-equilibrium effects in certain real applications. For this purpose, this chapter addresses the material properties of arc plasma under different conditions, such as LTE, non-thermodynamic equilibrium and non-chemical equilibrium. The calculation of radiation emission coefficient of arc plasma is also considered.

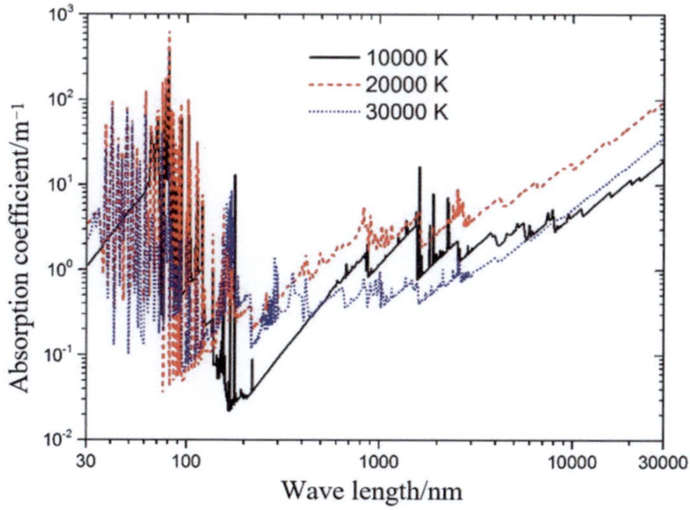

Fig. 3.2 Absorption coefficient of SF6-Cu arc plasma at different temperatures (wavelength range: 30–30,000 nm, calculated pressure: 1 bar, volume concentration of copper: 10%)

Table 3.1 P-1 wavelength bands used for radiation calculation in SF6-Cu arc plasma

Band number *i*	Wavelength range (nm)
1	30.0–44.1
2	44.1–73.1
3	73.1–149.9
4	149.9–299.8
5	299.8–3000.0
6	3000.0–30,000.0

References

1. Rong M Z, Zhong L L, Wang X H. Review of microscopic property calculation of equilibrium and non-equilibrium arc plasma. Transactions of China Electrotechnical Society, 2016, 31(19):54-65.
2. Guo Z Y. Arc and thermal plasma[M]. Beijing: Science Press, 1986.
3. Kuiken H K. An asymptotic treatment of the Elenbaas–Heller equation for a radiating wall-stabilized high-pressure gas-discharge arc[J]. Journal of Applied Physics, 1991, 70(10):5282.
4. Liu X M, Yuan D Y, Cao Y D. 2D arc modeling and simulation of SF6 circuit breaker based on magnetohydrodynamics[J]. Journal of Shenyang University of Technology, 2009, 31(3):257-261.
5. Wu Y, Rong M Z, Yang F. Introduction of 6-band P-1 radiation model for numerical analysis of three-dimensional air arc plasma[J]. Acta Physica Sinica, 2008, 57(9):5761-5767.
6. Gleizes A, Gonzalez J J, Freton P. Thermal plasma modelling[J]. Journal of Physics D: Applied Physics, 2005, 38(9).
7. Drellish K S, Aeschlim D P, Cambel A B. Partition functions and thermodynamic properties of nitrogen and oxygen plasmas[J]. Physics of Fluids, 1965, 8(9):1590.

8. Gordon S, McBride B J. Computer program for calculation of complex chemical equilibrium composition, rocket performance, incident and reflected shocks, and Chapman–Jouguet detonations[R]. NASA publication SP-273, 1971.
9. Murphy A B. Thermal plasmas in gas mixtures[J]. Journal of Physics D: Applied Physics, 2001, 34(20).
10. Coufal O. Composition and thermodynamic properties of thermal plasma up to 50 kK[J]. Journal of Physics D: Applied Physics, 2007, 40(11):3371-3385.
11. Fauchais P, Boulos M I, Pfender E. Thermal plasmas—fundamentals and applications Vol I[M]. New York: Plenum, 1994.
12. Kovitya P. Physical properties of high-pressure plasmas of hydrogen and copper in the temperature range 5000–60000 K[J]. IEEE Transactions on Plasma Science, 1985, 13(6):587-594.
13. Chapman S, Cowling T G. The mathematical theory of non-uniform gases[M]. 3rd edn. Cambridge: Cambridge University Press, 1970.
14. Ferziger J H, Kaper H G. Mathematical theory of transport processes in gases[M]. 2nd edn. Amsterdam: North-Holland, 1972.
15. Hirschfelder J O, Curtis C F, Bird R B. Molecular theory of gases and liquids[M]. 2nd edn. New York: Wiley, 1964.
16. Aubreton J, Elchinger M F, Hacala A, et al. Transport coefficients of typical biomass equimolar CO2–H2 plasma[J]. Journal of Physics D: Applied Physics, 2009, 42:0952069.
17. Sourd B, Aubreton J, Elchinger M F, et al. High temperature transport coefficients in e/C/H/N/O mixtures[J]. Journal of Physics D: Applied Physics, 2006, 39(6):1105-1119.
18. Aubreton J, Elchinger M F, Vinson J M. Transport coefficients in water plasma. part I: equilibrium plasma[J]. Plasma Chemistry and Plasma Processing, 2009, 29(2):149–171.
19. Andre P, Aubreton J, Clain S, et al. Transport coefficients in thermal plasma, applications to Mars and Titan atmospheres[J]. European Physical Journal D, 2010, 57(2):227-234.
20. Capitelli M, Cappelletti D, Colonna G, et al. On the possibility of using model potentials for collision integral calculations of interest for planetary atmospheres[J]. Chemical Physics, 2007, 338(1):62-68.
21. Laricchiuta A, Colonna G, Bruno D, et al. Classical transport collision integrals for a Lennard-Jones-like phenomenological model potential[J]. Chemical Physics Letters, 2007, 445(4-6):133-139.
22. Lombardi A, Palazzetti F. A comparison of interatomic potentials for rare gas nanoaggregates[J]. Journal of Molecular Structure: Theochem, 2008, 852(1-3):22-29.
23. Bruno D, Catalfamo C, Capitelli M, et al. Transport properties of high-temperature Jupiter atmosphere components[J]. Physics of Plasmas, 2010, 17:11231511.
24. Laricchiuta A, Bruno D, Capitelli M, et al. High temperature Mars atmosphere. Part I: transport cross sections[J]. European Physical Journal D, 2009, 54(3):607–612.
25. Chervy B, Gleizes A, Razafinimanana M. Thermodynamic properties and transport coefficients in SF6–Cu mixtures at temperatures of 300–30000 K and pressures of 0.1–1 MPa[J]. Journal of Physics D: Applied Physics, 1994, 27(6):1193.
26. Wang X H, Zhong L L, Cressault Y, et al. Thermophysical properties of SF6–Cu mixtures at temperatures of 300–30000 K and pressures of 0.01–1.0 MPa: part 2. Collision integrals and transport coefficients[J]. Journal of Physics D: Applied Physics, 2014, 47(49):495201.
27. Cressault Y, Gleizes A. Thermodynamic properties and transport coefficients in Ar–H2–Cu plasmas[J]. Journal of Physics D: Applied Physics, 2004, 37(4):560-572.
28. André P, Bussière W, Rochette D. Transport coefficients of Ag–SiO2 plasmas[J]. Plasma Chemistry and Plasma Processing, 2007, 27(4):381-403.
29. Devoto R S. Simplified expressions for the transport properties of ionized monatomic gases[J]. Physics of Fluids, 1967, 10(10):2105.
30. Rat V, Murphy A B, Aubreton J, et al. Treatment of non-equilibrium phenomena in thermal plasma flows[J]. Journal of Physics D: Applied Physics, 2008, 41(18):183001.
31. Rat V, André P, Aubreton J, et al. Two-temperature transport coefficients in argon–hydrogen plasmas—i: elastic processes and collision integrals[J]. Plasma Chemistry and Plasma Processing, 2002, 22(4):453-474.

32. Aubreton J, Elchinger M F, Rat V, et al. Two-temperature transport coefficients in argon–helium thermal plasmas[J]. Journal of Physics D: Applied Physics, 2004, 37(1):34-41.
33. Ghorui S, Heberlein J, Pfender E. Thermodynamic and transport properties of two-temperature nitrogen–oxygen plasma[J]. Plasma Chemistry and Plasma Processing, 2008, 28(4):553-582.
34. Colombo V, Ghedini E, Sanibondi P. Thermodynamic and transport properties in non-equilibrium argon, oxygen and nitrogen thermal plasmas[J]. Progress in Nuclear Energy, 2008, 50(8):921-933.
35. Colombo V, Ghedini E, Sanibondi P. Two-temperature thermodynamic and transport properties of argon–hydrogen and nitrogen–hydrogen plasmas[J]. Journal of Physics D: Applied Physics, 2009, 42:0552135.
36. Colombo V, Ghedini E, Sanibondi P. Two-temperature thermodynamic and transport properties of carbon–oxygen plasmas[J]. Plasma Sources Science and Technology, 2011, 20(3):0350033SI.
37. Wang W Z, Rong M Z, Yan J D, et al. Thermophysical properties of nitrogen plasmas under thermal equilibrium and non-equilibrium conditions[J]. Physics of Plasmas, 2011, 18:11350211.
38. Wu Y, Chen Z, Cressault Y, et al. Two-temperature thermodynamic and transport properties of SF6–Cu plasmas[J]. Journal of Physics D: Applied Physics, 2015, 48(41):415205.
39. Yang F, Chen Z, Wu Y, et al. Two-temperature transport coefficients of SF6–N2 plasma[J]. Physics of Plasmas, 2015, 22(10):103508.
40. Murphy A B. Transport coefficients of hydrogen and argon–hydrogen plasmas[J]. Plasma Chemistry and Plasma Processing, 2000, 20(3):279-297.
41. Prigogine I. Le domaine de validité de la thermodynamique des phénomènes irréversibles[J]. Physica, 1949, 15(1):272-284.
42. Potapov A V. Chemical equilibrium of multitemperature systems (law of mass action derivation of chemical equilibrium ionization of multitemperature system when electron and heavy particle temperature differ)[J]. Teplofizika Vysokikh Temperatur, 1966, 4:55-58.
43. Van de Sanden M C M, Schram P, Peeters A G, et al. Thermodynamic generalization of the Saha equation for a two-temperature plasma[J]. Physical Review A, 1989, 40(9):5273.
44. White W B, Johnson S M, Dantzig G B. Chemical equilibrium in complex mixtures[J]. The Journal of Chemical Physics, 1958, 28(5):751-755.
45. André P. Plasmas out of thermal equilibrium[J]. IEEE Transactions on Plasma Science, 1995, 23(3):453.
46. André P, Abbaoui M, Lefort A, et al. Numerical method and composition in multi-temperature plasmas: application to an Ar–H2 mixture[J]. Plasma Chemistry and Plasma Processing, 1996, 16(3):379-398.
47. Wang W Z, Rong M Z, Wu Y, et al. Thermodynamic and transport properties of two-temperature SF6 plasmas[J]. Physics of Plasmas, 2012, 19(8):083506.
48. Tanaka Y, Yokomizu Y, Ishikawa M, et al. Particle composition of high-pressure SF6 plasma with electron temperature greater than gas temperature[J]. IEEE Transactions on Plasma Science, 1997, 25(5):991-995.
49. Gleizes A, Chervy B, Gonzalez J J. Calculation of a two-temperature plasma composition: bases and application to SF6[J]. Journal of Physics D: Applied Physics, 1999, 32(16):2060.
50. André P, Aubreton J, Elchinger M F, et al. A new modified pseudoequilibrium calculation to determine the composition of hydrogen and nitrogen plasmas at atmospheric pressure[J]. Plasma Chemistry and Plasma Processing, 2001, 21(1):83-105.
51. Capitelli M, Colonna G, D'Angola A. Fundamental aspects of plasma chemical physics: thermodynamics[M]. Springer Science & Business Media, 2011.
52. Aubreton J, Elchinger M F, Fauchais P. New method to calculate thermodynamic and transport properties of a multi-temperature plasma: application to N2 plasma[J]. Plasma Chemistry and Plasma Processing, 1998, 18(1):1-27.
53. Trelles J P, Heberlein J V R, Pfender E. Non-equilibrium modelling of arc plasma torches[J]. Journal of Physics D: Applied Physics, 2007, 40(19):5937.

54. Devoto R S. Transport properties of ionized monatomic gases[J]. Physics of Fluids (1958–1988), 1966, 9(6):1230–1240.
55. Devoto R S. Third approximation to the viscosity of multicomponent mixtures[J]. Physics of Fluids (1958–1988), 1967, 10(12):2704–2706.
56. Devoto R S. Transport coefficients of partially ionized argon[J]. Physics of Fluids (1958–1988), 1967, 10(2):354–364.
57. Bonnefoi C. Limoges University, 1983.
58. Rat V, André P, Aubreton J, et al. Transport properties in a two-temperature plasma: theory and application[J]. Physical Review E, 2001, 64(0264092Part 2).
59. Colombo V, Ghedini E, Sanibondi P. Two-temperature thermodynamic and transport properties of argon-hydrogen and nitrogen-hydrogen plasmas[J]. Journal of Physics D: Applied Physics, 2009, 42(0552135).
60. Rat V, André P, Aubreton J, et al. Two-temperature transport coefficients in argon–hydrogen plasmas—II: Inelastic Processes and Influence of Composition[J]. Plasma Chemistry and Plasma Processing, 2002, 22(4):475-493.
61. Chen X, Li H P. The reactive thermal conductivity for a two-temperature plasma[J]. International Journal of Heat and Mass Transfer, 2003, 46(8):1443-1454.
62. Butler J N, Brokaw R S. Thermal conductivity of gas mixtures in chemical equilibrium[J]. Journal of Chemical Physics, 1957, 26(6):1636-1643.
63. Wang W Z, Rong M Z, Yan J D, et al. The reactive thermal conductivity of thermal equilibrium and nonequilibrium plasmas: application to nitrogen[J]. IEEE Transactions on Plasma Science, 2012, 40(4):980-989.
64. Tsai S T, Morishita H. Calculation of radiation and thermal transfer in thermal plasmas[J]. Journal of Physics D: Applied Physics, 2010, 43(24):245205.
65. Gleizes A, Gonzalez J, Liani B, et al. Calculation of net emission coefficient of thermal plasmas in mixtures of gas with metallic vapour[J]. Journal of Physics D: Applied Physics, 1993, 26(11):1921.
66. Lowke J. Predictions of arc temperature profiles using approximate emission coefficients for radiation losses[J]. Journal of Quantitative Spectroscopy and Radiative Transfer, 1974, 14(2):111-122.
67. Cressault Y, Gleizes A, Riquel G. Properties of air–aluminum thermal plasmas[J]. Journal of Physics D: Applied Physics, 2012, 45(26):265202.
68. Aubrecht V, Bartlova M, Coufal O. Radiative emission from air thermal plasmas with vapour of Cu or W[J]. Journal of Physics D: Applied Physics, 2010, 43(43):434007.
69. Billoux T, Cressault Y, Boretskij V F, et al. Net emission coefficient of CO2–Cu thermal plasmas: role of copper and molecules[J]. Journal of Physics: Conference Series, 2012, 406:012027.
70. Zhang J, Fang M, Newland D. Theoretical investigation of a 2 kA DC nitrogen arc in a supersonic nozzle[J]. Journal of Physics D: Applied Physics, 1987, 20(3):368.
71. Sevast'yanenko V. Radiation transfer in a real spectrum: Integration over frequency[J]. Journal of Engineering Physics and Thermophysics, 1979, 36(2):138-148.
72. Aubrecht V, Lowke J. Calculations of radiation transfer in SF6 plasmas using the method of partial characteristics[J]. Journal of Physics D: Applied Physics, 1994, 27(10):2066.
73. Eby S, Trepanier J, Zhang X. Modelling radiative transfer in circuit-breaker arcs with the P-1 approximation[J]. Journal of Physics D: Applied Physics, 1998, 31(13):1578.
74. Zhang Q, Yan J D, Fang M T C. The modelling of a SF6 arc in a supersonic nozzle: I. Cold flow features and DC arc characteristics[J]. Journal of Physics D: Applied Physics, 2014, 47(21):215201.
75. Liuxiaoming, Wangerzhi, Caoyundong. Study on dynamic arc model for high voltage SF6 circuit breaker[J]. Proceedings of the SCEE, 2004, 24(2):102–106.
76. Rong M Z, Ma Q, Wu Y, et al. The influence of electrode erosion on the air arc in a low-voltage circuit breaker[J]. Journal of Applied Physics, 2009, 106(2):023308.
77. Murphy A B. The effects of metal vapour in arc welding[J]. Journal of Physics D: Applied Physics, 2010, 43(43):434001.

78. Schnick M, Fuessel U, Hertel M, et al. Modelling of gas–metal arc welding taking into account metal vapour[J]. Journal of Physics D: Applied Physics, 2010, 43(43):434008.
79. Boulos M I, Fauchais P, Pfender E. Thermal Plasmas: Fundamentals and Applications[M]. Springer Science & Business Media, 1994.
80. Cressault Y, Gleizes A. Thermal plasma properties for Ar–Al, Ar–Fe and Ar–Cu mixtures used in welding plasmas processes: I. Net emission coefficients at atmospheric pressure[J]. Journal of Physics D: Applied Physics, 2013, 46(41):415206.
81. Cressault Y, Rouffet M E, Gleizes A, et al. Net emission of Ar–H2–He thermal plasmas at atmospheric pressure[J]. Journal of Physics D: Applied Physics, 2010, 43(33):335204.
82. Penghuimin. Radiation transport in plasma and radiation hydrodynamics [M]. Beijing: National Defence Industry Press, 2008.
83. Li Z Y. Welding plasma radiation spectrum theory and its industrial application [M]. Beijing: Peking University Press, 2013.
84. Bernath P F. Spectra of Atoms and Molecules[M]. Oxford University Press, 2005.
85. Walkup R, Stewart B, Pritchard D. Collisional line broadening due to van der Waals potentials[J]. Physical Review A, 1984, 29(1):169.
86. Gleizes A, Gongassian M, Rahmani B. Continuum absorption coefficient in SF6 and SF6–N2 mixture plasmas[J]. Journal of Physics D: Applied Physics, 1989, 22(1):83.
87. Liani B, Rahmouni M, Belbachir A, et al. Computation of net emission of CH4–H2 thermal plasmas[J]. Journal of Physics D: Applied Physics, 1997, 30(21):2964.
88. Hannachi R, Cressault Y, Teulet P, et al. Net emission of H2O–air–MgCl2/CaCl2/NaCl thermal plasmas[J]. Journal of Physics D: Applied Physics, 2008, 41(20):205212.
89. Naghizadeh-Kashani Y, Cressault Y, Gleizes A. Net emission coefficient of air thermal plasmas[J]. Journal of Physics D: Applied Physics, 2002, 35(22):2925.
90. Jan C, Cressault Y, Gleizes A, et al. Calculation of radiative properties of SF6–C2F4 thermal plasmas: application to radiative transfer in high-voltage circuit breakers modelling[J]. Journal of Physics D: Applied Physics, 2014, 47(1):015204.
91. Dixon C M, Yan J D, Fang M T C. A comparison of three radiation models for the calculation of nozzle arcs[J]. Journal of Physics D: Applied Physics, 2004, 37(23):3309-3318.
92. Bogatyreva N, Bartlova M, Aubrecht V. Mean absorption coefficients of air plasmas[C]. IOP Publishing, 2011:012009.
93. Randrianandraina H Z, Cressault Y, Gleizes A. Improvements of radiative transfer calculation for SF6 thermal plasmas[J]. Journal of Physics D: Applied Physics, 2011, 44(19):194012.
94. Bogatyreva N, Bartlova M, Aubrecht V, et al. P1-approximation for radiative transfer: application to SF6–Cu arc plasmas[J]. Open Chemistry, 2015, 13(1):502-508.

Chapter 4
Measurement of Plasma Parameters for Switching Arcs

4.1 Introduction

Switching arcs encountered in high voltage gas blast circuit breakers are basically of a transient nature with a gas temperature up to 30,000 K and a duration of typically 10–20 milliseconds, accompanied by strong light emission and possible shock waves in gas flow. In low voltage gas circuit breakers, the arc temperature is around 10, 000 K with a wider range of arc duration (3-20 ms), and the unique back-arc breakdown phenomenon is a fast transient process with a time scale of microsecond or shorter. Due to the very natures of high temperature, transient, and burning in an enclosure, the application of most diagnostic methods becomes challenging and applicable methods are rather limited. Additionally, the measurement of key parameters is extremely difficult and in fact many parameters, including but not limited to the temperature of heavy particle, electron temperature, electron density, cannot be directly measured. Computer simulation then becomes an important tool to obtain the details of the concerned parameters over space and time for the analysis of the arcing process and optimum design of switchgear products. However, computer prediction cannot be used alone at present without the support of experimental results. The establishment of a mathematical model for the arc needs knowledge from experiment; results from computer simulation also need experimental verification due to the approximations or assumptions introduced. Therefore, experiment and diagnosis is still an essential branch in the field of computing methodology of switchgears.

For switching arcs, physical quantities for measurement can be divided into two categories: macroscopic parameters (external) and microscopic parameters (internal) [1]. Macroscopic parameters include arc voltage, current, arc shape, etc. Microscopic parameters include plasma electron temperature, electron number density, etc. Measurement methods can be broadly divided into two types: intrusive and non-intrusive methods. As an example of the intrusive methods, a Langmuir probe is in direct contact with the plasma to be measured. The experimental equipment is generally simple, as is the principle (for certain plasmas), and it is one of the earliest

M. Rong et al., *Computing Methodology for Switchgear*,
https://doi.org/10.1007/978-981-95-2840-0_4

methods used in plasma diagnosis. It has been used in the diagnosis of switching arc plasma [2–4]. However, the insertion of a detector will inevitably disturb the plasma, resulting in an measurement error. In addition, the temperature at the centre of the switching arc is usually tens of thousands Kelvin, and the detector may risk melting quickly. So the life time of the probe is a concern to be addressed. The application of different types of Langmuir probes to vacuum arc measurement was reported previously [5], providing suggestions on the choice of suitable probe.

The non-intrusive methods can be further divided into passive and active measurements. Passive measurement refers to the method of obtaining microscopic parameters by analysing the radiation from the plasma, such as the emission spectroscopy. Active measurement requires an external light source, which is used as the probing light to interact with the plasma to be measured, to obtain the microscopic parameters of the plasma. Examples include absorption spectroscopy, laser Thomson scattering, laser interference imaging, etc. Compared with the intrusive methods, non-intrusive diagnostic methods have the following characteristics: less disturbance to the plasma (particularly the passive methods); no loss or damage of detectors, and reduced experimental cost. Different non-intrusive methods need to be chosen for the measurement of different microscopic parameters with sufficient spatial and temporal resolutions. The commonly used plasma diagnostic methods are listed in Table 4.1.

The principles of microscopic parameter measurement will be introduced in each section of this chapter, with examples given.

Table 4.1 Diagnostic methods for switching arc plasma

	Method	Characteristics
Contact methods	Langmuir probe	Measures arc temperature and electron density; the probe disturbs the plasma and can melt easily; measurement result is closely linked to the surface sheath; response can be slow
	Split-electrodes	Measures arc root current density
Non-contact methods	High-speed photography	Measures arc shape; mature theory and simple equipment requirements
	Thermal radiation detection	Measures arc radiation; mature theory, simple equipment requirements
	Emission spectrometry	Measures arc temperature and electron density; active measurement, mature theory, simple equipment requirements, accuracy may be low
	Absorption spectroscopy	Measures atomic number density; passive measurement, mature theory, simple equipment requirements, accuracy may be low
	Laser Thomson scattering	Measures electronic temperature and electronic number density; passive measurement, high spatial and temporal resolutions, no need for LTE assumption, difficult to set up, high accuracy

4.2 Measurements of Temperature and Electron Density by Atomic Emission Spectroscopy

The spectral diagnosis of arc plasma began in the 1950s, when the spectral measurements were used in astrophysics to infer the temperature ranges of several types of arcs from the intensity of spectral lines. In the 1960s, Griem, Lochte and Holtegreven systematically elaborated the theory of quantitative spectral diagnosis [6, 7], providing data on atomic spectral line strength, transition probability and broadening [8]. At present, atomic emission spectrometry is the most widely used method of plasma diagnosis due to its mature theory, simple equipment requirements and high sensitivity.

In the field of switch arc diagnosis, atomic emission spectrometry has been widely used. Hai Tao and Wang Qiping of Xi'an Jiaotong University [9] measured the radial distribution of temperature in the nozzle throat of the single-pressure SF_6 circuit breaker model by using Cu I 510.55 nm and 521.82 nm spectral lines. Airey et al. [10] measured the change of SF_6 arc temperature with current attenuation in high voltage circuit breakers by using F I 623.96 nm and S II 545.38 nm spectral lines. Jones et al. [11] measured the spectral line strength of copper metal vapour and tungsten metal vapour in SF_6 circuit breakers, and obtained the ratio of their concentrations. Methling et al. [12] obtained the electron number density of vacuum arc by using the Stark broadening method, and compared the simulation results with spectral lines, they found that the plasma in the near-pole region deviated from LTE.

4.2.1 *Principle of Measurement*

Atomic emission spectrometry can be divided into two categories from the perspective of analysing the measured spectral data, one is a measurement method based on spectral radiation intenszity, where the other is based on spectral line analysis.

1. Measurement method based on spectral radiation intensity analysis

According to Einstein's quantum theory of radiation, the net intensity of spectral line emission I_{ji} (Wm^{-3}) can be expressed as formula (4.1):

$$I_{ji} = N_j A_{ji} h\nu_{ji} + N_j B_{ji} \rho(\nu) h\nu_{ji} - N_i B_{ij} \rho(\nu) h\nu_{ji} \tag{4.1}$$

where the first term on the right hand side of the equation represents the intensity of spontaneous emission, the second term represents the intensity of stimulated emission, and the third term represents the contribution of stimulated absorption to the net intensity. h is the Planck constant, ν_{ji} is the frequency, N_i, N_j are the number densities of particles (atoms and ions) at energy levels i and j (m^{-3}), $\rho(\nu)$ is the energy spectrum density of the radiation field, A_{ji} is the Einstein coefficient of spontaneous emission from high energy level j to low energy level i, B_{ji} is Einstein's constant of stimulated radiation, B_{ij} is Einstein's constant of stimulated absorption. According to Planck's radiation law, the relationship between spontaneous emission, stimulated emission and stimulated absorption constant is as follows [13]:

$$B_{ij} = \frac{g_j}{g_i} B_{ji} \tag{4.2}$$

$$A_{ji} = \frac{8\pi h v^3}{c^3} B_{ji} \tag{4.3}$$

where c is the speed of light in vacuum. In the thermal radiation field, the spontaneous radiation in each mode is much larger than the stimulated radiation. When the self-absorption of the plasma to be measured is ignored, the net intensity of spectral line emission can be expressed as formula (4.4):

$$I_{ji} = N_j A_{ji} h v_{ji} \tag{4.4}$$

Under the thermodynamic equilibrium hypothesis, different energy levels of a particle have the same excitation temperature T_{ex} (K), and the total density of the particle (such as CuI) $N = \sum_j N_j$ and the number density of the particle N_j at energy level j follow the Maxwell-Boltzmann distribution.

$$N_j = N \frac{g_j}{Z} e^{(-E_j/k_B T_{ex})} \tag{4.5}$$

where $g_j = 2J_j + 1$ is the statistical weight factor, the degenerate sub-energy level number whose total angle quantum number is J_j. Energy level j corresponds to energy level E_j (J); the partition function is $Z = \sum_j g_j e^{(-E_j/k_B T_{ex})}$. By substituting Eq. 4.5 into Eq. 4.4, the net emission intensity of the spectral line can be obtained, as shown in Eq. 4.6, and then the emission coefficient, which is the emission intensity per unit solid angle, can be calculated as shown in Eq. 4.7.

$$I_{ji} = N A_{ji} h v_{ji} \frac{g_j}{Z} e^{(-E_j/k_B T_{ex})} \tag{4.6}$$

$$\varepsilon = \frac{I_{ji}}{4\pi} = N A_{ji} h v_{ji} \frac{g_j}{4\pi Z} e^{(-E_j/k_B T_{ex})} = A_{ji} h v_{ji} \frac{g_j}{4\pi Z} \frac{p_N}{k_B T_{ex}} e^{(-E_j/k_B T_{ex})} \tag{4.7}$$

where p_N is the partial pressure of the species. On the premise that the total pressure of the plasma is certain, the relationship between the partial pressure of the species and temperature can be obtained based on Gibbs minimum free energy method. For specific measured spectral lines, A_{ji}, v_{ji}, g_j, Z and E_j are constant values and can be obtained from relevant databases or literature, such as [14].

In the process of measuring arc microscopic parameters by atomic emission spectroscopy, the selection of spectral lines is very important. The selected spectral lines are required to be easy to distinguish, less affected by self-absorption, and with high accuracy of transition probability. There are many excited states of particles at high temperature, which will generate multiple characteristic spectral lines. For the arcing stage, the resonance lines of atoms are usually selected as the characteristic spectral lines. Resonance line is a spectral line generated by the transition of electrons in the outer layer of an atom from the excited state to the ground state. At the same time, the basic physical parameters of such spectral lines, such as wavelength, energy level and degenerate energy series, are easy to obtain, which is conducive to quantitative calculation. For the post-arc stage, Tanaka pointed out that the resonance line would generate self-absorption, leading to a large deviation in the measured temperature [15].

2. Measurement method based on spectral line analysis

Particles that are radiating outward in the plasma are disturbed by the micro level electric field of the surrounding charged particles. This interference is reflected in the line shape of the spectral line, which causes the spectral line to widen. This is called Stark broadening. This broadened line type belongs to the Lorentz type. The effect of ions will lead to asymmetric broadening; Electrons, in turn, cause symmetry broadening. Stark broadening is strongly related to the electron density of the plasma and has nothing to do with the plasma composition and whether the LTE assumption is satisfied. When the Stark spreading method is used to measure the electron number density, only the FWHM of the spectral line needs to be known, and the absolute intensity of the spectral line does not need to be known, so the absolute intensity calibration of the measuring instrument is not required. The Stark broadening method is a more accurate and direct method for measuring the number density of plasma electrons [8, 16].

Since Stark broadening is related to the type and atomic structure of the emitted particles, there is currently no universal analytical solution. Griem proposed a semi-empirical formula for Stark broadening and electron number density [6, 17]. The relationship between the broadening of hydrogen-like atoms (FWHM) and the electron number density [18, 19] is shown in formula (4.9). The relationship between Stark broadening (FWHM) of non-hydrogen-like atoms and the number density of electrons is shown in formula (4.10):

$$\Delta\lambda_{\mathrm{FWHM}} = 25\alpha N_e^{2/3} \tag{4.8}$$

$$\Delta\lambda_{\mathrm{FWHM}} = 2\times10^{-6} N_e \mathrm{w}\left[1 + 1.75\times10^{-4}\alpha N_e^{\frac{1}{4}}\left(1 - 0.068 N_e^{\frac{1}{6}} T_e^{-\frac{1}{2}}\right)\right] \tag{4.9}$$

where the unit of $\Delta\lambda_{\mathrm{FWHM}}$ is m, N_e is electron number density (cm^{-3}), T_e is the electronic temperature (K), w is the electron impact parameter, α is the static ion broadening parameter. w and α are available in related databases and literature, and are related to temperature and atomic species. In calculating the electron density using Eq. 4.10, it is necessary to know the approximate range of temperature, which can be obtained in advance by laser Thomson scattering, Boltzmann diagrams, or other measurements. More simply, ignoring the effect of ions, Eq. 4.10 becomes Eq. 4.11.

$$\Delta\lambda_{\mathrm{FWHM}} = 2 \times 10^{-6} N_e \mathrm{w} \quad (4.10)$$

The atomic emission spectra of arc plasma obtained by spectrometer are the result of a variety of broadening, among which instrument broadening, Doppler broadening and Stark broadening play major roles. Before calculating the electron density, we first need to separate the stark broadening from the measured spectral lines, that is, deconvolution of the measured emission spectral lines.

4.2.2 Examples of Application

1. Measurement of excitation temperature by absolute strength method

The absolute strength method is also known as the single line method, the absolute line method, and the classical Optical Emission Spectrometer (OES) technique. The excitation temperature can be obtained by analysing only one characteristic emission line of the measured plasma. However, the absolute strength of the spectral line measurement accuracy is difficult to ensure, the measured excitation temperature error is large. Bachmann et al. [20] used Ar I 696.54 nm and Ar II 487.98 nm lines respectively to measure the arc temperature of steady-state Tungsten Inert Gas Welding (TIG) arc. The specific steps are as follows:

(1) Calibrate the intensity measured by the spectrometer with an irradiation calibration light source (such as a mercury band lamp) to obtain the distribution of absolute intensity with wavelength $I(\lambda)$ ($\mathrm{Wm^{-2}nm^{-1}sr^{-1}}$).
(2) Fitting $I(\lambda)$ with corresponding line-shape or integrating $I(\lambda)$ directly in adjacent wavelengths to obtain radiation L ($\mathrm{Wm^{-2}sr^{-1}}$).
(3) The emission coefficient ε ($\mathrm{Wm^{-3}sr^{-1}}$) is obtained by Abel inverse transformation with the distribution of radiation in space.
(4) Given the components in the LTE state and the total pressure, the partial pressure of species can be obtained. According to the emission coefficient formula (4.7), the excitation temperature T_{ex} and the number density N of the particle can be obtained. Fig. 4.1 shows the radial temperature distribution curve under DC 200 A.

2. Measurement of excitation temperature by double-spectrum method

The double-spectrum method is also known as the two-line method or as the relative line intensity method. The ratio of emission coefficients between two different energy levels of the same particle is used to measure the excitation temperature. From formula (4.7), the emission coefficient ratio of the two spectral lines can be obtained as follows:

$$\frac{\varepsilon_1}{\varepsilon_2} = \frac{A_{ji,1}\nu_{ji,1}g_{j,1}}{A_{ji,2}\nu_{ji,2}g_{j,2}} \exp\left(-\frac{E_{j,1} - E_{j,2}}{k_B T_{ex}}\right) \quad (4.11)$$

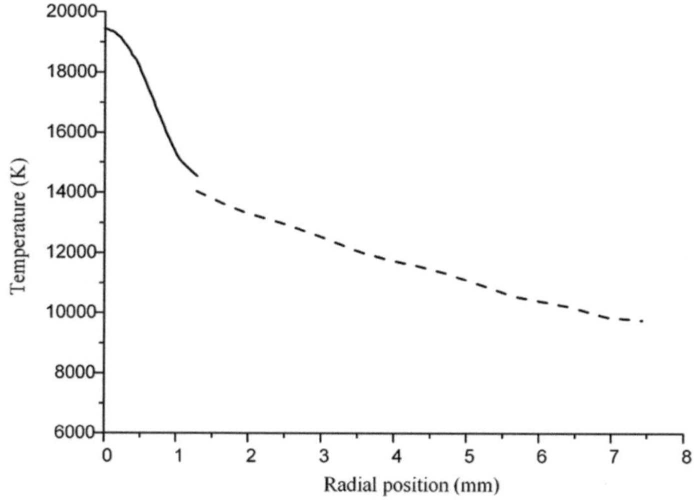

Fig. 4.1 Radial temperature distribution measured by absolute intensity method. The arcing DC current is 200 A [20]

With subscript 1 and 2 representing emission lines of two different energy levels. Taking the natural logarithm of both sides of Eq. 4.12 and rearranging gives:

$$\frac{1}{T_{ex}} = \frac{k_B}{E_{j,2} - E_{j,1}} \ln \frac{A_{ji,2} g_{j,2} \lambda_{ji,1} \varepsilon_1}{A_{ji,1} g_{j,1} \lambda_{ji,2} \varepsilon_2} \tag{4.12}$$

The difference between the energy levels of the two spectral lines has a great impact on the measurement of the excitation temperature. The larger the energy level difference $E_{j,2} - E_{j,1}$, the more accurate the calculated excitation temperature. It is generally recommended that the energy level difference is greater than 2 eV [7, 21]. If the unit of the energy level obtained from the database is eV, the above formula can be written as in formula (4.13). If instead, the unit of the excitation energy is cm^{-1}, the constant 5040 is changed to 0.625:

$$T_{ex} = 5040(E_{j,2} - E_{j,1}) / \log \frac{A_{ji,2} g_{j,2} \lambda_{ji,1} \varepsilon_1}{A_{ji,1} g_{j,1} \lambda_{ji,2} \varepsilon_2} \tag{4.13}$$

From Eq. 4.13, the excitation temperature of the atom can be calculated without calculating the atomic number density N and the partition function Z. Tanaka et al. [15] measured the spectral lines Fe I 426.0 nm and 442.7 nm with an optic fibre spectrometer with an iron electrode for the flat-plate SF_6 model circuit breaker, and obtained the decaying behaviour of the axial temperature distribution in the post-arc channel after current zero. The measured intensity of the two spectral lines and the temperature calculated by the double-spectrum method are shown in Fig. 4.2.

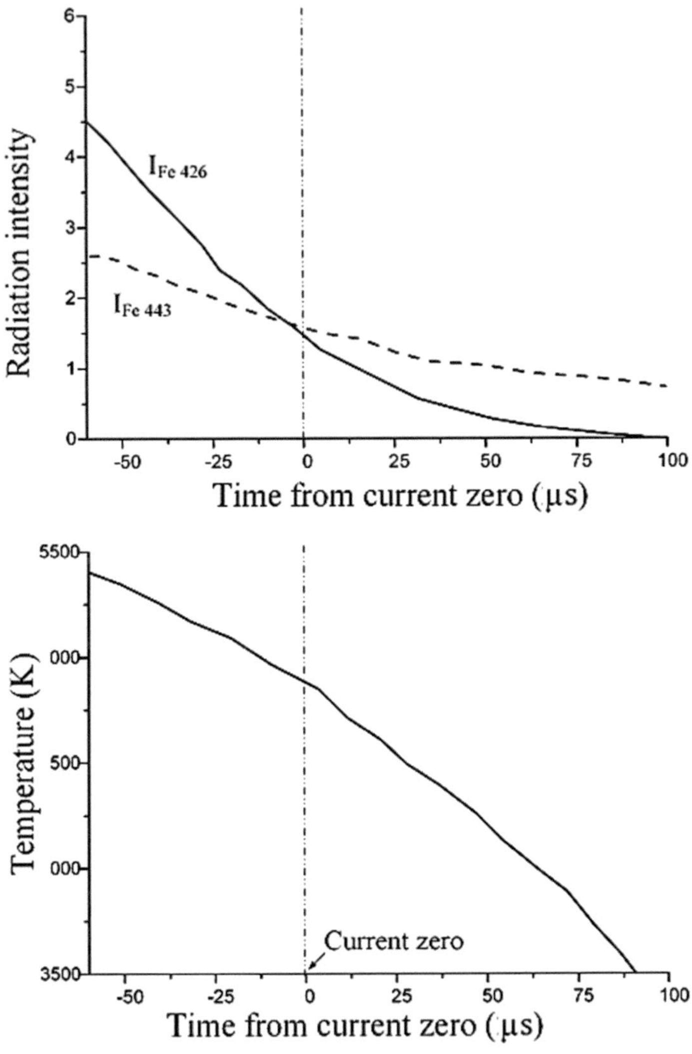

Fig. 4.2 Peak current 5 kA, emission intensity and temperature distribution in the zero-crossing phase

3. Measurement of excitation temperature by multi-line method

The multi-line method, also known as the Boltzmann diagram method, uses different excitation energy lines of the same atom to obtain the excitation temperature of the plasma to be measured. The double-spectrum method is therefore actually a special case of the multi-line method. Taking the natural logarithm of both sides of Eq. 4.7 generates Eq. 4.14:

$$\ln \frac{\varepsilon \lambda_{ji}}{A_{ji} g_j} = -\frac{1}{k_B T_{ex}} E_j + \ln \frac{Nhc}{4\pi Z} \tag{4.14}$$

Equation 4.14 is in the form of a linear equation ($y = ax + b$), where $b = \ln \frac{Nhc}{4\pi Z}$ is taken as the constant term, $y = \ln \frac{\varepsilon \lambda_{ji}}{A_{ji} g_j}$, and $x = E_j$. Using the data of multiple spectral lines, multiple points (x, y) are obtained, and the least squares fitting of these points is carried out to obtain the slope $a = -\frac{1}{k_B T_{ex}}$ of the linear equation, and then the excitation temperature T_{ex} of the atom is obtained.

The multi-line method uses the data of multiple characteristic spectral lines in its calculation and analysis, and the measurement results have high accuracy. According to Eq. 4.14, when the multi-line method is used for measurement, the difference in excitation energy of selected characteristic spectral lines is required to be as large as possible; smaller differences in excitation energy cause larger measurement error.

Vacquie et al. [22] selected 11 spectral lines of sulphur ion and measured the temperature of SF_6 cascade arc by using multi-line method. Fig. 4.3 is a Boltzmann diagram of a point on the axis of the cascade arc measured. Each point on the graph represents a spectral line. After linear fitting of 11 points on the graph, it can be obtained that the temperature of the measured region is 11,300 K.

4. Measurements of electron density and temperature by Stark broadening method

For some diagnostic spectral lines commonly used in plasma, such as Ar I 696.5 nm and Fe I 538.3 nm, Zielińska [23] gave more accurate expressions between Stark's broadening (FWHM, nm), electron number density (m^{-3}), and temperature (K) based on a large amount of experimental data.

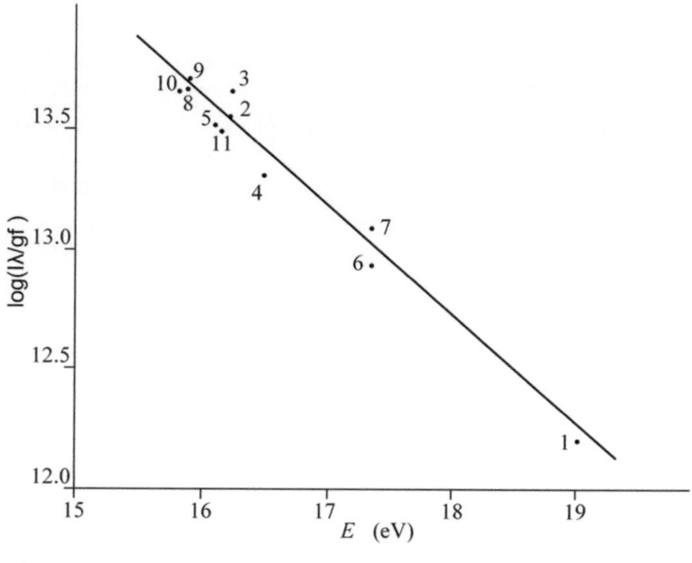

Fig. 4.3 Application of Boltzmann diagram in SF_6 arc

$$\Delta\lambda_{696.5}^{Ar} = 0.0814\frac{n_e}{10^{23}}(\frac{T_e}{13000})^{0.3685} \tag{4.15}$$

$$\Delta\lambda_{538.3}^{Fe} = 0.2648\frac{n_e}{10^{23}}(\frac{T_e}{13000})^{1.670} \tag{4.16}$$

The Stark broadening corresponding to these two spectral lines is measured experimentally, and then the electron number density and temperature can be obtained through Eqs. 4.15 and 4.16.

4.3 Measurement of Atom Number Density by Atomic Absorption Spectroscopy

In the 1970s, some scholars began to use absorption spectrum to diagnose switching arcs. Tuma et al. [24] measured the ablative products of cathode spots of copper electrodes under vacuum conditions. Jenkins et al. [25] studied the decay rate of metal atom vapour after a large current crossed zero in vacuum through absorption spectroscopy, and then demonstrated the generation mechanism of metal vapour.

In recent years, various devices (such as light source, ICCD, spectrometer, etc.) for absorption spectrum research have been made great progress, providing favourable conditions for the further development of arc plasma absorption spectrum diagnosis. Hong et al. [26] designed a high-brightness, continuous wavelength light source using the z-pinch effect, and measured the concentration of copper, silver and iron atoms in the air-switch arc plasma. Horvath [27] used a broadband LED light source to make a probe light, and measured the change of Cr atomic vapour number density with time after the arc crossed zero under air conditions. Takahashi [28] selected the characteristic spectral line of 327.4 nm of copper atoms, and measured the attenuation of copper atom number density after the vacuum arc crossed zero by the method of laser absorption spectrum, and verified it by the method of laser induced fluorescence (LIF).

Compared with the arc plasma emission spectrum diagnosis method, the most obvious difference with absorption spectroscopy is that it requires a high-intensity external light source (laser light source or white light source) as the light. In order to strictly exclude the influence of arc emission intensity on the measurement results, splitter and other devices should be added to the optical path setting to record arc radiation, so the optical path setup is a little more complicated.

4.3.1 Principle of Measurement

The energy state of an atom is determined by the position of the electron in the atomic orbital. The lowest energy state of an atom is called the ground state. The outer electrons of lower-energy atoms can be excited to higher-energy orbits by absorbing a photon whose energy is equal to the difference between two orbitals, a process called electron transition. When a beam of light passes through the object to be measured with a low-energy atom, the low-energy atom will absorb the light of a specific frequency and jump to the excited state, causing the change of incident light intensity and generating atomic absorption spectrum, as shown in Fig. 4.4.

Where, I_0 is the incident light intensity, and I_t is the transmitted light intensity after passing through the object to be tested. This absorption process satisfies:

$$hv = E_j - E_i \tag{4.17}$$

where, E_i and E_j are the energy levels of the atoms before and after the absorption of photons, and v is the frequency of the absorbed photons. An atom will jump to a higher energy level after absorbing photons of a certain frequency, but this excited state cannot be maintained. After a short period of time, the excited atom will return to a lower energy orbit by releasing photons. As the atom transitions to lower energy levels, the photons released are emitted in all directions. The observer will therefore see an intensity sag in the spectrum at a wavelength consistent with the wavelength being absorbed. Although photons of a specific wavelength corresponding to these intensity dips are re-emitted, the re-emitted photons are scattered in all directions, so the re-emission intensity observed on the detector is lower. The intensity sag in the spectrum is reflected in the spectral line function (the distribution of luminous energy or photon number with wavelength and frequency), that is, the absorption peak will appear at a specific wavelength (frequency) position. For information about spectral lines and their characteristic parameters, please refer to Section 6.7.

The absorption lines produced by the transition of the outer electrons from the ground state to the lower excited state are called resonance absorption lines. Different atoms absorb light at different frequencies due to their different structures. Therefore, the resonance absorption line of each atom has different frequency, which is called the characteristic absorption line of the atom. Generally speaking, the probability of the transition from the ground state to the low excited state is much larger than that of other transitions, and the atoms are mostly in the ground state, so the resonance absorption line is the most sensitive spectral line of all the spectral lines of the element.

According to Beer-Lambert law, assuming that the measured region is uniform, the theoretical transmittance can be expressed via formula (4.18):

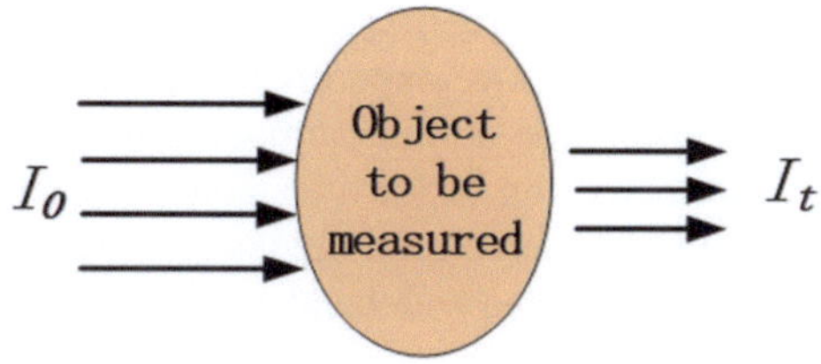

Fig. 4.4 Schematic diagram of absorption spectrum generation

$$t(\lambda) = \frac{I_t(\lambda)}{I_0(\lambda)} = e^{-\tau(\lambda)} = e^{-\alpha(\lambda)L} \tag{4.18}$$

where λ is the wavelength, is the optical depth, L is the absorption length, that is, the propagation distance of the probe light in the homogeneous medium to be measured, and $\alpha(\lambda)\tau$ is the absorption coefficient of the medium to be measured. The expression of the absorption coefficient is shown in Eq. 4.20 [13, 25]:

$$\alpha(\lambda) = \frac{\lambda^4 A_{ji}}{8\pi c} \frac{g_j}{g_i} \times S(\lambda, \lambda_0) N_1 \tag{4.19}$$

where, A_{ji} is the Einstein radiation coefficient of spontaneous emission, j is the higher energy level, i is the lower energy level, g_j and g_i are the statistical weights of energy levels j and i, $S(\lambda, \lambda_0)$ is the normalised shape function, λ_0 is the central wavelength, and N_1 is the number density of atoms in the ground state. It is assumed that the normalised shape function is of Lorentz line type affected by pressure broadening, and $\Delta\lambda_L$ is the FWHM of theoretical line type. The shape function can be expressed as formula (4.20):

$$S(\lambda, \lambda_0) = \frac{2}{\pi} \frac{\Delta\lambda_L}{4(\lambda - \lambda_0)^2 + \Delta\lambda_L^2} \tag{4.20}$$

Considering the instrument broadening caused by limited resolution of measuring equipment (such as spectrometer), the actual transmittance measured by spectrometer is the combination of theoretical transmittance and instrument broadening [26] [27]:

$$t_{\exp}(\lambda) = t(\lambda) * P_{inst}(\lambda) = \int_0^{\infty} t(\lambda - x) P_{inst}(x) dx \tag{4.21}$$

Through Eqs. 4.18 and 4.21, the function $t(\lambda) * P_{inst}(\lambda)$ with wavelength λ as independent variable and two parameters α_L and $N_1 L$ is obtained. Using the least squares method, the function is fitted with the transmittance measured by the spectrometer to obtain, and α_L and $N_1 L$. Given L, the ground state atomic number density of the object to be measured can be obtained.

4.3.2 Examples of Application

Based on the above experimental principles, an experimental device can be set up to determine the atomic number density of the ground state by atomic absorption spectrometry. Fig. 4.5 is a feasible optical path scheme.

The light source used here is broadband white light source. ENERGETIQ's EQ-99FC laser excited wideband white light source, for example, has a wavelength range of 170 nm–2100 nm, covering ultraviolet to infrared wavelengths. When the probe

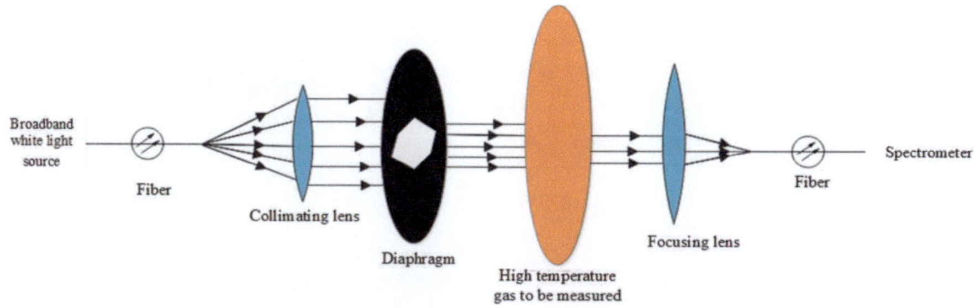

Fig. 4.5 Optical path of atomic absorption spectroscopy experiment

light source is a narrowband light source, in order to obtain an effective absorption spectrum of the gas atom, the incident light needs to be divided into two parts by a spectrometer: the reference beam and the probe beam. There are other factors that may cause the light intensity to weaken (such as particles, scattering of water droplets, etc.), and this causes the set up of the experimental device to become relatively complicated. Compared with a narrowband light source, when the external light source is a broadband light source, only a certain frequency of light will be absorbed by the particle to be measured, and light of other frequencies will pass through the high temperature gas to be measured to the same extent. In the subsequent processing of data, interference factors can be effectively eliminated. In this case, the spectrum measured by the spectrometer will show a series of absorption peaks, corresponding to the wavelength of the light absorbed by the atom as it transitions from the ground state to the excited state. There are two main functions of the collimator. One is to keep the light emitted by the broadband white light source parallel after emission. The other is to reduce the diameter and increase the brightness of the probe light, which is conducive to measurement.

In the setting of the spectrometer, the following points need to be explained. First, the spectral grating characterisation density of the spectrometer and the spectral band range displayed on the software interface are selected according to the known position of the absorption peak of the element to be measured, which can facilitate the processing of subsequent experimental data. The other is to track the change of the atom number density with time by means of absorption spectrum, and obtain the change of the atom number density during and after the arcing by means of time resolution.

4.3.3 Experimental Results

Using a broadband LED light source, Horvath [27] used three resonance absorption lines of Cr atoms at 425.43 nm, 427.48 nm and 428.97 nm to obtain the variation process of Cr atomic vapour outside the plate contact gap after current crossing zero, as shown in Fig. 4.6.

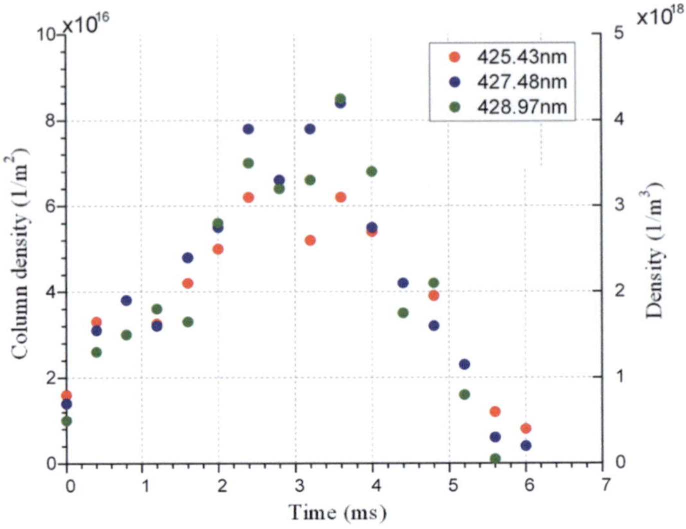

Fig. 4.6 Density of Cr atomic vapour variation with time

Hong et al. [26] obtained the number density of Cu I vapour in high temperature gas plasma by using the two resonance absorption lines of Cu I 324.75 nm and 327.40 nm, as shown in Fig. 4.7.

4.3.4 Limitations

The particle number density of arc plasma can be obtained by atomic absorption spectrometry, which is independent of the equilibrium hypothesis, thus bringing great flexibility to its application. However, this method also has some limitations. Firstly, in Eq. 4.19, it is necessary to assume that the plasma density to be measured is uniform, which is different from the actual situation of an arc. At the same time, this also makes the absorption spectroscopy not suitable for measuring the particle number density of the spatial distribution. Secondly, the value of the absorption length L cannot be accurately obtained, which will cause deviation of the calculated particle density. Thirdly, the arc itself emits light. During the arc burning phase, the emitted light of the arc is strong enough to "cancel" the probe light, which makes measurement difficult. In the arc extinguishing stage, the arc itself emits less light, but the accuracy of the measurement results will be affected to some extent.

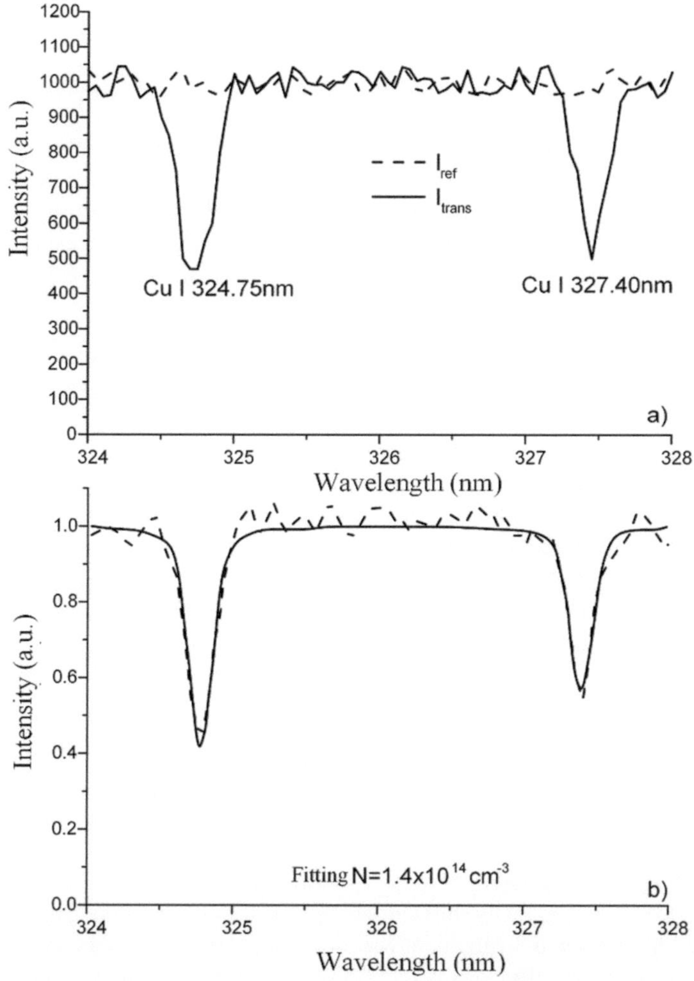

Fig. 4.7 The atomic vapour density of Cu obtained by absorption spectra

4.4 Measurements of Temperature and Electron Density by Laser Thomson Scattering

The arc current-zero period corresponds to the dielectric recovery process of arc plasma transforming from a conductive state to an insulative state. The attenuation of electron number is an important parameter used to judge the insulation recovery process, so the accurate measurement of electron number density in the current-zero period is very important.

However, residual high temperature gas between electrodes during the current-zero period tends to deviate from the LTE state, and most measurement methods based on atomic emission spectroscopy are based on the LTE assumption, which is not applicable to the measurement of micro level parameters in the arc current-zero period. Therefore, an arc plasma diagnosis method independent of the LTE hypothesis must be introduced.

Laser Thomson scattering was first applied to the measurement of plasma parameters in nuclear fusion in the 1960s. It is widely accepted as being able to accurately measure the electron density of plasma. In recent years, the Laser Thomson scattering method has been gradually applied to various plasma diagnostics such as welding arc [29], cold plasma [30] and switching arc [31, 32]. Laser Thomson scattering has the following advantages:

(1) LTE hypothesis is not required, and non-equilibrium plasma can be measured.
(2) The object to be measured is not required to be axisymmetric, and the measurement results directly reflect the physical information at the laser focal point. There is no need to carry out Abel transformation on the measurement results, which can be used to measure plasma of complex geometric shape.
(3) The measurement results are not affected by plasma components, such as metal vapour and gas-producing material vapour.

4.4.1 Principle of Measurement

The physical explanation of Thomson Scattering is that electrons in the incident wave field will generate scattering electromagnetic waves due to oscillation. Thomson Scattering is the elastic scattering between low-energy free electrons and low-energy photons. It is assumed that at a certain moment the motion velocity of the electron is v, the frequency of the scattered electromagnetic wave will be different from the frequency of the incident electromagnetic wave, which is called the Doppler frequency shift. The frequency shift can be expressed by the formula $\Delta\omega = k \cdot v$, and k is the difference between the incident wave vector and the scattered wave vector. The motion information of electrons is carried on the scattered electromagnetic waves, which is why Thomson Scattering can be used to diagnose plasma parameters [33].

In order to obtain the electron temperature and electron number density from the signal of Thomson Scattering, the experimental results must be fitted with the corresponding theoretical line shape.

The line shape of the scattering spectral line is determined by the shape factor $S(\mathbf{k}, \omega)$, where $\mathbf{k} = \mathbf{k_s} - \mathbf{k_i}$ is the difference between the wave vector of the scattered light and the incident light. When the scattering angle θ is fixed and the effect of relativity is not considered, $\mathbf{k}$ can be approximated as a constant, $\mathbf{k} = |k| = \frac{4\pi}{\lambda_0}\sin(\frac{\theta}{2})$, λ_0 is the wavelength of the incident light, and $\omega = \omega_s - \omega_i$ is the difference between the angular frequency of the scattered light and the incident light.

For the shape factor of high-density arc plasma, there are generally two methods for theoretical calculation in literature: RPA(Random Phase Approximation) method, also known as Salpeter approximation [34], which ignores collisions between particles; and the MFF(Memory Function Formalism) method, considering the collision between particles, should be used when the results of the collision cannot be ignored.

The applicability of these two methods depends on the coupling degree of the plasma, which can be evaluated by the number of electrons Λ in a Debye sphere, $\Lambda = 4\pi\lambda_D^3 n_e/3$, where $\lambda_D = \sqrt{(\varepsilon_0 k_B T_e)/(n_e e^2)}$ is Debye radius, ε_0 is vacuum dielectric constant, k_B is Boltzmann constant, n_e is electron number density, and e is electron charge. The RPA method is suitable for an ideal plasma ($\Lambda \geq 1$), which has been verified by Thomson Scattering experiments in many applications. However, in non-ideal plasma ($\Lambda = 2 - 4$), such as plasma spraying, the RPA method produces overly high results. For such applications in the Thomson Scattering experiment, researchers have proposed the MFF method. However, the MFF method needs an additional parameter in fitting, which will lead to the decrease of accuracy and instability of the calculation [35].

This part focuses on the introduction of the RPA method. The shape factor of RPA scattering spectral line is shown in Eq. 4.22:

$$S(\mathbf{k}, \omega) = \frac{2n_e\pi}{k}|1 - \frac{\chi_e}{1 + \chi_e + \chi_i}|^2 f_{ek}(\frac{\omega}{k}) + \frac{2n_e\pi Z}{k}|\frac{\chi_e}{1 + \chi_e + \chi_i}|^2 f_{ik}(\frac{\omega}{k}) \quad (4.22)$$

The spectral lines of Thomson Scattering are composed of two parts: the electron characteristic part and the ion characteristic part. The former is caused by the fluctuation of the number density of the free electrons, while the latter is due to the electrostatic influence of the density fluctuations of the ions on the free electrons. The first term on the right side of the Eq. 4.22 is the electronic characteristic part, from which the relevant information of electron temperature and electron number density can be obtained; the second term on the right side is the ion characteristic part, from which the relevant information of ion temperature and ion number density can be obtained. The characteristic part of the ion is usually stronger, but the broadening of spectral line is narrow. It is represented as a peak near the central wavelength of incident light. The characteristic part of the ion is also called the ion peak.

The broadening of the ion characteristic spectral line is very narrow (< 0.02 nm), and is often covered by the central non-scattering part in the actual measurement, so it is difficult to measure effectively. Some literature has used the serial Fabry-Pérot interferometer with high resolution (0.0006 nm) to measure the spectral line of the ion characteristic part under specific conditions [36].

In Eq. 4.22, Z is the ionic charge, $f_{ek}(\frac{\omega}{k})$ and $f_{ik}(\frac{\omega}{k})$ are the velocity distributions of electrons and ions in the $\mathbf{k}$ direction, and χ_e and χ_i are the susceptibility (also called magnetisation factor) of the electrons and ions:

$$\chi_j = \frac{1}{(k\lambda_D)^2}(\frac{Z_j^2 n_j T_e}{n_e T_j})W(\xi_j) \quad (4.23)$$

$$W(\xi) = 1 - 2\xi e^{-\xi^2}\int_0^{\xi} e^{\varsigma^2} d\varsigma + i\sqrt{\pi}\xi e^{-\xi^2} \quad (4.24)$$

where, $\xi_j = \frac{\omega}{kv_{tj}}$, v_{tj} is the most probable speed of particles, $v_{tj} = \sqrt{\frac{2k_B T_j}{m_j}}$.

In most literature that uses laser Thomson scattering to measure the micro level parameters of arc plasma, only the relevant parameters of the electron are concerned, that is, the shape function of the characteristic part of the ion is ignored.

Assuming that electrons follow the Maxwellian distribution, an approximate expression of the shape factor can be obtained, as shown in Eq. 4.25:

$$S(k,\omega) = n_e \frac{1}{|1+\alpha^2 W(\xi)|^2} \frac{e^{-\xi^2}}{v_{te}} \tag{4.25}$$

where, $|1+\alpha^2 W(\xi)|^2$ is the linear correction term. $\alpha = \frac{1}{k\lambda_D}$ (also known as Salpeter number), which has a significant influence on the shape factor, and further affects the shape of the scattering spectral line, which is called the scattering parameter. It can be seen from the above formula that the spectral line of Thomson scattering is affected by the Gaussian line pattern of Doppler broadening and a correction term caused by Landau damping. When the Salpeter number is small, Gaussian line dominates the scattering line. When the Salpeter number is large, the influence of Landau damping correction term on the linear shape of spectral lines becomes important.

In order to obtain the electron temperature and electron number density from the Thomson scattering signal, the measured spectral results must be fitted with corresponding theoretical line shapes. Two parameters, electron density and electron temperature, can be obtained quantitatively and uniquely from the shape curve of one measurement.

The thin grey line below is the actual measurement line. Given the electron temperature, Thomson scattering spectral lines corresponding to different electron temperatures and electron number density can be obtained by using formula (4.25), as shown in Fig. 4.8. It can be seen that the higher the electron number density, the greater the distance between the two saddle-shaped peaks, the more obvious the effect of changing the electron number density on the crest, and the lower the trough. The curve obtained under the condition that the electron number density is 4.5×10^{23} m^{-3} is the closest to the experimental curve, as shown by the solid red line in Fig. 4.8.

4.4.2 Examples of Applications

Based on the above basic principles, our research group set up a laser Thomson scattering experiment platform to measure the electron density of arc plasma after current crossing zero. The optical path of the laser Thomson scattering experiment is shown in Fig. 4.9. The optical path consists of the following components. Nd: YAG pulse laser, where the central wavelength is 532 nm and the pulse width is 10 ns, and the pulse laser energy is set to 50 mJ; the laser is focused on the arc plasma to be measured through a convex lens with a focal length of 300 mm. The outgoing light is in the same direction as the incident light. The outgoing light enters the horn mirror

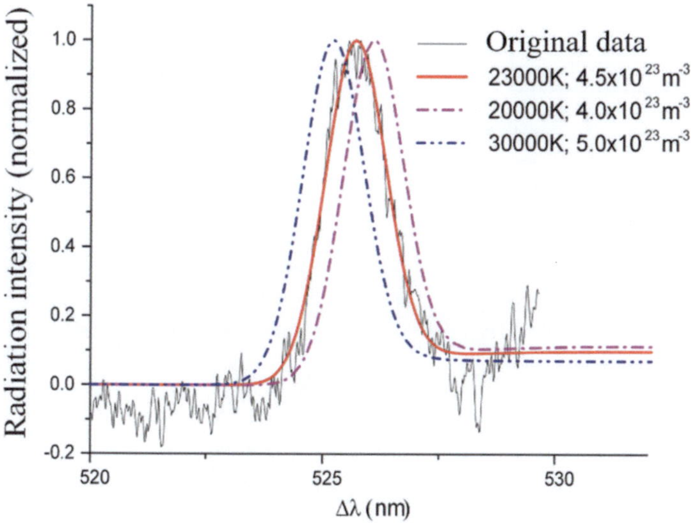

Fig. 4.8 Experimental results of typical laser Thomson scattering and fitting shape factor

through the observation window in the corresponding direction and is collected, so as to avoid a large amount of stray light interference caused by the scattering of outgoing light and to protect the safety of the experimenter. The scattered light is collected at a position perpendicular to the incident laser. It is collected by the lens array and focused on the incident slit of the spectrometer. The Intensified CCD (ICCD) is connected to the spectrometer at the back end for imaging. The exposure time of ICCD in this experiment was set as 2 μs. Since the laser emitted by the Nd: YAG pulse laser used in this paper is a pulsed laser, the emission frequency is only 10–30 Hz, and the post-arc process time is only about tens or even hundreds of microseconds, so a current-zero period can only be used for a single laser Thomson scattering test.

In order to measure points at different time in the current-zero period, precise timing control is needed. A digital delay generator is used to synchronise the main circuit with the output of Nd:YAG laser and ICCD exposure. The switch break of the main circuit is connected in parallel with a group of IGBT (Insulated Gate Bipolar Transistor), and the digital delay generator sends a trigger signal to the driver of IGBT, and the current through the break crosses zero. After a pre-set time, the Q-switch of Nd:YAG is triggered to turn on, and the laser is emitted; At the same time, the ICCD exposure is controlled to record the scattering spectra. The delay time can be controlled to measure the arc plasma at different moments after the current crosses zero.

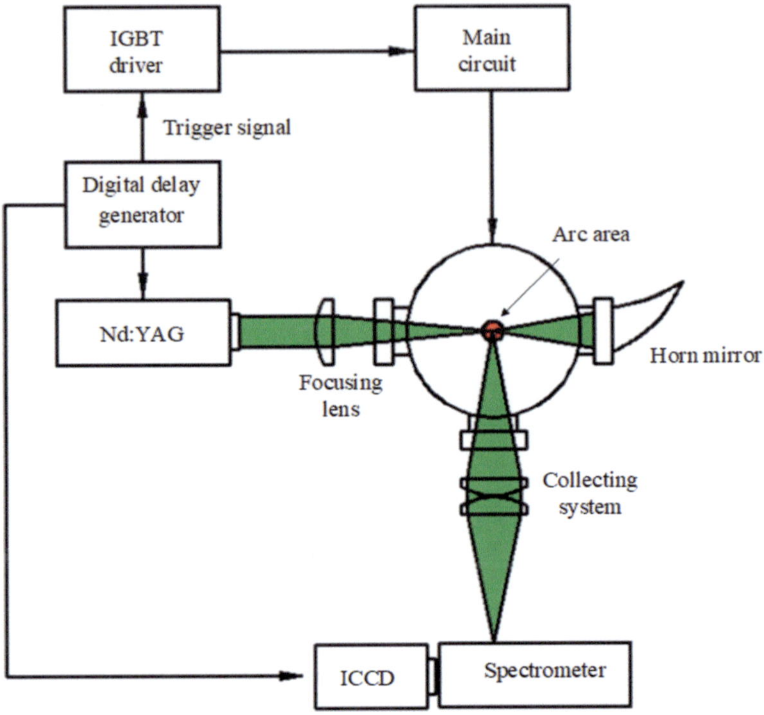

Fig. 4.9 Schematic diagram of optical path for laser Thomson scattering measurement

4.4.3 Limitations

In the laser Thomson scattering experiment, the incident probe laser focuses on the arc plasma, which inevitably causes the plasma energy at the focal point to rise, leading to a deviation of the relevant measurement results. Since the time for the heavy particles to reach equilibrium (μs) is much longer than the time for laser pulses (tens of ns), the probe laser has little effect on the measurement of ion temperature. However, the laser pulse time is much longer than the electron collision relaxation time (ps), and the probe laser energy has a great influence on the electron temperature measurement results. The main reason for the measurement deviation is the electron heating effect caused by inverse bremsstrahlung, which makes the measured electron temperature higher.

To measure the electron temperature accurately, the energy density of the probe laser must be gradually reduced, so that the corresponding electron temperature when the laser energy is zero can be derived. It should be pointed out that previous studies generally believed that the electron temperature and the laser pulse energy show a linear relationship, but that recent research results show that the electron temperature and the laser pulse energy do not have a linear relationship, especially when the laser energy is small, deviation from the linear fitting curve is large.

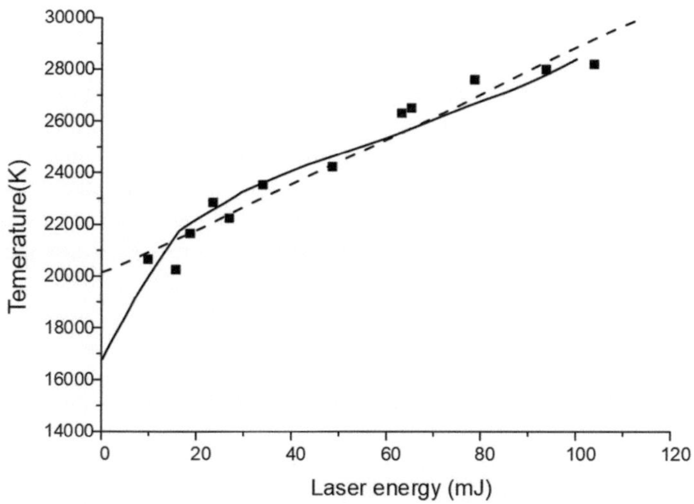

Fig. 4.10 Effect of incident laser energy on electron temperature

In Fig. 4.10, the solid boxes are experimental measurement results from [37], the dotted line is the fitting curve of the experimental result, and the solid line is the variation of electron temperature with pulsed laser energy simulated by radiative MHD [38].

For the above reasons, some researchers believe that the measurement of electron temperature by Thomson scattering method is unreliable [39]. Fortunately, the incident pulse laser energy has little effect on the measured electron density, which is believed to be accurate and reliable by the Thomson scattering method.

4.5 Measurement Relating to Arc Shape

By studying the shape of the switching arc under different factors and the movement process in time and space through experimental means, we can obtain the law of the occurrence, development and movement of the arc, observe the stagnation, reversal and any sudden changes in the arc movement, and analyse the phenomenon of back commutation during the switching arc movement and the process of the arc being cut by the splitter plates. In addition, the wall erosion and electrode ablation phenomena will change the particle composition in arc zone, thus changing the physical parameters of arc and making an impact on arc shape, so the study of the switching arc motion under the condition of wall erosion and electrode ablation is helpful in understanding the micro level plasma characteristics of the arc.

High-speed photography technology is a relatively mature means of testing the arc shape, which can achieve rapid, multiple sampling of the arc in a very short time period, and record and store the time and space information of the high-speed transient process of the arc by computer. When projected at a normal speed, the process of the arc appears clearly and slowly in front of our eyes. High speed photography technology has the outstanding advantages of real-time target capture, fast image recording, instant playback, and clear imaging.

The advantage of the high speed camera is that the image obtained is clear and complete and easy to analyse. The measurement requires that one side of the arc area of the tested product must use a fully transparent shell, so in some cases, due to the limitation of the switch's internal arc extinguishing chamber structure, the inability to fully expose the arc area will cause measurement difficulties.

4.5.1 Principle of High-Speed Photography

The high-speed camera converts the light signal of the arc into an electric signal through a high-speed CCD (Charge-coupled Device) light-receiving element, and then performs high-speed parallel storage, which can capture the image that changes during the high-speed movement of the arc. Its spatial resolution and temporal resolution are very high, so it has been widely used in the field of arc research.

4.5.2 Experiment and Result Analysis

1. Experimental set up

The experimental set up of the switching arc is shown in Fig. 4.11. The test product is a simplified arc extinguishing chamber, including two electrode arc runners, a splitter plate and an insulated wall bracket. One side of the insulated wall is made of transparent plexi-glass in order to capture the dynamic characteristics of the arc in the arc chamber with a high speed camera.

The test product, namely the model arc extinguishing chamber, is connected to the LC oscillation circuit as shown in the Fig 4.11. The high voltage probe Tektronix P6015 was used to measure the arc voltage of the sample, and the circuit current was measured through the Hall sensor. The measured signal was collected and recorded by a Tektronix TDS460A oscilloscope.

During the experiment, a PHANTOM V10 high-speed camera was used to shoot the dynamic characteristics of the arc, and the high-speed camera and oscilloscope were synchronously recorded through the control circuit. The sample rate was 20,000 frames per second, and the image depth was 8-bit 256 colours.

2. Results and analysis

Figure 4.12 shows the arc motion and cutting process recorded by a high-speed camera. It can be seen that after the arc ignites, it starts to move towards the splitter plate. The cathode arc roots are more concentrated and move more slowly than the

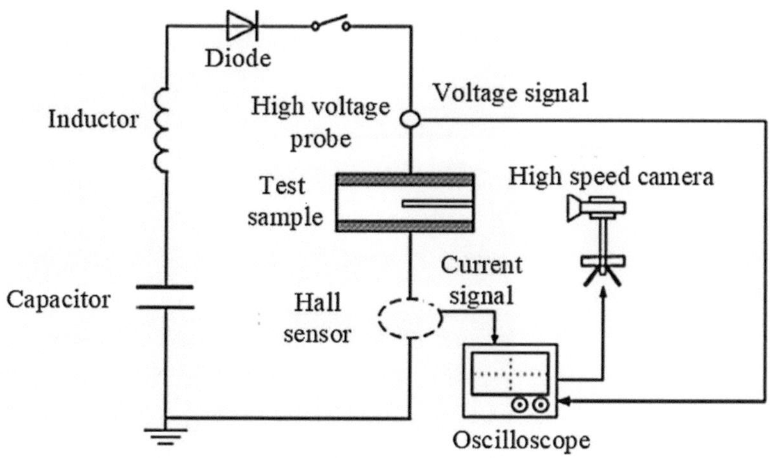

Fig. 4.11 Schematic diagram of experimental set up

anode arc roots in the lower running path. When it reaches the splitter plate, it is obvious that the arc root of the cathode on the upper surface of splitter plate goes in front of the arc root of the anode on the lower surface of the splitter plate. And when the arc is completely divided into two segments, the motion speed of the upper segment is obviously faster than that of the lower segment.

4.6 Measurement of Radiation from Arc

4.6.1 Principle of Measurement

Since the radiation emitted by a high temperature arc has a very complicated wavelength composition, it is usually necessary to choose a thermal detector to measure the arc radiation energy. This is because thermal detectors, such as thermocouples and thermal radiometers, have flat wavelength response characteristics in a wide wave band. However, the response time of such heat detectors is very slow, usually in the tens of ms or s level, which means they may only be suitable for observing light radiation varying over tens of ms or longer. The pursuit of faster response times (1ms or less) usually comes at the expense of the detector's flat wavelength response. For example, photodetectors (photomultiplier tubes, photodiodes, etc.) have extremely fast response speeds, ranging from a few ns to several hundred ns in general, which can realise effective time resolution in the transient process of optical radiation. At the same time, it has a very high sensitivity and can measure very weak light radiation power. However, photodetectors have different response capacities to optical signals with different wavelengths, so the corresponding photodetectors should be selected according to the wavelength range used in the application to ensure the highest

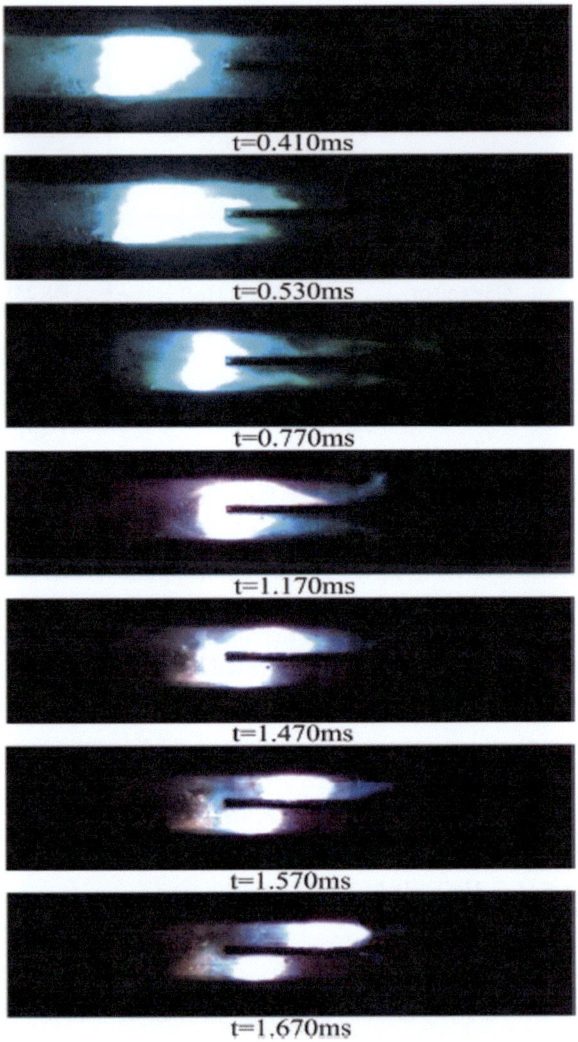

Fig. 4.12 Arc images recorded by a high speed camera

sensitivity and accuracy of the measured data in this band. When it is necessary to measure the radiation intensity of different wavelengths quantitatively or relatively, the response ability of photodetectors to different wavelengths should also be known. For an arc with a complex wavelength distribution, the response of the light detector to different wavelengths also becomes quite complex and difficult to measure, which means that it is unrealistic to use photodetectors to obtain the transient process of radiation in all wavelengths of the arc.

Due to the duration of the switching arc being very short, the light emission of the switching arc is a transient process with rapid changes. Considering the mutually restricting factors of arc radiation measured by the above detectors, our research group at Xi'an Jiaotong University proposed a method which combines thermopile and photomultiplier for measuring switch arc radiation, and realised the measurement of absolute arc radiation energy by utilising the characteristic that a thermopile has similar response ability to different wavelengths of light. However, the total absolute radiation energy accumulated during the arc burning period is measured by this method, which has no time resolution. In order to obtain the transient change of switching arc radiation, we choose the photomultiplier tube with super fast response. Although it is not possible to know the different response of each wavelength to the total light intensity in the experiment, the transient change of arc relative radiation can be measured through the photomultiplier tube instead of quantitative measurement. Then, the relative radiation obtained by the photomultiplier is calibrated by the total absolute radiant energy measured by the thermopile, and the transient process of the absolute radiant energy emitted by the switching arc can be calculated.

4.6.2 Examples of Application

Figure 4.13 shows the experimental schematic diagram of radiation energy measurement of fault arc. Arc radiation was transmitted through the optical fibre to the photomultiplier tube (PMTH-S1-PR928) for measuring the transient relative radiation power, and the total absolute radiation energy of the arc was measured by using the 12 A thermoelectric reactor probe of Ophir, Israel. The spectral measurement range was 190–20,000 nm and the energy range was 1–30,000 mJ. It is also equipped with a Nova II power meter head to display the measurement results. In the experiment, the thermopile probe was placed 4 m away from the arc outside the chamber, and the arc radiation was detected through the quartz glass on the chamber wall. Generally, the spectral range of arc radiation is UV-visible band, and the selected JGS1 type quartz glass (180–4,000 nm) has a transmittance of over 97% in UV-visible band, which ensures that the energy of arc radiation is collected as comprehensively as possible.

The probe was placed at a distance of 4 m from the arc so that the arc could be treated as a point light source, and the arc radiated a spherical intensity into the surrounding space. The arc radiant energy received by the hot reactor probe was the energy received by a small receiving area ($\varphi = 16$ mm in diameter) on a sphere 4 m away from the arc. Thus, there was a corresponding relationship between the actual radiation energy X of the arc and the energy Y measured by the probe as follows:

$$\frac{4\pi \times r^2}{\pi \times (\phi/2)^2} = \frac{X}{Y} \tag{4.26}$$

According to formula (4.26) and the data displayed by the power meter, the absolute radiation energy actually emitted by the arc can be calculated. By calibrating the relative radiant power measured by the photomultiplier tube with the absolute

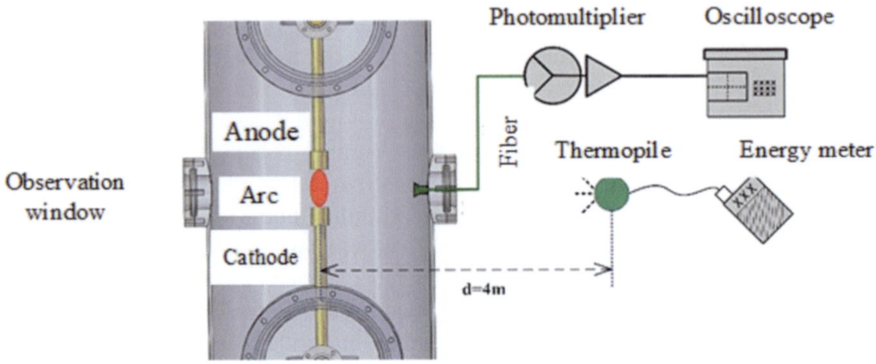

Fig. 4.13 Schematic diagram of arc radiation energy measurement

radiant energy, the transient process of the absolute radiant energy emitted by the arc can be calculated.

Figure 4.14 shows the schematic diagram of the experiment. The location of the ignition arc was the centre of the cavity, and the fault arc was generated by the copper wire. The container was filled with 0.1 MPa air, the electrode spacing is 30 mm, and the electrode material was pure copper. The current through the arc column was 1–20 kA dc current, and the arc burning time was 0.1 s. PCB 102B16 piezoelectric sensors were installed at two positions (p1 and p2) on the inner wall of the cavity to measure the pressure rise at different positions in the cavity. P1 was 0.475 m away from the upper cover plate of the cavity, and p2 was at the same horizontal plane as the arc centre.

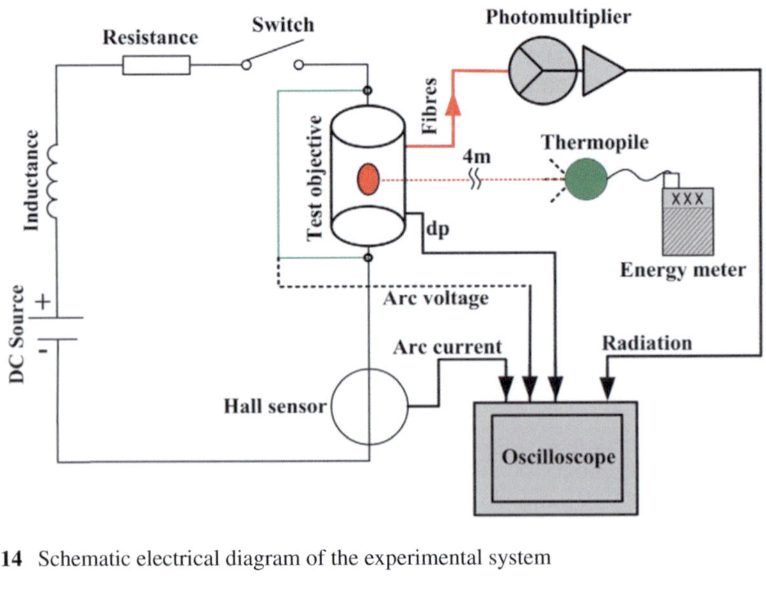

Fig. 4.14 Schematic electrical diagram of the experimental system

Arc voltage was measured by a high voltage probe (Tektronix-P6015A), arc current was collected and measured by CS20000HC series Hall current sensor, arc relative radiation power was measured by photomultiplier tube, and the absolute radiant energy of the arc was displayed by the thermal reactor probe. The pressure signal, arc current and arc relative radiation power were collected and recorded by an oscilloscope (Tektronix DPO4032). In addition, the motion of the arc was photographed using KODAK's high-speed camera, which was stored and observed on a computer at a maximum speed of 12,000 frames per second.

Figure 4.15 shows the typical experimental waveforms under the condition of arc current of 10 kA. Within the first few ms, the arc voltage rose rapidly due to the rapid melting and evaporation of the copper wire. At about 3 ms, the copper wire completely evaporates and the arc voltage shows a steep ignition peak. At the same time, the arc begins to emit bright arc light outward, and the cathode of the photomultiplier tube receives the radiation light from the arc, inducing the photocurrent inside.

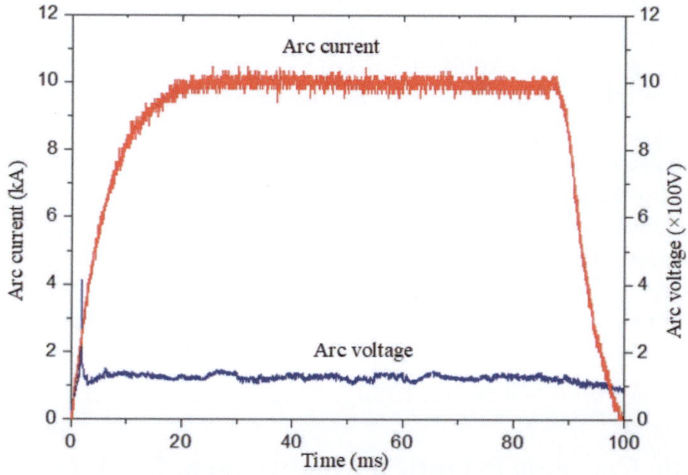

(a)

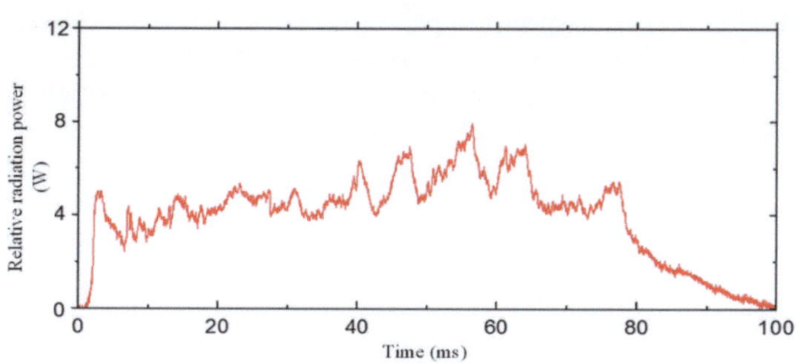

(b)

Fig. 4.15 Typical experimental waveforms for an arc current of 10kA

After that, the arc burns steadily, and arc voltage tends to be stable. At 15 ms, the current reaches its maximum. With the continuous injection of arc energy, the arc expands rapidly, and more and more high temperature areas are covered. On the one hand, arc energy is transferred to the surrounding area through conduction, convection and radiation, heating the surrounding gas, causing the increase of gas internal energy, which leads to the rapid increase of internal pressure. On the other hand, the accumulation of arc energy and the injection of a large amount of copper vapour cause the arc radiation to increase, the radiation power tends to rise, and continuous fluctuations occur in a certain range. At $t = 100$ ms, the current crosses zero, the arc is extinguished, no more arc energy is injected, and the maximum pressure rise occurs. Meanwhile, the arc radiation power is reduced to zero. As mentioned above, the transient process of the absolute radiant energy emitted by the arc can be obtained by calibrating the curve of the relative radiant power with the absolute radiant energy measured by the power meter.

Figure 4.16 shows the changes of absolute arc radiation energy measured under different current. At the first few ms of the experiment, the copper wire was heated to the evaporation temperature (about 4,000 K) and an arc was formed. The radiant energy began to increase. At about 3 ms, the copper wire was completely melted and evaporated. From 3 ms to 15 ms, the arc current rose rapidly, the joule heat increased rapidly, and the arc radiation intensity gradually increased, resulting in a slow increase in radiation energy. After 15 ms, the current reached its peak, and intense electrode ablation caused a large amount of copper vapour to diffuse into the arc column, rapidly increasing the arc radiation. Compared with the previous stage, the rising speed of radiation energy was obviously faster, which indicated that the injected copper vapour made the arc radiation effect significantly enhanced. After 85 ms, with the decrease of arc current, arc energy increased more slowly and arc radiation increased more gradually.

4.7 Basic Concepts in Optical Measurement for Arcs

The absorption and emission spectra of plasma are not strictly monochromatic. In the vicinity of the central frequency ν_0, it can be observed that there exists a distribution function $I(\nu)$ of spectral line intensity around the central frequency, which is called line shape. As shown in Fig. 4.17, the part between ν_1 and ν_2 is called the kernel of the line, and the rest is called the line wings. $I(\nu_1) = I(\nu_2) = I_m/2$, the frequency difference $\delta\nu = \nu_2 - \nu_1$ is called the FWHM of the spectral line. At the same time, the HWHM (half-width at half maximum) of the spectral line is often used, that is, half of the FWHM, which needs to be distinguished in calculation. Angular frequency and wavelength are sometimes used as variables to describe the line shape of spectral lines.

To compare different line shapes, a normalised strength line shape is defined $g(\nu - \nu_0) = I(\nu - \nu_0)/I_0$, where $g(\nu - \nu_0) = I(\nu - \nu_0)/I_0$. Common spectral shapes are Lorentzian, Gaussian, and Voigt.

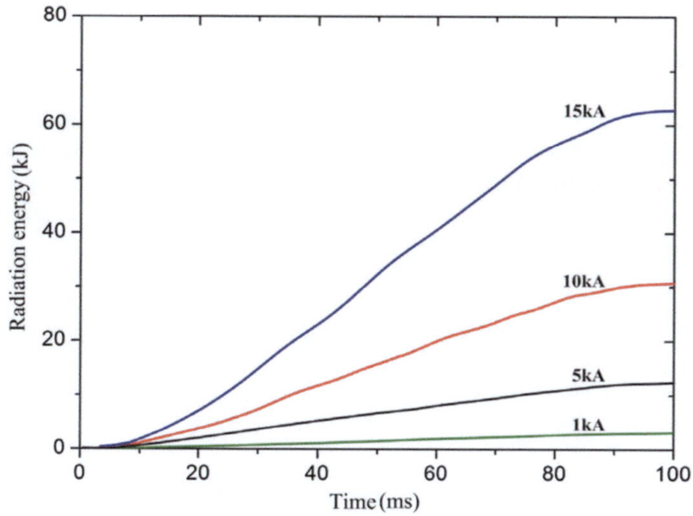

Fig. 4.16 Arc radiation energy measurement by experiment

Fig. 4.17 Line shape of a spectral line

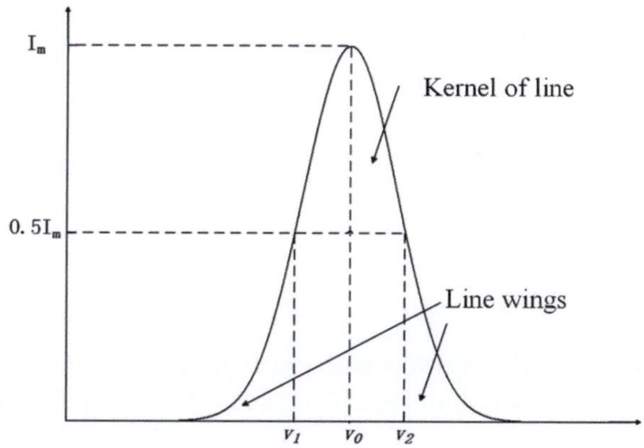

1. Lorentzian type

The line shapes of nature broadening or the intrinsic broadening, pressure broadening belongs to Lorentzian type.

Since the lifetime of the excited state energy level is limited, according to Heisenberg's uncertainty principle $\Delta E \Delta t \geq \hbar/2$(where $\hbar$ is the reduced Planck's constant), the uncertainty of the upper and lower energy levels leads to natural broadening [40]. Natural broadening is relatively narrow in the visible light range and can be ignored compared to other types of broadening.

Photon-emitting particles interact with the surrounding environment to cause pressure broadening, which is a uniform broadening. Resonance broadening is caused by the interaction between the same kinds of neutral particles; van der Waals broadening is caused by the interaction between different kinds of neutral particles; the broadening caused by charged particles is the Stark broadening.

The normalised Lorentzian type expression is shown in Eqs. 4.27 and 4.28.Where, α_L is HWHM of Lorentzian type in frequency form. $\Delta\lambda_L$ is FWHM of Lorentzian type in wavelength form. Lorentz line has symmetry about v_0, when $v = v_0$ there is a maximum, $g_{L,v,\max} = 1/\pi\alpha_L$; Lorentz line has symmetry about λ_0, when $\lambda = \lambda_0$ there is a maximum, $g_{L,\lambda,\max} = 2/\pi\Delta\lambda_L$.

$$g_L(v, v_0) = \frac{1}{\pi}\frac{\alpha_L}{(v - v_0)^2 + (\alpha_L)^2} = g_{L,v,\max}\frac{\alpha_L^2}{(v - v_0)^2 + (\alpha_L)^2} \tag{4.27}$$

$$g_L(\lambda, \lambda_0) = \frac{2}{\pi}\frac{\Delta\lambda_L}{4(\lambda - \lambda_0)^2 + (\Delta\lambda_L)^2} = g_{L,\lambda,\max}\frac{\Delta\lambda_L^2}{4(\lambda - \lambda_0)^2 + (\Delta\lambda_L)^2} \tag{4.28}$$

The FWHM of the convolution of two Lorentzian types is the sum of the two FWHMs of the two Lorentzian types [40], that is $\Delta\lambda_L = \Delta\lambda_{L1} + \Delta\lambda_{L2}$.

2. Gaussian type

The line shapes of instrumental broadening, Doppler broadening belong to Gaussian type. The normalised Gaussian line shape is shown in Eqs. 4.29 and 4.30:

$$g_G(v, v_0) = \sqrt{\frac{\ln 2}{\pi}}\frac{1}{\alpha_G}e^{-[\frac{\ln 2}{(\alpha_G)^2}(v - v_0)^2]} = g_{G,v,\max}\exp[-\frac{\ln 2}{(\alpha_G)^2}(v - v_0)^2] \tag{4.29}$$

$$g_G(\lambda, \lambda_0) = \sqrt{\frac{\ln 2}{\pi}}\frac{2}{\Delta\lambda_G}e^{-[\frac{4\ln 2}{(\Delta\lambda_G)^2}(\lambda - \lambda_0)^2]} = g_{G,\lambda,\max}\exp[-\frac{4\ln 2}{(\Delta\lambda_G)^2}(\lambda - \lambda_0)^2] \tag{4.30}$$

where α_G is HWHM of Gaussian type in frequency form. $\Delta\lambda_G$ is FWHM of Gaussian type in wavelength form. Gaussian type has symmetry about v_0, when $v = v_0$ there is a maximum, $g_{G,v,\max} = \sqrt{\frac{\ln 2}{\pi}}\frac{1}{\alpha_G}$; Gaussian type has symmetry about λ_0, when $\lambda = \lambda_0$ there is a maximum, $g_{G,\lambda,\max} = \sqrt{\frac{\ln 2}{\pi}}\frac{2}{\Delta\lambda_G}$.

Instrument broadening is the broadening of spectral lines due to the limited inherent resolution of spectral instruments. The actual spectrum is the combination of the theoretical spectral line with the instrument line (usually approximately the spectral line of the monochromatic incident light of the mercury lamp used for spectral instrument detection and calibration).

Doppler broadening is one of the main sources of spectral line broadening in low-pressure gas, which is caused by the thermal motion of particles emitting or absorbing photons. Doppler broadening (HWHM, s^{-1}) can be expressed as $\alpha_D = \frac{v_0}{c}(\frac{2k_BT}{m}ln2)^{1/2}$, m is the mass of the particle emitting or absorbing photons (kg). Thus it can be seen that the maximum and HWHM of Doppler broadening are both functions of the central frequency. Using the relationship between the gas constant,

Avogadro constant and Boltzmann constant, $R = N_A k_B$, M is the relative atomic mass (molar mass, g/mol), and the Doppler broadening (FWHM, s^{-1}) can be written as follows:

$$2\alpha_D = \frac{2\nu_0}{c}\sqrt{\frac{2\ln 2RT}{M}} \tag{4.31}$$

More commonly, Doppler broadening in wavelength form [17, 41] (FWHM) is used, as shown in Eq. 4.32:

$$\Delta\lambda_D = 2 \times \frac{\lambda_0^2}{c}\alpha_D = \frac{\lambda_0}{c}(\frac{8k_BT}{m}ln2)^{1/2} \approx 7.16 \times 10^{-7}\lambda_0\sqrt{\frac{T}{M}} \tag{4.32}$$

Doppler broadening increases linearly as the central frequency increases and is proportional to $\sqrt{T/M}$, and the higher the temperature, the more obvious the Doppler broadening becomes.

Under general experimental conditions, the Gaussian part of the measured spectral line is the joint effect of instrument broadening and Doppler broadening, and the broadening of the three has the following relationship, $\Delta\lambda_G = \sqrt{\Delta\lambda_{In}^2 + \Delta\lambda_D^2}$.

3. Voigt type

In the actual measured spectral lines, there are contributions of both uniform and non-uniform broadening, and the actual measured spectral lines are usually neither Lorentzian nor Gaussian type. Such an integrated line is called a Voigt type. For example, since the theoretical spectral line and instrument broadening co-exist in the measured results, the monochromatic incident light is usually first detected with the spectral instrument to obtain instrument broadening, and then deconvolution is carried out to the actual measured spectral line to remove the influence of instrument broadening.

It is assumed that there is only uniform spreading, which radiates energy $dI = I_0 g_L(\nu_1, \nu_0)d\nu_1$ in the frequency range $(\nu_1, \nu_1 + d\nu_1)$. On this basis, considering the instrument spreading, the radiant energy dI will be extended to the frequency range $(-\infty, +\infty)$. At this time, the central frequency of the instrument spreading line shape function is ν_1, and the energy extended from the radiant energy dI to the frequency range $(\nu, \nu + d\nu)$ is as in Eq. 4.33 [42]:

$$dI \cdot g_G(\nu, \nu_1)d\nu = I_0 g_L(\nu_1, \nu_0)d\nu_1 \cdot g_G(\nu, \nu_1)d\nu \tag{4.33}$$

The distribution of total radiation energy I_0 in the frequency range $(\nu, \nu + d\nu)$ is as follows:

$$I_0[\int_{-\infty}^{+\infty} g_L(\nu_1, \nu_0)g_G(\nu, \nu_1)d\nu_1]d\nu \tag{4.34}$$

Then the expression of the Voigt broadened line shape is as follows:

$$\begin{aligned} g_V(v, v_0) &= \int_{-\infty}^{+\infty} g_L(v_1, v_0) g_G(v, v_1) dv_1 \\ &= \sqrt{\frac{\ln 2}{\pi^3}} \int_{-\infty}^{+\infty} \frac{\alpha_L}{(v_1 - v_0)^2 + \alpha_L^2} \frac{1}{\alpha_G} e^{-[\frac{\ln 2}{\alpha_G^2}(v - v_1)^2]} dv_1 \end{aligned} \tag{4.35}$$

If $\alpha_G(v_1)$ is approximated as a constant independent of v_1, then

$$g_V(v, v_0) \approx V(v, v_0) = \frac{1}{\alpha_G} \sqrt{\frac{\ln 2}{\pi}} (\frac{\mu}{\pi} \int_{-\infty}^{+\infty} \frac{1}{(\xi - t)^2 + \mu^2} e^{-t^2} dt) \tag{4.36}$$

where, $t = \frac{\sqrt{\ln 2}}{\alpha_G}(v - v_1)$, $\mu = \sqrt{\ln 2} \frac{\alpha_L}{\alpha_G}$, $\xi = \sqrt{\ln 2} \frac{v - v_0}{\alpha_G}$.

Through strict derivation, it can be known that [42], the Voigt line shape satisfies the normalisation condition and is strictly symmetrical about v_0.

The maximum value V_{max} of the Voigt line shape is determined by the FWHM of the Lorentzian and Gaussian line shapes and is smaller than the maximum value of both Lorentzian and Gaussian line shapes.

There is no exact analytical relationship between the FWHM of the three types of lines. In practical use, empirical formula (4.37) [43] can be used:

$$\alpha_V = (1.0692\alpha_L + \sqrt{0.86639\alpha_L^2 + 4\alpha_G^2})/2 \tag{4.37}$$

4.8 Summary

This chapter provides an overview of measurement and diagnostic methods for the physical parameters and shapes of arcs in power switchgear equipment. These parameters include excitation temperature, electron density, atomic number density, arc column dynamics, and radiation power. We categorize non-intrusive measurement methods applicable to this chapter into two types: passive measurements and active measurements.

Passive measurement methods involve analyzing the plasma's radiation to obtain microscale parameters. Examples include emission spectroscopy for measuring electron density and electron temperature in steady-state arcs, high-speed photography for capturing arc morphology, and thermal detectors for measuring high-temperature radiation. Active measurements, on the other hand, require an external light source to interact with the plasma of interest, yielding its microscale parameters. Specific techniques include atomic absorption spectroscopy for measuring atomic number density and Thomson scattering for assessing electron density and temperature in non-steady-state arcs.

This chapter provides detailed explanations of the fundamental principles and practical applications of each method, along with discussions of their advantages and limitations. By understanding these concepts, researchers and practitioners can select the most suitable method for specific application scenarios.

References

1. Wang J M. Theoretical study and test of vacuum arc [M]. Xi'an: Xi'an Jiao Tong University Press,1993.
2. Schneider A V, et al. Anode temperature and plasma sheath dynamics of high current vacuum arc after current zero[J]. IEEE Transactions on Plasma Science, 2013, 41(8): 2022-2028.
3. Van Lanen E P A, et al. Current-zero characteristics of a vacuum circuit breaker at short-circuit current interruption[C]. 2006 International Symposium on Discharges and Electrical Insulation in Vacuum. IEEE, 2006, 1: 212-215.
4. Ivanov V A, et al. Time-resolved measurements of the parameters of arc cathode plasmas in vacuum[J]. IEEE Transactions on Plasma Science, 1985, 13(5): 334-336.
5. Klajn A. Langmuir probes in switching vacuum arc measurements[J]. IEEE Transactions on Plasma Science, 2005, 33(5): 1611-1617.
6. Griem H R. Plasma spectroscopy[M]. New York: McGraw-Hill, 1964.
7. Lochte-Holtegreven W. Plasma-diagnostics[M]. Amsterdam: North-Holland Publication Co., 1968.
8. Griem H R. Spectral line broadening by plasmas[J]. IEEE Ttanzacton, on Plasma Science, 1974, 3(4): 227.
9. Hai T, Wang Q P. Measurement of arc temperature in SF_6 arc extinguishing chamber model[J]. High Voltage Apparatus, 1991.
10. Airey D R, et al. Time-resolved radial temperature profiles for 10 kA SF_6 arcs[J]. Journal of Physics D: Applied Physics, 1975, 8(16): 1982.
11. Jones G R, et al. A spectroscopic investigation of electrode material entrained into the extinguishing arc of SF6 circuit breaker[J]. Journal of Physics D: Applied Physics, 1988, 21(9): 1414.
12. Methling R, et al. Spectroscopic investigation of a Cu-Cr vacuum arc[J]. IEEE Transactions on Plasma Science, 2015, 43(8): 2303-2309.
13. Demtröder W. Laser spectroscopy: basic concepts and instrumentation[M]. Springer Science & Business Media, 2013.
14. Sansonetti J E. Handbook of basic atomic spectroscopic data[J]. Journal of Physical and Chemical Reference Data, 2005, 34(4): 1559.
15. Tanaka Y, et al. Transient behaviour of axial temperature distribution in post-arc channel after current zero around nozzle throat in flat-type SF6 gas-blast quenching chamber[J]. Journal of Physics D: Applied Physics, 1995, 28(10): 2095.
16. Ling L G. Research of spectral information from laser-pulse MIG hybrid arc plasma[D]. Tian Jin University, 2010.
17. Pellerin S, et al. Stark width of - Ar I transition (696.543 nm) [J]. Journal of Physics B: Atomic, Molecular and Optical Physics, 1996, 29(17): 3911-3924.
18. Ionascut-Nedelcescu A, et al. Calculation of the gas temperature in a throughflow atmospheric pressure dielectric barrier discharge torch by spectral line shape analysis[J]. Journal of Applied Physics, 2008, 103(6).
19. Vogman G V, Shumlak U. Deconvolution of Stark broadened spectra for multi-point density measurements in a flow Z-pinch[J]. Review of Scientific Instruments, 2011, 82(10): 103504.
20. Bachmann B, et al. High-speed three-dimensional plasma temperature determination of axially symmetric free-burning arcs[J]. Journal of Physics D: Applied Physics, 2013, 46(12): 125203.
21. Hegazy H. Oxygen spectral lines for diagnostics of atmospheric laser-induced plasmas[J]. Applied Physics B: Lasers and Optics, 2010, 98(2-3): 601-606.

22. Vacquie S, et al. Measurements of electron density in a SF6 arc plasma[J]. Journal of Physics D: Applied Physics, 1985, 18(11): 2193.
23. Zielińska S, et al. Investigations of GMAW plasma by optical emission spectroscopy[J]. Plasma Sources Science and Technology, 2007, 16(4): 832.
24. Tuma D T, et al. Erosion products from the cathode spot region of a copper vacuum arc[J]. Journal of Applied Physics, 1978, 49(7): 3821-3831.
25. Jenkins J E, et al. Measurement of the neutral vapour density decay following the extinction of a high-current vacuum arc between copper electrodes[J]. Journal of Physics D: Applied Physics, 1975, 8(12): L139.
26. Hong D, et al. A new optical technique for investigations of low-voltage circuit breakers[J]. IEEE Transactions on Plasma Science, 2005, 33(2): 976-981.
27. Horvath B, Lamara T. Time-resolved optical resonant absorption spectroscopy of Cr metallic vapor in air using a broadband LED light source[J]. Plasma Sources Science and Technology, 2013, 22(4): 45006.
28. Takahashi S, et al. Laser measurement of copper vapor density after a high-current vacuum arc discharge in an axial magnetic field[J]. IEEE Transactions on Plasma Science, 2005, 33(5): 1519-1526.
29. Yang G, Heberlein J V. The anode region of high intensity arcs with cold cross flow[J]. Journal of Physics D: Applied Physics, 2007, 40(18): 5649.
30. Sonoda Y, et al. Laser Thomson scattering diagnostics of dielectric barrier discharge plasmas[J]. J. Plasma Fusion Res, 2009, 8: 696.
31. Tomita K, et al. Thomson scattering diagnostics of decay processes of Ar/SF_6 gas-blast arcs confined by a nozzle[J]. Journal of Physics D: Applied Physics, 2013, 46(38): 382001.
32. Snyder S C, et al. Direct evidence of departure from local thermodynamic equilibrium in a free-burning arc-discharge plasma[J]. Physical Review E, 1993, 48(5): 4124.
33. Lu Weitao. The research and design of the SG-III Thomson scattering diagnostic imaging system[D]. Graduate school of Chinese academy of sciences (Xi 'an institute of optics and precision mechanics), 2011.
34. Froula D H, et al. Plasma scattering of electromagnetic radiation[M]. Elsevier, 2011.
35. Yang G, Heberlein J. Anode attachment modes and their formation in a high intensity argon arc[J]. Plasma Sources Science and Technology, 2007, 16(3): 529.
36. Tanaka M, Ushio M. Plasma state in free-burning argon arc and its effect on anode heat transfer[J]. Journal of Physics D: Applied Physics, 1999, 32(10): 1153.
37. Bentley R E. Shape parameters for the Thomson electron feature as a means of interpreting laser scattering data[J]. Measurement Science and Technology, 1997, 8(10): 1139.
38. Murphy A B, et al. Electron heating in Thomson scattering measurements of plasma temperature: are thermal plasmas thermal[C]. PLASMA PHYSICS: 11th International Congress on Plasma Physics: ICPP2002. 2003, 669(1): 757-761.
39. Murphy A B. The effects of metal vapour in arc welding[J]. Journal of Physics D: Applied Physics, 2010, 43(43): 434001.
40. Kunze H-J. Introduction to plasma spectroscopy[M]. Springer Science & Business Media, 2009, 56.
41. Pan Chenggang. Diagnosis the Physical Characteristics of Pulsed MIG Welding Arc by Spectrum [D]. Shanghai Jiao Tong University, 2013.
42. Wu Chen. Research on spectral profile function and its fitting method[D]. North China Electric Power University, 2013.
43. Olivero J J, Longbothum R L. Empirical fits to the Voigt line width: A brief review[J]. Journal of Quantitative Spectroscopy and Radiative Transfer, 1977, 17(2): 233-236.

Chapter 5
Calculation and Analysis of Lorentz Force on Current Carrying Components

5.1 Introduction

The force between two current-carrying conductors or between a current-carrying conductor and a ferromagnetic component is called the Lorentz force. It is sometimes also referred to as electrodynamic force in other languages. On the positive side, the Lorentz force in electrical apparatuses can be used in current-limiting circuit breakers to rapidly separate electrical contacts and drive arc motion, electromagnetic cannons and electromagnetic ejection systems. On the opposite side, the Lorentz force can also cause a strong mechanical impact on electrical equipment, form unwanted arcs, and promote the welding of electrical contacts in switchgear.

To either maximise or suppress it, calculation and analysis of the Lorentz force is crucial. Electrical equipment should be able to withstand the Lorentz force due to short-circuit current without suffering structural damage or permanent deformation. For electrical contacts in switchgear, they should have the ability to overcome the repulsive effect or avoid welding due to the action of the Lorentz force. This ability defines the dynamic stability of electrical apparatuses, which is another key design consideration in addition to the thermal stability. The former is often quantified by the peak value of the impulse current during a short-circuit fault. This chapter is devoted to the calculation and analysis methods of the Lorentz force in power equipment.

5.2 Classification and Calculation of Short-Circuit Current in Power Systems

Short-circuit is the most serious fault in a power system, severely impacting the safe and reliable operation of the latter. There are four basic types of short-circuit faults in a three-phase system, i.e. three-phase, two-phase, two-phase grounding, and single-phase grounding short circuits. In a three-phase short circuit fault, the phases are

M. Rong et al., *Computing Methodology for Switchgear*,
https://doi.org/10.1007/978-981-95-2840-0_5

still symmetric and the fault is also called a symmetric short circuit. In contrast, the other three types of faults are asymmetric. In a network grounded at the neutral point, signal-phase grounding short circuit fault is the most common, accounting for about 90% of all the faults. In networks where the neutral point is not directly grounded, short-circuit faults are mainly between phases.

As the power systems continue to increase in complexity and capacity, the short-circuit current level has become a worrying issue in network planning and operation. The increased current level imposes more stringent requirements on electrical equipment in the grid, in addition to those on transmission capability and reliability of the main grid [1]. Short-circuit current calculation and analysis allows one to quantitatively evaluate the severity of the network faults, to verify the structure, wiring, and short-circuit current capacity of the network as a whole, and to provide baseline information for the selection of the interruption capacity of circuit breakers. There are different types of short-circuit current limiting measures, but they have different application scenarios, effectiveness in current limiting, and economic value [2]. Their impacts on power system operation are also different [2].

5.2.1 Classification of Short-Circuit Current in Power Systems

There are different causes for the occurrence of short circuits between phases of current-carrying conductors or between phase conductors and the ground, all leading to the short circuit of load impedance. Examples include ageing and damage of insulating materials, accidents short of human intervention, birds or animals bridging current-carrying components, natural events such as wind and snow, and accidents due to human negligence, etc.

The three-phase power system in China can be divided into two categories according to the neutral point grounding method used. The first one is a high grounding current system (or effective grounding system), including solid grounding and low reactance grounding. The other category is the low grounding current system (or non-effective grounding system), including non-grounded neutral, resonance grounding and grounding through high resistance. The method for neutral point grounding in a power system depends on a number of factors including reliability of the power supply, safety and protection level, equipment safety, insulation level, and over-voltage level.

Existing power systems of 220 kV and above across the globe make use of the direct (effective) grounding method while systems below 220 kV use the appropriate grounding method based on the system configuration [3]. In China, for electrical grids of 60 kV and below, the neutral point is not grounded when the single-phase grounding current (capacitive current) is less than 10 A). When the capacitive current is large, it is grounded through an arc suppression coil. In systems with a nominal voltage between 110 and 220 kV, partial neutral point grounding is adopted due to requirements from relay protection, system stability, limiting of single-phase short-circuit current, and other aspects. In high-voltage and ultra-high-voltage systems in China, neutral point direct grounding or grounding by reactor is adopted. Step-up transformers at 500 kV have been grounded via reactor.

For systems with neutral point grounded, the short-circuit types are: single-phase grounding, two-phase grounding, two-phase ungrounded, and three-phase short circuits. However, for a system with an ungrounded neutral point, single-phase grounding does not constitute a short circuit as the three-phase system is not disrupted and there is no short-circuit current. So for neutral point ungrounded systems, the short-circuit types are: two-phase grounded, two-phase ungrounded, and three-phase short circuits. In addition, short circuit may occur at multiple locations concurrently, which is called multiple short circuits or a compound fault.

5.2.2 Calculation Methods of Short-Circuit Current

Calculation of short-circuit current is an essential need in power system planning and design. It is the basis for the selection of electrical equipment ratings and setting of protection systems. It also provides an effective means for simulating and analysing fault accidents in the planning, design and operation of power systems. When a short circuit occurs, the short-circuit current can reach dozens or even hundreds of times the normal load current, which has a significant impact on the operation of the low-voltage system. During this transient process, the short-circuit current varies in a complicated manner. The maximum instantaneous value of the short-circuit current is called the inrush current. It can generate strong Lorentz forces in the electrical equipment. The Lorentz force generated in this way can be used to examine the dynamic stability of the mechanical structure of the electrical equipment when a short circuit occurs.

In the following section, the calculation of the three-phase short-circuit current of a power system with an infinite power source is given as an example for analysis.

The term 'infinite power source" refers to a power supply that can maintain constant terminal voltage and frequency when disturbances occur outside of the power supply. Based on the circuit in Fig. 5.1, the analysis and calculation of the transient process of the power system can be simplified. It then becomes relatively easy to obtain the various components of the short-circuit current, the effective current, and the impulse current during the short-circuit transient process. A true infinite power source does not exist in an actual power system. However, when multiple generators with limited capacities operate in parallel, or the electrical distance between the power source and the short-circuit point is very long, the equivalent power source can be regarded as an infinite power source.

Figure 5.1 is a simple three-phase circuit powered by an infinite power source. It is in a normal steady state before the short circuit occurs. The resistance and inductance of each phase are $R + R'$ and $L + L'$, respectively. Taking Phase A as an example, its voltage and current can be expressed as:

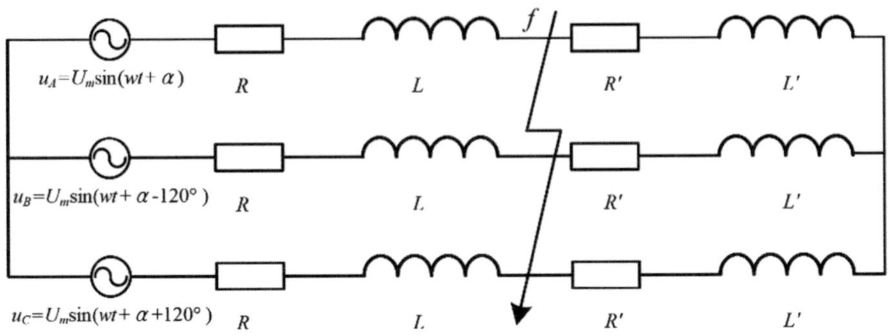

Fig. 5.1 Three-phase short circuit fault in a power system powered by an infinite power source

$$u_A = U_m \sin(\omega t + \theta) \tag{5.1}$$

$$i_A = I_{m|0|} \sin(\omega t + \theta - \varphi_{|0|}) \tag{5.2}$$

$$I_{m0} = \frac{U_m}{\left[(R + R')^2 + (L + L')^2\right]^{1/2}} \tag{5.3}$$

$$\varphi_{|0|} = \arctan \frac{\omega(L + L')}{R + R'} \tag{5.4}$$

where, u_A and i_A are the instantaneous values of Phase A's voltage and current, respectively; $\varphi_{|0|}$ is the current amplitude before the short circuit; $\varphi_{|0|}$ is the voltage amplitude of the source; $\varphi_{|0|}$ is *the* impedance angle of the circuit before the short circuit; θ is the initial phase angle of the source voltage; and ω is the angular frequency of the source voltage.

When the system experiences a three-phase short circuit fault at location f, the circuit is divided into two independent parts by the short-circuit point. The left part from the short-circuit point is still connected to the power source with an impedance for each phase, R + jωL. At the same time, the current in the circuit also gradually transitions from the original steady-state load current to the steady-state short-circuit current. The right side from the short-circuit point can be regarded as a passive network as there is no power source. The current in the passive network will decay exponentially, from the steady state load current to zero. In this process, the energy stored in the circuit will be converted into heat by the resistors. Therefore, the transient processes associated with the three-phase short circuit fault in the original circuit can be studied by examining only the electromagnetic transient processes of the circuit on the left-hand side of the short circuit point.

Close to the short circuit point, the impedance of each phase of the left-hand side circuit connected to the power supply becomes R + jωL, and the current changes from the previous steady-state value before the fault to a new steady-state value which is determined by the impedance $R + \mathrm{j}\omega L$. Assuming that a three-phase short-circuit fault occurs at t = 0, it is possible to study just a one-phase circuit instead of a three-phase circuit since the circuit on the left-hand side is a three-phase symmetrical circuit. For Phase A, the differential equation corresponding to the electromagnetic transient process is

$$U_m \sin(\omega t + \theta) = R \cdot i_A + L\frac{di_A}{dt} \tag{5.5}$$

The above formula is a first-order, non-homogeneous linear differential equation with constant coefficients, and the corresponding solution is the short-circuit current. Its solution consists of two parts. The first part is the special solution of Eq. 5.5, which characterises the periodic component of the short-circuit current. The second part is the general solution of the homogeneous equation corresponding to Eq. 5.5, which characterises the aperiodic component of the short-circuit current. For A, the short circuit current is

$$\begin{aligned} i_A &= I_{pm} \sin(\omega t + \theta - \varphi_k) + Ce^{-t/Ta} \\ &= i_p + i_{ap} \end{aligned} \tag{5.6}$$

In Eq. (5.6), $I_{pm} = U_m \Big/ \sqrt{R^2 + (wL)^2}$ is the periodic component of the short-circuit current, and $I_{pm} = U_m \Big/ \sqrt{R^2 + (wL)^2}$ is the aperiodic component of the short-circuit current. $I_{pm} = \frac{U_m}{\sqrt{R^2+(wL)^2}}$ is the amplitude of the periodic component of the short-circuit current and $\varphi_k = arctan\left(\frac{wL}{R}\right)$ is the short-circuit impedance angle. According to Lenz's law, the current of an inductive circuit cannot be changed suddenly, which means that the current in the circuit immediately after the short-circuiting is equal to the current before it. Therefore, the value of the constant C can be obtained from the solution. When t = 0, there is:

$$C = I_m \sin(\theta - \varphi) - I_{pm} \sin(\theta - \varphi_k) \tag{5.7}$$

From the above formula:

$$C = I_m \sin(\theta - \varphi) - I_{pm} \sin(\theta - \varphi_k) \tag{5.8}$$

Substituting Eq. 5.8 into Eq. 5.6, the Phase A short-circuit current can be expressed as:

$$i_A = I_{pm} \sin(\omega t + \theta - \varphi_k) + [I_m \sin(\theta - \varphi) - I_{pm} \sin(\theta - \varphi_k)]e^{-\frac{t}{Ta}} \tag{5.9}$$

Equation 5.9 shows that during the transient process following a sudden occurrence of a three-phase short circuit in a three-phase circuit powered by an infinite power source, the short-circuit current includes two components. One is the periodic component, or AC component, which is a steady-state short-circuit current which features a forced current. Its amplitude remains unchanged during the transient process and depends on the magnitude of the power supply voltage and circuit parameters. This is because the voltage amplitude of the infinite power supply is constant and the circuit parameters are not changed. The other is an aperiodic component or DC component and it is generated to prevent the current in the circuit from an abrupt change. Due to the lack of external power support and the presence of resistance in the circuit, it decays exponentially to zero with a time constant. When the aperiodic current decays to zero, indicating that the transient process is over, the circuit enters a stable short-circuit state. Since the three-phase circuit is still symmetrical (assuming $(\theta + 120°)$ and $(\theta + 120°)$ to replace the θ in Eq. 5.9), the short-circuit current expressions of Phases B and C can be determined, shown in Eqs. 5.10 and 5.11:

$$\begin{aligned} i_B = {} & I_{pm}\sin(\omega t + \theta - 120° - \varphi_k) + [I_m \sin(\theta - 120° - \varphi) \\ & - I_{pm}\sin(\theta - 120° - \varphi_k)]e^{-\frac{t}{Ta}} \end{aligned} \tag{5.10}$$

$$\begin{aligned} i_C = {} & I_{pm}\sin(\omega t + \theta + 120° - \varphi_k) + [I_m \sin(\theta + 120° - \varphi) \\ & - I_{pm}\sin(\theta + 120° - \varphi_k)]e^{-\frac{t}{Ta}} \end{aligned} \tag{5.11}$$

Figure 5.2 is a waveform diagram of the three-phase short-circuit current where i_{pA}, i_{pB} and i_{pC} are the periodic components of the short-circuit currents of Phases A, B, and C. i_{aA}, i_{aB} and i_{aC} are the aperiodic components of the short-circuit current decaying exponentially. Due to the presence of the DC components, the maximum instantaneous value of the short-circuit current that is greater than the amplitude of the AC component of the short-circuit current in each phase will occur. It is called the short-circuit impulse current and will generate a Lorentz force in the electrical equipment that is proportional to the square of the short-circuit impulse current. Therefore, to verify the dynamic stability of the equipment, the short-circuit impulse current must be calculated.

From Eq. 5.9, when the circuit parameters are constant (i.e. the power supply voltage amplitude and the short-circuit loop impedance are constant), the amplitude of the periodic component current is constant during the transient process. The aperiodic current component needs to have a maximum initial value to produce a maximum peak value (the impulse current) in the overall short-circuit current. The initial value of the aperiodic component is related to the initial angle of the power supply voltage and the value of the current in the circuit before the short circuit. From the derivation of Eq. 5.9, the initial value of the aperiodic current component of the short-circuit current in Phase A is:

$$i_{aA} = I_m \sin(\theta - \varphi) - I_{pm}\sin(\theta - \varphi_k) \tag{5.12}$$

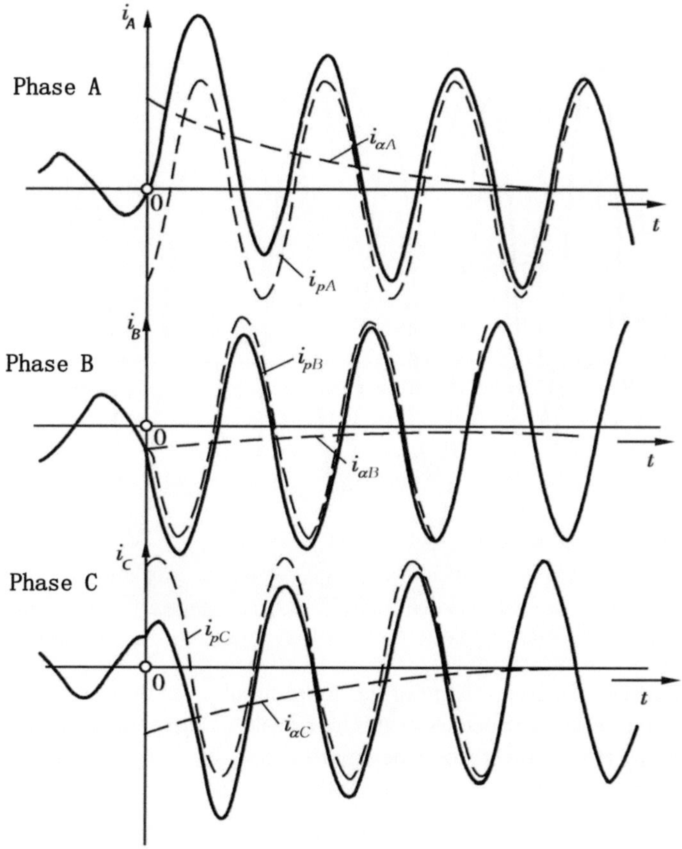

Fig. 5.2 Three-phase short-circuit current waveform

Evidently, the initial value of the aperiodic current component of Phase A is the difference between the normal load current immediately before the short circuit and the periodic component of the short-circuit current immediately after the onset of the short circuit. Because the amplitude of the current before the short circuit is much smaller than the amplitude of the periodic component of the short circuit current, the case with $I_m = 0$ and $\sin(\theta - \varphi_k) = 1$ means that there is no load current before the short circuit and i_{aA} will be at its maximum value shortly after the short circuit starts. A short circuit that meets these conditions is usually called a severe closing short circuit on a no-load line. Taking the above conditions into Eq. 5.9, we get:

$$i_A = -I_{pm}\cos(\omega t) + I_{pm}e^{-\frac{t}{T_a}} \tag{5.13}$$

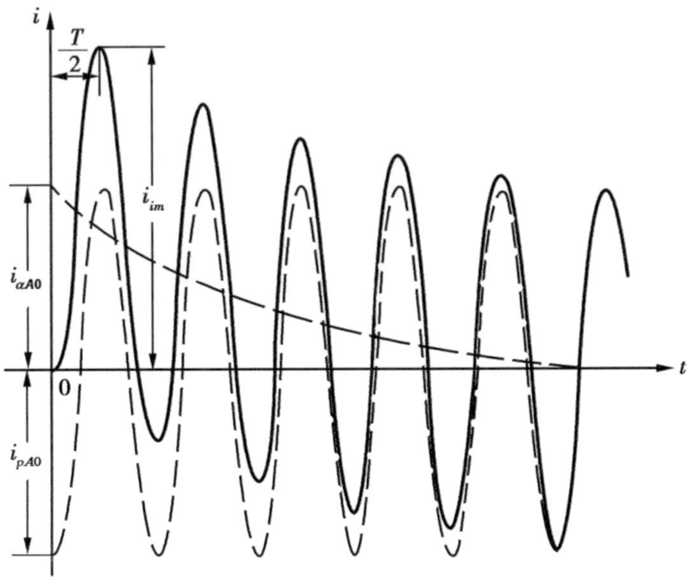

Fig. 5.3 Waveform of short circuit current under the worst conditions

Figure 5.3 shows the waveform of the short circuit current in the worst situation where the impulse current occurs at the half cycle of the periodic component, i.e. when t = 0.01 s (for a power frequency of 50 Hz):

$$\begin{aligned} i_M &= I_{pm} + I_{pm} e^{-\frac{0.01}{T_a}} \\ &= I_{pm}(1 + e^{-\frac{0.01}{T_a}}) \\ &= I_{pm} \cdot K_M \end{aligned} \tag{5.14}$$

In Eq. 5.14, K_M is called the impulse coefficient, which represents the ratio of the impulse current to the amplitude of the periodic component of the short-circuit current. It can be seen that K_M depends on the time constant, T_a, which is a parameter of the circuit. When the time constant varies from zero to infinity, the K_M ranges from 1 to 2. In practical calculations, when a short circuit occurs on the generator bus, K_M equals 1.9; when a short circuit occurs on the high-voltage bus of a power plant, K_M equals 1.85; when a short circuit occurs in other locations, K_M equals 1.8.

The impulse current is mainly used to check the mechanical stability of the electrical equipment especially the current-carrying conductors due to Lorentz force at high current. It can be seen from the above analysis that to determine the impulse current and the aperiodic component of the short-circuit current, the periodic component must be calculated. The sole task of short-circuit calculation in most cases is to calculate the periodic component of the short-circuit current. When the power supply potential is given, the calculation of the periodic component of the short-circuit current only involves the solution of a steady-state sinusoidal AC circuit. In practical engineering problems, due to the complexity of the actual power grid and the difficulty of calculating the short-circuit current, it is generally only necessary to calculate the periodic component and its initial value at the instant of short circuiting. In addition, for asymmetrical short circuits in power systems, symmetrical analysis is generally used. The symmetric component method breaks down an asymmetric set of three-phase phasors into three sets of symmetric phasors, which have been widely used in the analysis and calculation of asymmetric three-phase circuits. However, the symmetric component method also has limitations. When the network parameters (such as line impedance) of a three-phase circuit are not symmetric, the method is no longer applicable.

5.3 Principles and Methods of Lorentz Force Calculation

When the moving and stationary contacts of a circuit breaker are in the closed position, the Lorentz force on the moving contact can be divided into two parts. One is the Lorentz force, F_L, of the moving contact rod as a current-carrying conductor existing in the magnetic field generated by the conductive circuit. The other is the Holm force F_H induced by current contraction on the contact surface between the moving and stationary contact. The current carried by the contact rod has to pass the contact surface through multiple spots. The Lorentz force and the Holm force will be introduced briefly within this chapter.

5.3.1 Calculation Method of Lorentz Force

The Lorentz force is essentially the force that a current-carrying conductor receives in a magnetic field. Therefore, it is necessary to accurately calculate the current density distribution and the magnetic field generated by the current. When a circuit breaker carries a stable current, the current density inside the conductive path of the contact can be obtained by solving the Maxwell equations shown in Eqs. 5.15 and 5.16:

$$\begin{cases} \nabla^2 \varphi = 0 \\ \vec{E} = -\nabla \varphi \\ \vec{J} = \vec{E} \cdot \sigma \\ \nabla \cdot \vec{J} = 0 \end{cases} \tag{5.15}$$

$$I = \oint_s \vec{J} \cdot dS \tag{5.16}$$

where φ is the electric potential, $\vec{E}$ is the electric field, $\vec{J}$ is the current density, σ is the electrical conductivity and I is the total current flowing through the conductor.

After calculating the distribution of the current density, the magnetic field distribution can be calculated according to Eq. 5.17:

$$\begin{cases} \nabla \times \dfrac{1}{\mu} \nabla \times \overrightarrow{A} = \overrightarrow{J} \\ \overrightarrow{B} = \nabla \times \overrightarrow{A} \end{cases} \tag{5.17}$$

The Lorentz force on the electrode rod can then be calculated according to Ampere's law. For a contact structure involving rotational movement, the resultant moment on the rotating shaft, $\vec{M}$, received by the moving contact rod needs to be calculated first, and then the repulsive force, $\vec{F}$, on the contact surface is calculated. The rotating moment and repulsive force can be calculated by Eq. 5.18:

$$\begin{cases} \overrightarrow{M} = \int_V \vec{r} \times (\vec{J} \times \vec{B}) dv \\ \overrightarrow{F} = \overrightarrow{M} / l \end{cases} \tag{5.18}$$

where l is the distance from the centre of the contact surface to the rotating shaft (pivot) of the moving contact rod.

5.3.2 Calculation of Holm Force

The principle of generation of the Holm force is shown in Fig. 5.4. When the moving and stationary contacts come in touch with each other, only the protruding points (spots) of the seemingly smooth contact surfaces make the electrical connection and only a small fraction of the mating surfaces conduct electricity. When the current passes through these spots, the current density lines converge into the spots, forcing the current to reverse in the direction parallel to the surfaces, thereby generating a repulsive Lorenz force.

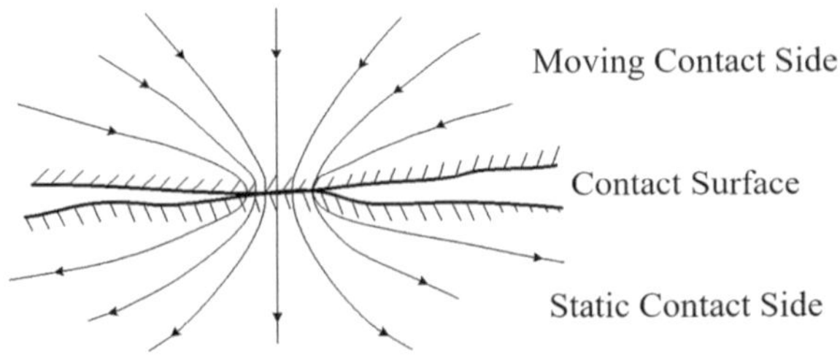

Fig. 5.4 Diagram showing the principle of Holm force generation at the interface of two electrical contacts

According to electromagnetism theory, an analytical expression for the Holm force, F_H, is given by Eq. 5.19 [4]:

$$F_H = \frac{\mu_0}{4\pi} I^2 \ln \sqrt{\frac{\xi H A}{F}} \tag{5.19}$$

where H is the Brinell hardness of the contact material and for 15% silver cadmium oxide it is between 700 and 1250 HBW. A is the apparent contact area of the two surfaces. Since the two contacts physically touch each other, A is taken as the overlapping area of the two surfaces. F is the final contacting force between the two contacts. ξ is related to the smoothness and microscopic deformation of the surfaces. With plastic deformation, the smoother the surface is, the closer ξ is to 1. The rougher the contact surface is, the smaller ξ is and the deformation is elastic. To adequately consider the effect of the Lorentz force on the contact system to ensure its operational stability, ξ is recommended to be set to 0.6.

The force, F, in Eq. 5.19 is equal to the difference between the pressing force, F_k, and two repulsive Lorentz forces is also considered, that is:

$$F = F_k - F_L - F_H \tag{5.20}$$

where F_k is the contact terminate pressing force, which involves the contact pressure after the contact is closed. This force is measured when the moving contact is no longer moving forward, that is, when the contact is completely closed. F_L is the Lorenz force induced on the bulk of the moving contact. It is therefore a body force. Then the expression for Holm force becomes:

$$F_H = \frac{\mu_0}{4\pi} I^2 \ln \sqrt{\frac{\xi H A}{F_k - F_L - F_H}} \tag{5.21}$$

which can be rewritten as:

$$f(F_H) = F_H - \frac{\mu_0}{4\pi} I^2 \ln \sqrt{\frac{\xi H A}{F_k - F_L - F_H}} = 0 \tag{5.22}$$

With known values of the final contacting force onto the contact surface, the current and apparent contact area, the Holm force between the moving and stationary contacts can be obtained by solving the above equation. Since Eq. 5.22 is a nonlinear equation, it can be solved by a numerical method such as the dichotomy method. In addition, Eq. 5.22 can also be used to optimise the mechanical force from the operating mechanism onto contact surface.

The calculation procedure is shown in Fig. 5.5. First, choose a small initial value for F_k such as zero. This initial value will not be involved in the solution loop. Second, F_k will be varied by a small step value before the existence of a solution to Eq. 5.22 is checked by obtaining the maximum value of $f(F_H)$ via setting its derivative with

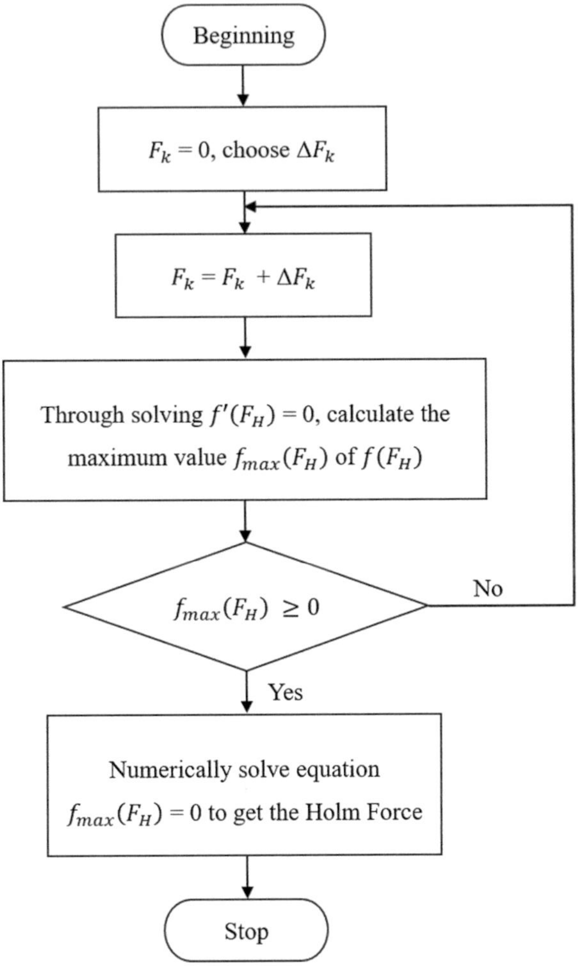

Fig. 5.5 Procedure to obtain optimal electrode force

respect to F_H to zero. If Eq. 5.22 has no solution (maximum of $F(F_H) < 0$), then the above process is repeated with an updated value of F_k until a solution exists. The evaluation loop is then terminated and a solution to Eq. 5.22 is obtained. It should be noted that the compressing force (F_k) calculated at this stage is the minimum value to ensure that the moving contact is not repelled by current I, in theory. The F_{H} value obtained by solving Eq. 5.22 is the Holm force experienced by the two contacts corresponding to the minimum pressing force from the operating mechanisms (F_k).

Analysis of the dynamic stability of the contact system in a circuit breaker requires consideration of more details. Taking the contact system of a low-voltage large-capacity DC circuit breaker as an example, in the closed state, the force of the moving contact rod is shown in Fig. 5.6a. Since the position of the moving contact rod is fixed at this time, the horizontal force received by the moving contact rod

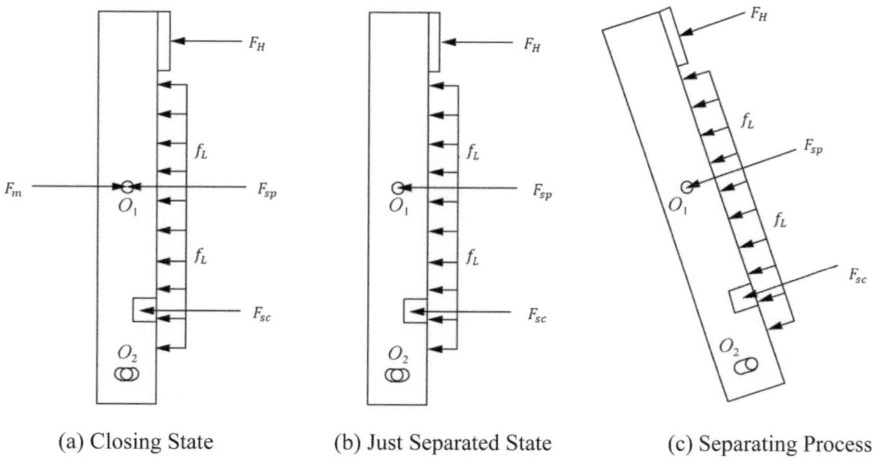

Fig. 5.6 Working principle of the electrode system

should be zero. Therefore, the mechanical force, F_m that is supplied by the moving contact rod, is:

$$F_m = F_H + F_{sc} + \int_V f_L dV + F_{sp} \tag{5.23}$$

where F_H is the Holm force, f_L is the bulk Lorentz force on the conductors which is assumed as a repulsive force, F_{sp} is the opening spring force in the operating mechanism, F_{sc} is the contact spring force and F_g is the gas aerodynamic force.

The forces depicted in Fig. 5.6 may need the use of a software package to obtain their values, such as ANSYS. The finite element method based on a mesh system can be chosen to solve for the current density and magnetic field distribution involving the contact system. To reduce the total number of elements in the mesh and the computational cost, simplifications are often needed. The arc running track, mechanical spring, insulating bracket and other components that have unimportant effects on the current density and magnetic field distributions do not need to be considered. Only the upper busbars, moving and stationary contacts, moving contact rods, soft electrical connections and lower busbars, that act as current-carrying components, are included. The actual shape of the conductive parts is modelled as closely as possible, including the round hole in the central shaft (O_1), the slotted hole for the bottom shaft (O_2) on the moving contact rod, and the part going through the middle of the closing rod. This helps ensure the accuracy of the calculated Lorentz force acting on the contact system.

5.4 Modelling Method for Lorenz Force in Low-Voltage Electrical Apparatuses

There are many types of low-voltage electrical apparatuses. This chapter takes low-voltage air circuit breakers (ACB) and moulded case circuit breakers (MCCB) as the main objects, and ANSYS as the main calculation tool [5–7] to introduce the relevant knowledge of electric power calculation in detail, supplemented by case studies [8].

5.4.1 Calculation of Short-Circuit Lorentz Force in Moulded Case Circuit Breaker

1. Geometric structure modelling

The calculation in a dual-breakpoint moulded case circuit breaker (MCCB) is taken as an example [9, 10]. MCCB is a commonly used electrical switchgear that has symmetry in its structure. To start with, factors and details that have little effect on the calculation results are usually simplified or removed to minimise the computational effort, so due to the symmetry in structure, only half of the unit needs to be included in the calculation and analysis. Second, a three-dime, nsional solid model including the filling air is established in the ANSYS workspace. Due to the existence of the ferromagnetic sheets, the Edge Element Method based on SOLID117 is used as an example. SOLID117 has a total of 20 nodes, of which 8 nodes at the vertices have a degree of freedom of VOLT, and 12 nodes at the midpoint of the elemental edge have a degree of freedom of an edge flux AZ. The SOLID117 element is then directly coupled to solve the current density and magnetic field distributions simultaneously. The material properties and element types defined and assigned in ANSYS are shown in Table 5.1. Finally, discretisation to the geometric model is carried out using the free mesh method [11]. The model is shown in Fig. 5.7.

Table 5.1 List of ANSYS simulation analysis parameters

Structure	Elemental types	Material property
Conductive part (moving and stationary conductive rod)	SOLID117 (KEYOPT(1) = 5)	$\mu_1 = 1, \rho = 1.75 \times e^{-8}$ (copper)
Conductive part (moving and stationary contact)	SOLID117 (KEYOPT(1) = 5)	$\mu_1 = 1,\ \rho = 3.4 \times e^{-8}$ (Silver tungsten carbide)
Ferromagnetic area (ferromagnetic sheet and magnetizing block)	SOLID117 (KEYOPT(1) = 0)	Read B-H curve from material library
Cylindrical air	SOLID117 (KEYOPT(1) = 0)	$\mu_0 = 1$

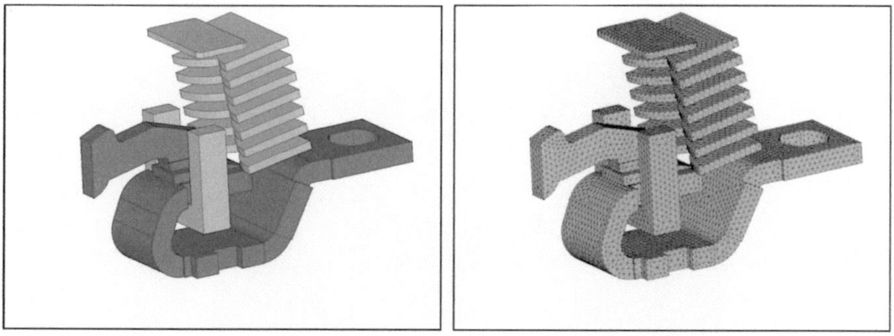

(a) Half geometry model (b) Half finite element model

Fig. 5.7 Lorentz force calculation model for moulded case circuit breakers

2. Calculation and analysis of Lorentz force of moulded case circuit breakers

Applying correct boundary conditions to the solution of the electric current density and the magnetic flux density is essential when they are coupled in the solution process. In solving the current density distribution in a current-carrying conductor, three steps are followed to set the boundary conditions and specify the load when ANSYS is used:

(1) Force the VOLT DOF, i.e. the degree of freedom, of the nodes where current enters the domain (current inlet nodes) to 0;
(2) Couple the VOLT DOF of the current inlet nodes properly so they form an equipotential surface;
(3) A node on the equipotential surface is selected to impose half of the total current, *i*, (due to the symmetry of the unit).

To satisfy the condition that the divergence of the current density is zero for DC or low-frequency AC systems, the current inlet plane and current outlet plane on current-carrying conductors must reside on the outer boundary of the entire domain. For example, using the Element Edge Method based on SOLID117 in ANSYS, the far-field elements cannot be employed for this purpose. Instead, forced boundary conditions have to be adopted. The computational domain that contains conductors, as well as air as insulation, needs to be chosen in such a way that the location of imposing the boundary conditions will not significantly affect the results while minimising the total number of elements in the model.

Boundary conditions are also needed on the symmetry plane. Magnetic lines on the symmetry plane need to be perpendicular to the latter (the symmetry plane) while parallel on the other boundary surfaces. In ANSYS, the perpendicular boundary condition is assumed by default when the Element Edge Method is used. The parallel boundary condition can be activated by setting the parameter AZ (magnetic vector potential) to 0. Figure 5.7 shows the half-unit domain together with the mesh system generated by ANSYS following the implementation of the magnetic and electric current boundary conditions.

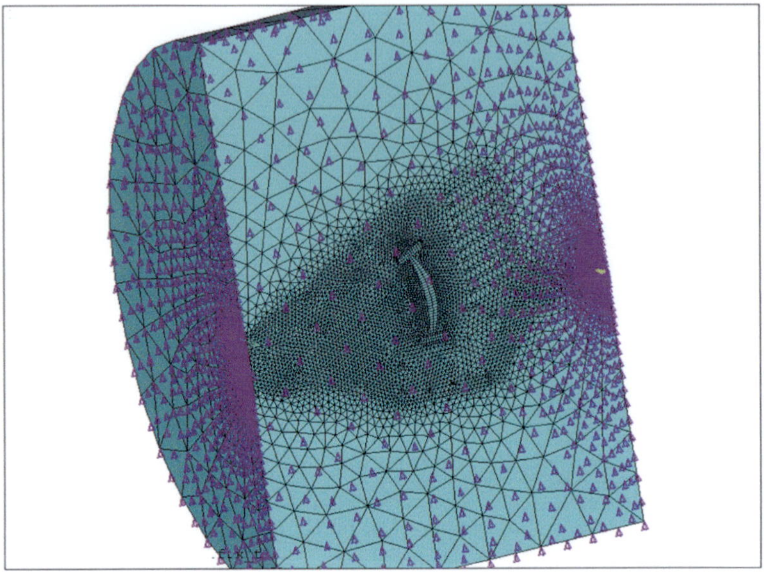

Fig. 5.8 Computational domain and mesh system based on the symmetric half of a moulded case circuit breaker with electric current and magnetic boundary conditions (purple triangles) implemented in ANSYS

Post-processing tools are available in commercial software packages. Figure 5.8 shows an example of the current density vector and magnetic induction intensity vector obtained with ANSYS at 10 kA. In this calculation, the arc column was considered as a conducting route with a very simple shape as in Fig. 5.8a. The ferromagnetic sheets play a significant role in shaping the magnetic field, which can be seen in the lower part of Fig. 5.8b.

After obtaining the current density and magnetic fields, the results in the arcing unit of J_{i}, B_{i}, and V_{i} which represent components of different directions of J, B and V can be exported to a table (by using the ETABLE command), and cross-product of the vectors carried out (by VCROSS). Finally, the Lorentz force on the arc is obtained by integration (using SMULT and SSUM in ANSYS). The actual Lorentz force should be twice that obtained for the symmetric half of the unit. Based on the direction of the arc motion, a component of the Lorentz force (F_y in the y direction) acts as the blowing force to drive the arc into the arc chutes [22, 23]. As an example, the arc-blowing force is 60.7 N at a current of 10 kA. By changing the current, a curve of the arc-blowing force as a function of the current is obtained, as shown in Fig. 5.9.

Further, the Holm force can be calculated substituting parameters of $F_L = 1.23$ N, $A = 20$ mm^2, $H = 500$ N/mm^2, $\xi = 0.45$ and $i = 2{,}400$ A into Eq. 5.21 gives:

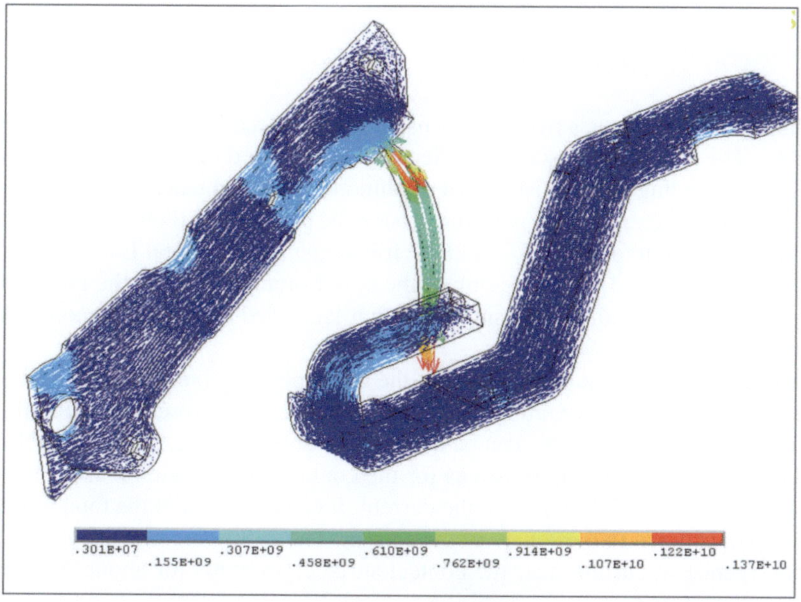

(a) Current density distribution

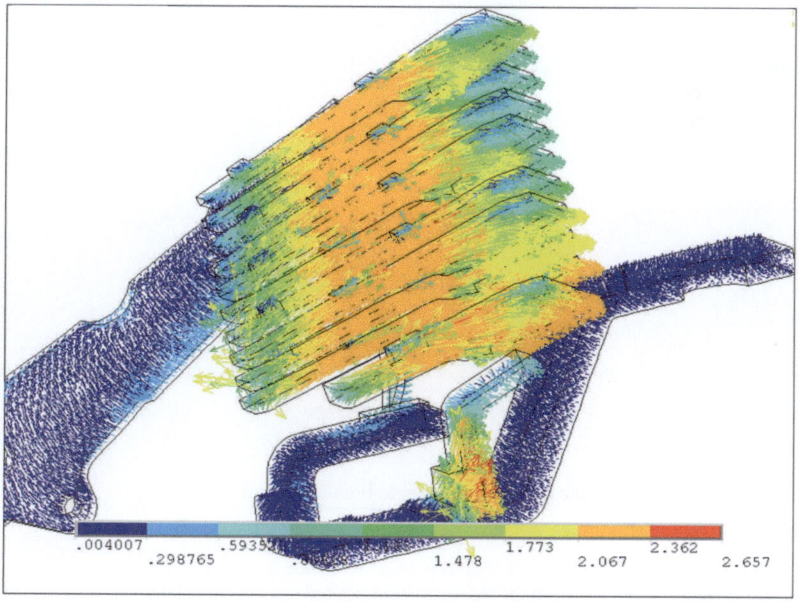

(b) Magnetic induction intensity distribution

Fig. 5.9 Electric current density and magnetic fields in a moulded case circuit breaker with a current load of 10 kA. Results obtained by ANSYS

$$F_h = \frac{\mu_0}{4\pi} I^2 \ln \sqrt{\frac{\xi HA}{F_k - F_L - F_h}} = 2.5 \ln \frac{0.45 * 500 * 20}{F_k - 1.23 - F_h} \tag{5.24}$$

The above equation has two unknowns, the Holm force F_H and the total pressing force F_k. The procedure described in Sect. 3.2 is followed to obtain a critical value of F_k corresponding to the existence of a solution to Eq. 5.24 when F_k is varied. This value of F_k represents the optimal value. Using the principle of dichotomy, the above equation is solved to obtain $F_k = 4.26$ N. It is larger than the total Lorentz force on the contacts which is 3.98 N, the sum of the bulk Lorentz force and the Holm force. In theory, F_k is the optimal pressing force that is needed to ensure the contacts are not repelled apart at 2,400 A.

By changing the value of the current, the corresponding body Lorentz force F_L, Holm force F_H, the total Lorentz force (sum of body Lorentz force and Holm force), and the total pressing force F_k can be obtained, as shown in Fig. 5.10 and Table 5.2.

The change in the repulsive forces for the contacts of a single breaking point is shown in Fig. 5.11 as a function of the current. It can be seen that the total repulsive force is proportional to the square of the current. The Holm force, which is an effect of convergence of current near the contact surface, accounts for about 70% of the total repulsive force.

It can also be seen from Fig. 5.9 that the magnetic flux density (or magnetic induction intensity) is strong on the magnetic boosting blocks on both sides of the

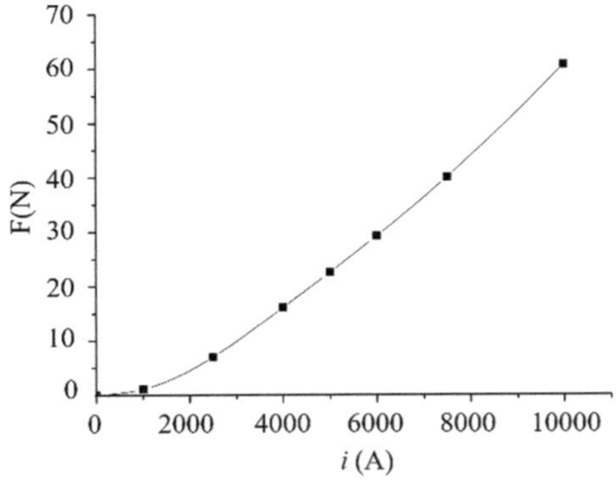

Fig. 5.10 Relationship between the arc blowing force and the current

Table 5.2 Forces acting on the contacts of a single breaking point

Current/kA	1	2	3	4	5	6	7	8	9	10
F_L/N	0.237	0.95	2.00	3.21	4.65	6.33	8.26	10.43	12.81	15.43
F_H/N	0.552	1.98	4.17	6.91	10.35	14.16	18.50	23.36	28.64	34.22
Repulsive force/N	0.789	2.92	6.17	10.12	15.00	20.49	26.77	33.78	41.45	49.65
F_k/N	0.839	3.1	6.62	10.92	16.25	22.29	29.22	36.98	45.50	54.65

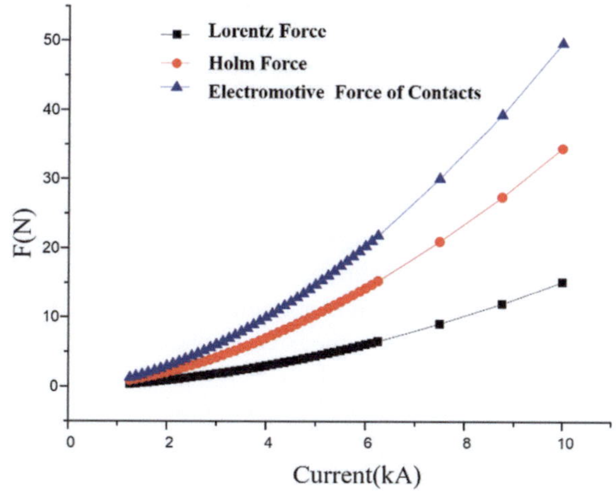

Fig. 5.11 Variation of different forces acting on a single-side contact as a function of current

contact. To clarify the influence of the magnetic boosting blocks, calculation of the forces can be carried out with the magnetic boosting blocks removed and a comparison is given in Fig. 5.12 in terms of the net contacting force on the contacts.

It can be seen from Fig. 5.12 that the total repulsive Lorenz force increases with the use of magnetic boosting blocks. Since the ferromagnetic material can effectively enhance the magnetic field, the shape of the block will be an important factor. For example, when a current flows in the stationary conductive rod, the U-shaped magnetic boosting blocks will generate a circular magnetic field around the rod, increasing the repulsive Lorentz force acting on the contact. Therefore, when the pressing force on the contact is determined during the design process, the influence of the magnetic boosting blocks should be taken into account.

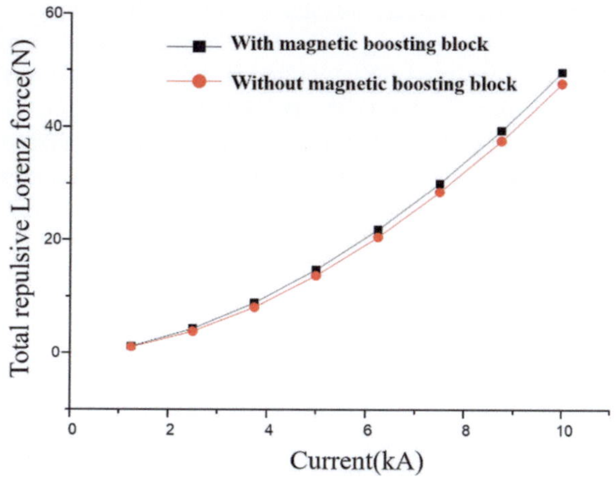

Fig. 5.12 A comparison of the influence of the magnetic boosting blocks on the total repulsive Lorentz force on the contacts

5.5 Modelling and Analysis of Short-Circuit Current Lorentz Force in High-Voltage Appliances

When a surge current occurs in the closing process or a short circuit develops, there will be strong repulsive Lorentz force caused by current contraction near the surface of flat-type contacts in high-voltage circuit breakers. This repulsive force can greatly reduce the net contacting force between the contacts, leading to contact bouncing and subsequent arcing, melting, welding or evaporation of the contact material. All of these can affect the normal operation of the contacts and may trigger a major accident. The greater the current, the stronger the repulsive Lorentz force. The circuit breaker operating mechanism needs to operate at a sufficiently faster speed under short-circuit conditions, which demands high operational reliability and mechanical strength of the mechanism. Therefore, it is important to consider the effect of the Lorentz force caused by short-circuit current.

This section takes the high-voltage vacuum circuit breakers as the object of study and considers the calculation of the Lorentz force during its closing or breaking operation under short circuit conditions. The force calculation model is first introduced before a dynamic simulation model of the vacuum circuit breaker's closing or breaking operation is developed. For the closing process of three phases under short-circuit conditions, the effects of simultaneous and non-simultaneous pre-breakdown on the effective value of the Lorentz force and the angular displacement curve of the main shaft (of the drivetrain) are assessed.

5.5.1 Calculation of Short-Circuit Lorentz Force in High-Voltage Appliances

When a circuit breaker closes or opens with a short-circuit current, the Lorentz force exerted on the moving contact is composed of two parts [12, 13]. The first is due to the attractive force F_A generated by the current under the action of a magnetic field, and the other is due to the repulsive force F_R generated by the contraction of the conduction path of the current near the contact surface.

The chosen contact system in this section is a cup-shaped axial magnetic field contact structure (from a Type VS1 vacuum circuit breaker), as shown in Fig. 5.13a. The mechanisms of the generation of attractive and repulsive forces between the two contacts are different. To establish a model for Lorentz force calculation, the following assumptions are made:

(1) The current passes across the contact surface through a thin path, that is, the contact surface area has no influence on the Lorentz force.
(2) When the moving and stationary contacts are in the closed position, they only have electrical contact within a small circle at the contact centre.

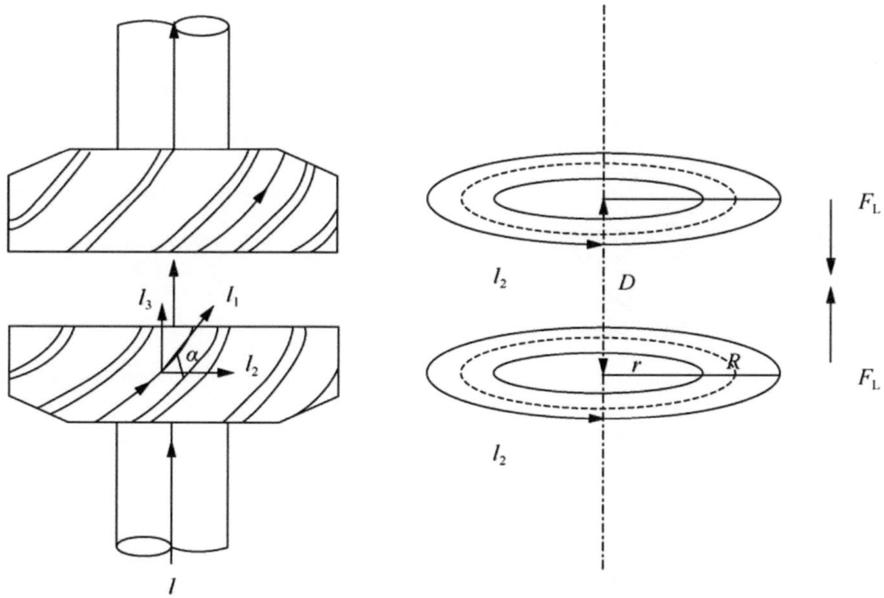

(a) Contact structure (b) Attractive force generated between the contacts

Fig. 5.13 Contact structure and generation of attractive force between the two contacts. Explain the meaning of symbols

(3) There is only a single arc column and a single arc root when an arc is developed in the contact gap.

In Fig. 5.13a, the current I_1 can be decomposed into a tangential component I_2, which is parallel to the contact surface, and an axial component I_3. $I_1 = I/6$ (each contact has 6 slots). In Fig. 5.13b, the magnitude of the two current loops is I_2. $I_2 = I_1\cos\alpha = I/6\cdot\cos\alpha$. In this example, $\alpha = 27°$ and $I/6\cdot\cos 27° \approx 0.15I$. Effectively, these two current loops are located in the cross-sectional plane half to their axial thickness. The axial distance, D, between them, varies with the contact gap length. When the contacts are in the closed position, $D = D_0$ is the sum of the half thicknesses of each contact. In this example, $D_0 = 22$ mm, the outer radius, R, of the current loops is 35 mm (same as the contact radius), and the inner radius of the current loops, r, is 30 mm.

In Fig. 5.13b, assuming that the current is evenly distributed between the inner and outer radii of the current loops, the middle current loop (the dotted circle in Fig. 5.13b) is set to be the equivalent current loop for calculating the attractive Lorentz force. Following [14, 15], the attractive Lorentz force generated by the two equivalent current loops on the moving contact is:

$$F_1 = \mu_0 I_2^2 \left(\frac{R + r}{2D}\right) \tag{5.25}$$

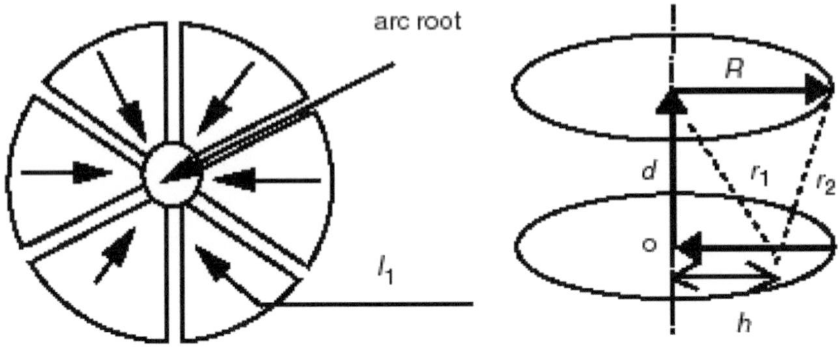

(a) Current distribution on the contact surface (b) Calculation of the repulsive force

Fig. 5.14 Generation of repulsive Holm force due to the contract of the current path near the contact surfaces

where μ_0 is the permeability of the vacuum.

When pre-breakdown takes place between the moving and stationary contacts, a current flows across the contact gap and contraction of the current path appears inside the contact body. As a result, a repulsive force is generated and acting on the moving contact, designated as F_2. For the model shown in Fig. 5.14b, Biot-Savart's law is used to calculate the distributed repulsive force between the two contacts:

$$f = \frac{\mu_0}{4\pi} I_1^2 \frac{1}{d}\left(\frac{h}{r_1} + \frac{R-h}{r_2}\right) \tag{5.26}$$

For convenience of simulating the dynamic process of the moving contact, a point force F_2 acting at the centre of the contact, O, collates the overall effect of the distributed force defined in Eq. 5.26, so that:

$$\begin{aligned} F_2 &= \int_0^R f dh = \frac{\mu_0 I_1^2}{4\pi d} \int_0^R \left(\frac{h}{r_1} + \frac{R-h}{r_2}\right) dh \\ &= \frac{\mu_0 I_1^2}{4\pi d} \int_0^R \left(\frac{h}{\sqrt{d^2+h^2}} + \frac{R-h}{d^2 + (R-h^2)}\right) dh \\ &= \frac{\mu_0 I_1^2}{12\pi d}\left(\sqrt{d^2+R^2} - d\right) \end{aligned} \tag{5.27}$$

When the moving and stationary contacts physically touch each other, the current path shrinks near the contact surface, generating a repulsive Holm force, F_H, whose calculation [16] is based on the following approximations:

(a) There is only one conductive spot located at the centre of the contact surfaces, i.e. all small conductive spots are concentrated at the centre to form a single conductive spot.
(b) Within the conductive spot, the contact surface is not an ideal flat surface but a superconductive ball of equipotential across its body.

In summary, during the process of a circuit breaker closing with the presence of a short-circuit current, the net Lorentz force, F_T, is the difference between the attractive force, F_1, and the repulsive forces, F_2 and F_H. The attractive force lasts over the whole closing process, but the two repulsive forces are complementary, i.e. only one of them exists at any time. Suppose the acting time of F_2 is $[t_{e1}, t_{e2}]$, and let:

$$g(t) = \begin{cases} 1 & t_{e1} \le t \le t_{e2} \\ 0 & t < t_{e1} \text{ or } t > t_{e2} \end{cases} \qquad h(t) = \begin{cases} 1 & t_{e2} \le t \\ 0 & t < t_{e2} \end{cases} \tag{5.28}$$

Then the following equation is obtained:

$$F_T = F_2 g(t) + F_H h(t) - F_1 = \frac{\mu_0 I^2}{12\pi(s+4)}\left(\sqrt{(s+4)^2 + 35^2} - (s+4)\right)g(t) + \frac{\mu_0 I^2}{4\pi}(6.38 - \ln F)h(t) - 0.0225\mu_0 I^2 \frac{32.5}{s+22} \tag{5.29}$$

where I is the effective value of the total short-circuit current and s is the distance between the moving and stationary contacts.

The vacuum circuit breaker under study in this section has a three-pole contact structure, and the magnetic fields generated by the other two phases currents have an insignificant effect on the Lorentz force generated by the phase under focus.

5.5.2 *Short-Circuit Lorentz Force in High-Voltage Electrical Apparatuses*

The medium-voltage system in many countries, such as China, has an ungrounded neutral point. In this section, the influences of the Lorentz force on the dynamic characteristics of the circuit breakers under different closing conditions will be studied when they close a neutral-ungrounded three-phase system under short circuit conditions.

1. Simultaneous pre-breakdown in three-phase vacuum circuit breakers

When a three-phase ungrounded short-circuit event occurs in the power system, the instantaneous values of the short-circuit currents are:

$$
\begin{aligned}
i_A &= \sqrt{2}I_r[\sin(wt+\psi-\varphi)-\sin(\psi-\varphi)e^{-\frac{t}{\tau}}]\\
i_B &= \sqrt{2}I_r[\sin(wt+\psi-\varphi-\frac{2}{3}\pi)-\sin(\psi-\varphi-\frac{2}{3}\pi)e^{-\frac{t}{\tau}}]\\
i_C &= \sqrt{2}I_r[\sin(wt+\psi-\varphi-\frac{4}{3}\pi)-\sin(\psi-\varphi-\frac{4}{3}\pi)e^{-\frac{t}{\tau}}]
\end{aligned}
\tag{5.30}
$$

where I_r is the effective value of the periodic components of the short-circuit currents; ψ is the phase angle of a voltage at the instant of the short-circuiting; φ is the phase angle of a current lagging its corresponding voltage, and τ s the time constant of the circuit.

The influence of the arc voltage (resistance) on the short-circuit current is not taken into account in Eq. 5.30. Under the action of an axial magnetic field, the movement of the arc column and the outward diffusion of the plasma is slowed down [14, 17], thereby reducing the radial loss of ions and electrons. The arc voltage is much smaller than the system-rated voltage and it has virtually no current limiting effect on the first half wave of the short-circuit currents.

With the onset of pre-breakdown, the sudden change in current is related to the phase angle of the voltage when pre-breakdown takes place. For the three-phase circuit breaker under study, the short-circuit currents of the three-phase synchronous fault can still be expressed by Eq. 5.30. Ignoring the influence of the circuit resistance, φ can be assumed to be equal to $\pi/2$. Equation 5.30 is then simplified into:

$$
\begin{aligned}
i_A &= \sqrt{2}I_r[\sin(wt+\psi-\frac{\pi}{2})-\sin(\psi-\frac{\pi}{2})e^{-\frac{t}{\tau}}]\\
i_B &= \sqrt{2}I_r[\sin(wt+\psi-\frac{7}{6}\pi)-\sin(\psi-\frac{7}{6}\pi)e^{-\frac{t}{\tau}}]\\
i_C &= \sqrt{2}I_r[\sin(wt+\psi-\frac{11}{6}\pi)-\sin(\psi-\frac{11}{6}\pi)e^{-\frac{t}{\tau}}]
\end{aligned}
\tag{5.31}
$$

when I_r is given a representative value of 31.5 kA. Taking $\tau = 45$ ms, the short-circuit currents can be calculated, as shown in Fig. 5.15. In obtaining the results, it is assumed that the circuit breaker contacts of all three phases physically close at the same time.

2. Asynchronous pre-breakdown in three-phase vacuum circuit breakers

In the practical operation of circuit breakers, the three pairs of contacts do not necessarily physically close simultaneously during the closing process of the circuit. Even if the contacts of the three phases physically close at the same time, the condition in the form of Eq. 5.32 is still not satisfied because the phase difference of the voltages is 120° and the closing speed of the moving contact of the circuit breaker used in the example (VS1) does not exceed 2 m/s:

$$
v \geq \omega l_u U_m \tag{5.32}
$$

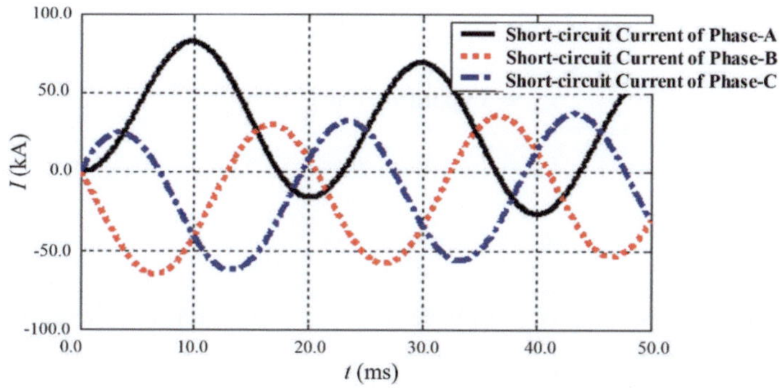

Fig. 5.15 Short-circuit current waveform calculated by Eq. 5.39 for three-phase synchronous fault

Fig. 5.16 Three-phase short circuiting where Phase A and B first experience pre-breakdown

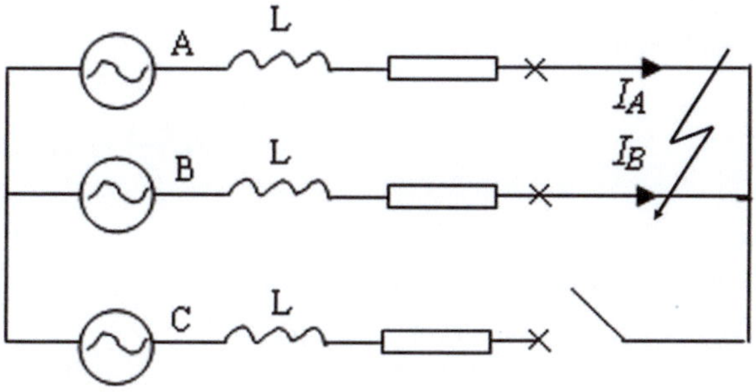

where ω is the angular frequency of the power supply voltage, l_u is the pre-breakdown distance of unit voltage and U_m is the amplitude of the power supply voltage. The phase angle of the power supply voltage at the instant of pre-breakdown is mostly less than 90°, so it is impossible for the three phases to have pre-breakdown at the same time.

The asynchronous pre-breakdown process of the three-phase contacts of the circuit breaker needs to be simulated to study the influence of the Lorentz force on the closing process of the breaker under different closing conditions. In a neutral-ungrounded three-phase system, as is the case for most medium-voltage networks including that in China, large short-circuit current occurs only when pre-breakdown takes place on at least two phases in the closing process. This applies to the three-phase-to-ground fault where the pre-breakdown instants across the three contact gaps are all different, or to the case of a three-phase fault (ungrounded) where the pre-breakdown instant of two phases is different from the other phase.

Assuming that a three-phase fault pre-breakdown first appears on Phases A and B, at this instant it is a two-phase short-circuit, as shown in Fig. 5.16. The currents through Phase A and Phase B will no longer follow Eq. 5.31. This situation will be analysed below.

Because the three-phase power supply voltage is:

$$
\begin{aligned}
u_A &= U_m \sin(\omega t + \psi) \\
u_B &= U_m \sin(\omega t + \psi - \frac{2}{3}\pi) \\
u_C &= U_m \sin(\omega t + \psi - \frac{4}{3}\pi)
\end{aligned} \tag{5.33}
$$

The instantaneous value of the phase-to-phase line voltage after pre-breakdown on Phases A and B is:

$$
u_{AB} = u_A - u_B = \sqrt{3} U_m \sin(\omega t + \psi + \frac{\pi}{6}) \tag{5.34}
$$

Meanwhile, after the pre-breakdown of the two phases, the following differential equation holds:

$$
u_{AB} = -(2L\frac{di_A}{dt} + 2i_A R) \tag{5.35}
$$

where R is the circuit resistance. Applying the vector method [18] to the circuit to solve this differential equation yields:

$$
i_A = -\frac{\sqrt{3} U_m \sin(\omega t + \psi + \frac{\pi}{6} - \varphi)}{2\sqrt{(\omega L)^2 + R^2}} + \frac{\sqrt{3} U_m \sin(\psi + \frac{\pi}{6} - \varphi)}{2\sqrt{(\omega L)^2 + R^2}} e^{\frac{-t}{\tau}} \tag{5.36}
$$

where φ is the power factor angle and ψ the phase angle of the pre-breakdown voltage.

As stated previously, the magnitude of the circuit resistance is much smaller than the reactance, so $\varphi = \pi/2$. t in Eq. 5.36 is set to zero at the pre-breakdown instant (t_{kI}) of Phases A and B. Following the onset of pre-breakdown, the short-circuit current in Phases A and B is given in Eq. 5.37 based on an arbitrary time zero. The short-circuit current in Phase C is still zero at this time:

$$
\begin{aligned}
i_{A1} &= -\frac{\sqrt{3}}{2} I_r \sin[\omega(t - t_{k1}) + \psi - \frac{\pi}{3}] + \frac{\sqrt{3} I_r \sin(\psi - \frac{\pi}{3})}{2} e^{\frac{-(t-t_{k1})}{\tau}} \\
i_{B1} &= \frac{\sqrt{3}}{2} I_r \sin[\omega(t - t_{k1}) + \psi - \frac{\pi}{3}] - \frac{\sqrt{3} I_r \sin(\psi - \frac{\pi}{3})}{2} e^{\frac{-(t-t_{k1})}{\tau}}
\end{aligned} \tag{5.37}
$$

Under the same conditions, pre-breakdown on Phase C happens at time t_{k2} and the short-circuit current changes to three-phase short-circuit current. From Eq. 5.31, the three-phase short-circuit currents are:

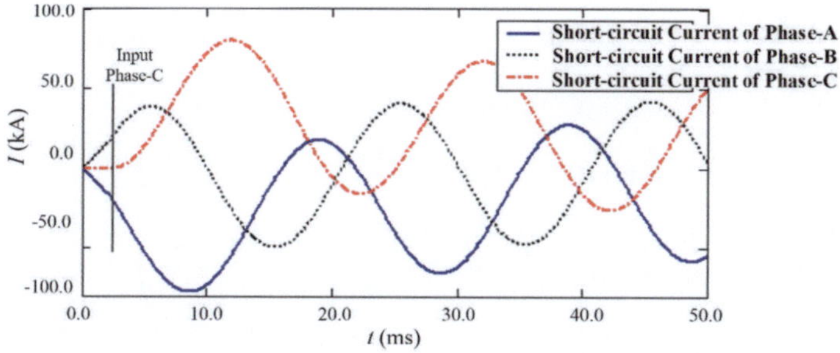

Fig. 5.17 Short-circuit current waveform in the vacuum circuit breaker when pre-breakdown occurs asynchronously in the three-phase system

$$
\begin{aligned}
i_{A2} &= \sqrt{2}I_r \sin(\omega(t-t_{k2})+\psi-\frac{7}{6}\pi)-[\sqrt{2}I_r\sin(\psi-\frac{7}{6}\pi)-i_{A1}(t_{k2}))]e^{-\frac{(t-t_{k2})}{\tau}}]\\
i_{B2} &= \sqrt{2}I_r \sin(\omega(t-t_{k2})+\psi-\frac{11}{6}\pi)-[\sqrt{2}I_r\sin(\psi-\frac{11}{6}\pi)-i_{B1}(t_{k2}))]e^{-\frac{(t-t_{k2})}{\tau}}]\\
i_{C2} &= \sqrt{2}I_r[\sin(\omega(t-t_{k2})+\psi-\frac{3}{6}\pi)-\sin(\psi-\frac{3}{6}\pi)e^{-\frac{(t-t_{k2})}{\tau}}]
\end{aligned}
\tag{5.38}
$$

where $i_{A1}(t_{k2})$ is the current in Phase A at the instant of pre-breakdown in Phase C and $i_{B1}(t_{k2})$ is that for Phase B.

Figure 5.17 shows the short-circuit current waveforms calculated by Eqs. 5.37 and 5.38 when the system has a peak short-circuit current of 80 kA. The pattern of the short-circuit current experiences an apparent change when pre-breakdown in Phase C occurs. Before this instant, the currents in Phases A and B are 180° out of phase, whereas after this instant the phase difference in the short-circuit currents of Phases A, B and C becomes 120°, and the magnitude of the currents also changes.

3. Lorentz force due to short-circuit current on the contacts of vacuum circuit breakers during closing

Substituting 5.31 into 5.29, the maximum Lorentz force generated by the short-circuit current can be obtained, corresponding to the occurrence of pre-breakdown on all three phases, as given below:

$$
\begin{aligned}
F_T a = {} & 2\times 10^{-7} I_r^2 \times [\sin(100\pi t+\psi-\frac{\pi}{2})\\
& -\sin(\psi-\frac{\pi}{2})e^{-\frac{t}{\tau}}]^2 \times \left(\frac{\sqrt{(s+4)^2+35^2}-(s+4)}{3(s+4)}\right) f(t)\\
& +2\times 10^{-7} I_r^2 \times [\sin(100\pi t+\psi-\frac{\pi}{2})\\
& -\sin(\psi-\frac{\pi}{2})e^{-\frac{t}{\tau}}]^2 \times (6.38-\ln\sqrt{F})g(t)
\end{aligned}
$$

$$-\frac{9.18}{s+22}\times 2\times 10^{-7}I_r^2\times[\sin(100\pi t+\psi-\frac{\pi}{2})-\sin(\psi-\frac{\pi}{2})e^{-\frac{t}{\tau}}]^2 \tag{5.39}$$

$$\begin{aligned}F_Tb = {} & 2\times 10^{-7}I_r^2\times[\sin(100\pi t+\psi-\frac{7\pi}{6}) \\ & -\sin(\psi-\frac{7\pi}{6})e^{-\frac{t}{\tau}}]^2\times\left(\frac{\sqrt{(s+4)^2+35^2}-(s+4)}{3(s+4)}\right)f(t) \\ & +2\times 10^{-7}I_r^2\times[\sin(100\pi t+\psi-\frac{7\pi}{6}) \\ & -\sin(\psi-\frac{7\pi}{6})e^{-\frac{t}{\tau}}]^2\times(6.38-\ln\sqrt{F})g(t) \\ & -\frac{9.18}{s+22}\times 2\times 10^{-7}I_r^2\times[\sin(100\pi t+\psi-\frac{7\pi}{6})-\sin(\psi-\frac{7\pi}{6})e^{-\frac{t}{\tau}}]^2\end{aligned} \tag{5.40}$$

$$\begin{aligned}F_Tc = {} & 2\times 10^{-7}I_r^2\times[\sin(100\pi t+\psi-\frac{11\pi}{6}) \\ & -\sin(\psi-\frac{11\pi}{6})e^{-\frac{t}{\tau}}]^2\times\left(\frac{\sqrt{(s+4)^2+35^2}-(s+4)}{3(s+4)}\right)f(t) \\ & +2\times 10^{-7}I_r^2\times[\sin(100\pi t+\psi-\frac{11\pi}{6}) \\ & -\sin(\psi-\frac{11\pi}{6})e^{-\frac{t}{\tau}}]^2\times(6.38-\ln\sqrt{F})g(t) \\ & -\frac{9.18}{s+22}\times 2\times 10^{-7}I_r^2\times[\sin(100\pi t+\psi-\frac{11\pi}{6})-\sin(\psi-\frac{11\pi}{6})e^{-\frac{t}{\tau}}]^2\end{aligned} \tag{5.41}$$

Figure 5.18 shows the timeline of different events during the closing process of the circuit breaker under short-circuit conditions, used to help select the relevant time parameters in simulation. The trip time of the holding shaft of the energy storage mechanism of the breaker is set as time zero ($t_0 = 0$) in the simulation and the duration of simulation is $t_3 = 50$ ms. Prior to this, the circuit breaker is in open position, held by the latch. It is assumed that the short-circuit current still exists at the end of the simulation.

Pre-breakdown in the closing process has a certain randomness. The maximum pre-breakdown time of a 12 kV vacuum circuit breaker is typically 1.0 ms to 2.5 ms [19]. Considering the most severe case, the pre-breakdown time is assumed to be 2.5 ms in the simulation. Pre-breakdown depends on the contact gap length instead of time. To set up the simulation, the pre-breakdown time needs to be converted into the corresponding contact gap length. As the circuit breaker used as an example in this section has a closing speed (average speed within 2 mm before the contacts physically touch each other) of about 0.65 m/s under no-load condition, this gives a contact gap length of 1.5 mm.

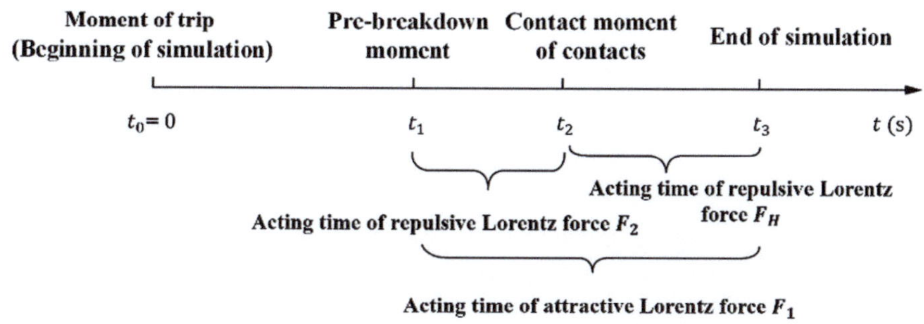

Fig. 5.18 Timeline for different events during circuit breaker closing operation under short circuit conditions

In a medium-voltage transmission network, the maximum time constant that can occur during a three-phase short circuit is 45 ms [20], so the time constant $\tau = 45$ ms is used for simulation. Since the simulated vacuum circuit breaker takes about 40 ms from the instant of tripping to reach its final position for the moving contact when no Lorentz force is in action, the simulation time is set to 50 ms, that is, $t_3 - t_0 = 50$ ms when a margin is taken into account. According to Fig. 5.18, the parameter of t_{e1} for *f(t)* in Eq. 5.28 should be t_1 in the figure.

The dynamic simulation is performed in ADAMS where the relevant expressions for the Lorentz force are implemented in the simulation model of the circuit breaker as forces that act on the centre of the contact surface. The movement of the moving contact of the circuit breaker during closing is then simulated under short circuit conditions. Figure 5.19 shows the variation of the Lorentz force with time.

Results show that after the pre-breakdown in the contact gaps occurs at about 25 ms, the movement of the contact starts to be affected by the Lorentz force induced by the short circuit current: the closing speed is reduced and the closing process is prolonged. This is clearly shown in Fig. 5.20 where a comparison of the angular

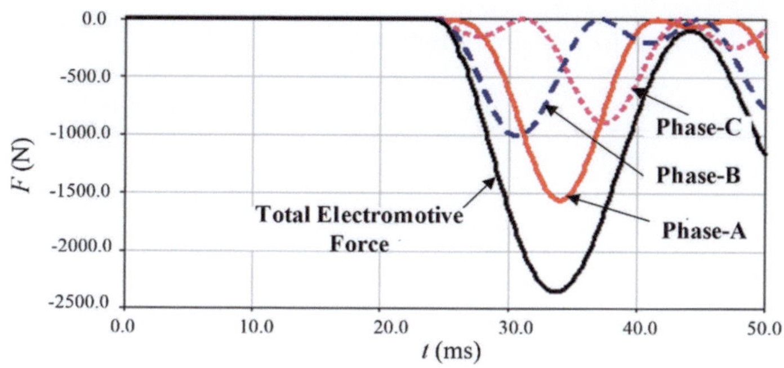

Fig. 5.19 Lorentz forces acting on the phase contacts for the relatively extreme case of simultaneous pre-breakdown of the three phases

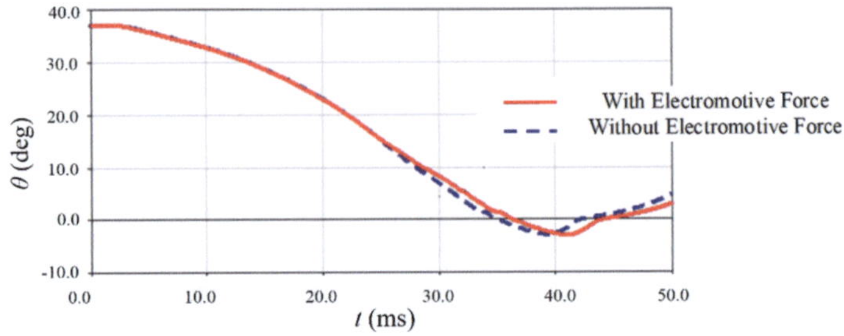

Fig. 5.20 Effect of short-circuit current Lorentz force on the movement of the contact, expressed in the angle of the main shaft

displacement of the main shaft is made for cases with and without Lorentz force included. The closing time is longer with Lorentz force.

In reality, during the closing process of a three-phase circuit breaker, each phase may have a different pre-breakdown time due to variations in motion characteristics as a result of manufacturing and due to differences in the circuit for each phase. The expression of the action time of the repulsive forces in Phase A and Phase B, F_2 and F_H, is given in Eq. 5.42, and the expression of the action time of the repulsive force in Phase C, F_2 and F_H, in Eq. 5.43:

$$f_1(t) = \begin{cases} 1 \; t_1 \le t \le t_3 \\ 0 \; t < t_1 \end{cases} \quad g_1(t) = \begin{cases} 1 \; t_3 \le t \\ 0 \; t < t_3 \end{cases} \tag{5.42}$$

$$f_2(t) = \begin{cases} 1 \; t_2 \le t \le t_3 \\ 0 \; t < t_2 \end{cases} \quad g_2(t) = \begin{cases} 1 \; t_3 \le t \\ 0 \; t < t_3 \end{cases} \tag{5.43}$$

When pre-breakdown occurs on only two phases, I in Eq. 5.29 is replaced by i_A and i_B in 31 and the short-circuit current induced Lorentz force can be calculated by:

$$\begin{aligned} F_T a1 &= 10^{-7} I_r^2 \times [-\frac{\sqrt{3}}{2}\sin[100\pi(t - t_{k1}) + \psi - \frac{\pi}{3}] \\ &+ \frac{\sqrt{3}\sin(\psi - \frac{\pi}{3})}{2} e^{\frac{-(t-t_{k1})}{\tau}}]^2 \times \left(\frac{\sqrt{(s+4)^2 + 35^2} - (s+4)}{3(s+4)}\right) f_1(t) \\ &+ 10^{-7} I_r^2 \times [-\frac{\sqrt{3}}{2}\sin[100\pi(t - t_{k1}) + \psi - \frac{\pi}{3}] \\ &+ \frac{\sqrt{3}\sin(\psi - \frac{\pi}{3})}{2} e^{\frac{-(t-t_{k1})}{\tau}}]^2 \times (6.38 - \ln\sqrt{F}) g_1(t) \\ &- \frac{9.18}{s+22} \times 10^{-7} I_r^2 \times [-\frac{\sqrt{3}}{2}\sin[100\pi(t - t_{k1}) \\ &+ \psi - \frac{\pi}{3}] + \frac{\sqrt{3}\sin(\psi - \frac{\pi}{3})}{2} e^{\frac{-(t-t_{k1})}{\tau}}]^2 \end{aligned} \tag{5.44}$$

$$
\begin{aligned}
F_T b1 = {} & 10^{-7} I_r^2 \times [\frac{\sqrt{3}}{2}\sin[100\pi(t-t_{k1})+\psi-\frac{\pi}{3}] \\
& - \frac{\sqrt{3}\sin(\psi-\frac{\pi}{3})}{2} e^{\frac{-(t-t_{k1})}{T}}]^2 \times \left(\frac{\sqrt{(s+4)^2+35^2}-(s+4)}{3(s+4)}\right) f_1(t) \\
& + 10^{-7} I_r^2 \times [\frac{\sqrt{3}}{2}\sin[100\pi(t-t_{k1})+\psi-\frac{\pi}{3}] \\
& - \frac{\sqrt{3}\sin(\psi-\frac{\pi}{3})}{2} e^{\frac{-(t-t_{k1})}{T}}]^2 \times (6.38-\ln\sqrt{F}) g_1(t) \\
& - \frac{9.18}{s+22} \times 10^{-7} I_r^2 \times [\frac{\sqrt{3}}{2}\sin[100\pi(t-t_{k1})+\psi-\frac{\pi}{3}] \\
& - \frac{\sqrt{3}\sin(\psi-\frac{\pi}{3})}{2} e^{\frac{-(t-t_{k1})}{T}}]^2
\end{aligned} \tag{5.45}
$$

$$
F_T c1 = 0 \tag{5.46}
$$

When the third phase experiences also the pre-breakdown, the short-circuit current expressed in Eq. 5.28 is substituted into Eq. 5.29 to yield:

$$
\begin{aligned}
F_T a2 = {} & 2 \times 10^{-7} I_{A2}^2 \times \left(\frac{\sqrt{(s+4)^2+35^2}-(s+4)}{3(s+4)}\right) f_2(t) \\
& + 2 \times 10^{-7} I_{A2}^2 \times (6.38-\ln\sqrt{F}) g_2(t) - \frac{9.18}{s+22} \times 2 \times 10^{-7} I_{A2}^2
\end{aligned} \tag{5.47}
$$

$$
\begin{aligned}
F_T b2 = {} & 2 \times 10^{-7} I_{B2}^2 \times \left(\frac{\sqrt{(s+4)^2+35^2}-(s+4)}{3(s+4)}\right) f_2(t) \\
& + 2 \times 10^{-7} I_{B2}^2 \times (6.38-\ln\sqrt{F}) g_2(t) - \frac{9.18}{s+22} \times 2 \times 10^{-7} I_{B2}^2
\end{aligned} \tag{5.48}
$$

$$
\begin{aligned}
F_T c2 = {} & 2 \times 10^{-7} I_{C2}^2 \times \left(\frac{\sqrt{(s+4)^2+35^2}-(s+4)}{3(s+4)}\right) f_2(t) \\
& + 2 \times 10^{-7} I_{C2}^2 \times (6.38-\ln\sqrt{F}) g_2(t) - \frac{9.18}{s+22} \times 2 \times 10^{-7} I_{C2}^2
\end{aligned} \tag{5.49}
$$

Figures 5.21 and 5.22 show the instant occurrence of each event in the process of closing with a short-circuit current. These figures are used to determine the relevant time parameters in the simulation. The trip time of the holding shaft of the energy storage mechanism of the breaker is set as time zero ($t_0 = 0$) in the simulation and the duration of simulation is $t_3 = 50$ ms. Prior to this, the circuit breaker is in open position, held by the latch. It is assumed that the short-circuit current still exists at the end of the simulation.

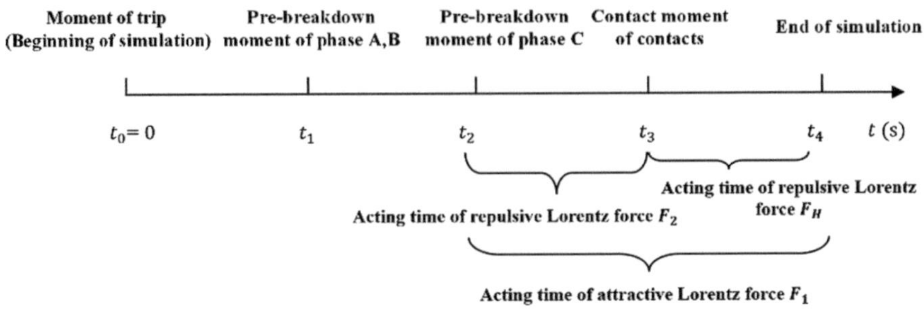

Fig. 5.21 Timeline for different events in Phase A and Phase B during circuit breaker closing operation under short circuit conditions

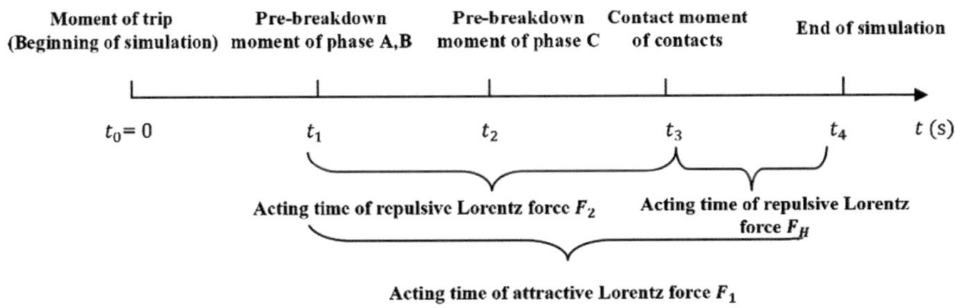

Fig. 5.22 Timeline for events in Phase C during circuit breaker closing operation under short circuit conditions

According to China's power industry standard (DL 403–91), the differences in the closing time of the three phases in vacuum circuit breakers under study are no more than 2 ms. In such a situation, the DC component of the two-phase short-circuit current has no time to undergo a significant decay before pre-breakdown occurs in Phase C. When setting up a simulation of the dynamic process of a circuit breaker under short circuit conditions, such as in ADAMS, the relevant expressions of the Lorentz forces are implemented as point forces. Quantitative results can then be obtained on the influences of the asynchronization of the three phases over the different Lorentz forces when closing with short circuit currents, and on the four characteristic quantities that describe the angular displacement of the main shaft of the mechanical system. It is assumed that the pre-breakdown of Phase A and Phase B occurs first, and the pre-breakdown of Phase C then occurs after 2 ms. The expression of the above electromotive force is added to the simulation model of the circuit breaker in a forced manner, and the motion process of the circuit breaker closing the short-circuit current is simulated. Figure 5.23 shows the Lorentz force waveform when pre-breakdown first occurs on Phase A, Phase B and Phase C experiences pre-breakdown 2 ms later.

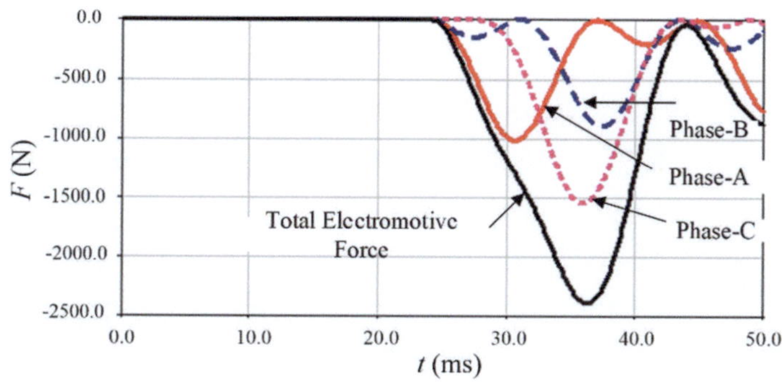

Fig. 5.23 Variation of Lorentz forces under short circuit conditions with asynchronous pre-breakdown between the three phases

4. Lorentz force in vacuum circuit breakers during short circuit current interruption process

In the opening operation of a three-phase circuit breaker, the current in different phases will not pass their zero points concurrently due to the phase difference, which means that the arcs for the three phases cannot be extinguished simultaneously. For medium-voltage power systems with an ungrounded neutral point, such as those in China, a typical feature is that the arc in the first opened phase of a three-phase circuit breaker is extinguished first, and the arcs in the other two phases are then extinguished at the same time. The variation in the three-phase short circuit currents during opening can be calculated by applying considering the process in a way similar to the closing process. When Phase A opens and clears first, the short-circuit current flowing in the other two phases can be calculated by taking $\varphi = \pi/2$ [21] in the following equations:

$$
\begin{aligned}
i_B &= -\frac{\sqrt{3}}{2}I_r\sin[\omega(t-t_k)+\psi-\frac{\pi}{3}]+(\frac{\sqrt{3}I_r\sin(\psi-\frac{\pi}{3})}{2}+i_{B1}(t_k))e^{\frac{-(t-t_k)}{T}}\\
i_C &= \frac{\sqrt{3}}{2}I_r\sin[\omega(t-t_k)+\psi-\frac{\pi}{3}]-(\frac{\sqrt{3}I_r\sin(\psi-\frac{\pi}{3})}{2}+i_{C1}(t_k))e^{\frac{-(t-t_k)}{T}}
\end{aligned}
\tag{5.50}
$$

where i_{B1} and i_{C1} are, respectively, the instantaneous values of current in Phase B and Phase C when Phase A passes its current zero and clears at time t_k.

The currents in the circuit when all three phases in the short circuit are given by Eq. 5.31. The currents in the other two phases after Phase A clears are given by Eq. 5.50. Using both, the short circuit current waveforms over the complete interruption process are shown in Fig. 5.24 where the three sets of moving and stationary contacts corresponding to the three phases separate at the same time.

Substituting I in Eq. 5.29 by iA in Eq. 5.31 and assigning appropriate values to the former allows the calculation of the Lorentz force during the circuit breaker breaking process:

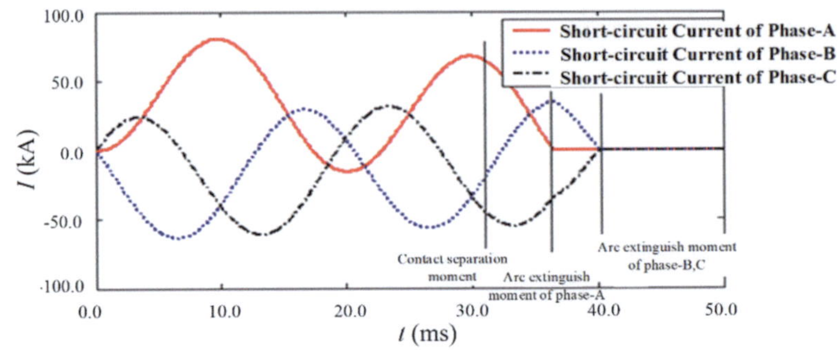

Fig. 5.24 Waveforms of the three-phase short-circuit currents when the three sets of arc contacts separate concurrently and Phase A clears first

$$F_{Ta} = 2 \times 10^{-7} I_r^2 \times [\sin(314t - \frac{\pi}{2}) + e^{\frac{-t}{T}}]^2 \times ((\frac{\sqrt{(s+4)^2 + 35^2} - (s+4)}{3(s+4)})f(t) + (6.38 - \ln\sqrt{F})g(t) - \frac{9.18}{s+22}) \quad (5.51)$$

where $f(t)$ is a function defining the active period of the repulsive force F_2 and $g(t)$ for the repulsive force F_H. The coefficient ξ can be assumed to take a maximum value of 1. I_r is the effective short circuit current for the circuit breaker.

The Lorentz forces that Phase B and Phase C experienced before Phase A clears can be obtained in a similar manner through the use of Eq. 5.31. After Phase A clears, the short-circuit currents in Phase B and Phase C start to follow Eq. 5.50. Thus, the following equations can be used to calculate the maximum short-circuit Lorentz forces for Phase B and Phase C:

$$\begin{aligned} F_{Tb} &= 1.5 \times 10^{-7} I_r^2 \times ([\sin(314(t - t_k) \\ &- \frac{\pi}{2}) + e^{\frac{-(t-t_k)}{\tau}}]^2 \times ((\frac{\sqrt{(s+4)^2 + 35^2} - (s+4)}{3(s+4)})f(t - t_k) \\ &+ (6.38 - \ln\sqrt{F})g(t - t_k) - \frac{9.18}{s+22}) \end{aligned} \quad (5.52)$$

$$\begin{aligned} F_{Tc} &= 1.5 \times 10^{-7} I_r^2 \times ([\sin(314(t - t_k) - \frac{\pi}{2}) \\ &- e^{\frac{-(t-t_k)}{\tau}}]^2 \times ((\frac{\sqrt{(s+4)^2 + 35^2} - (s+4)}{3(s+4)})f(t - t_k) \\ &+ (6.38 - \ln\sqrt{F})g(t - t_k) - \frac{9.18}{s+22}) \end{aligned} \quad (5.53)$$

The Lorentz forces experienced by the arc contacts of each phase under short circuit conditions can be obtained by implementing the relevant expressions for the Lorentz forces in the circuit breaker dynamic simulation model, as shown in Fig. 5.25.

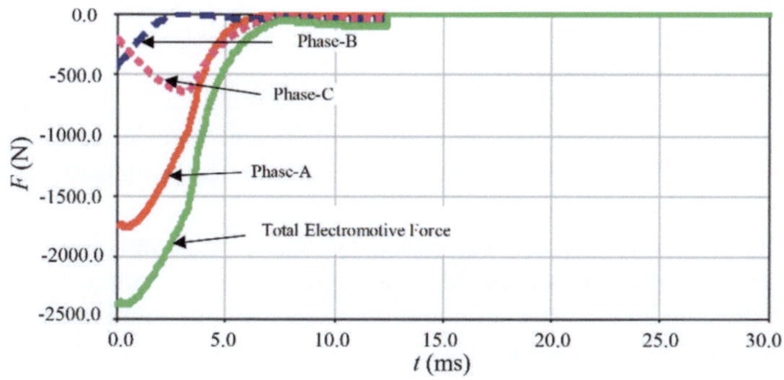

Fig. 5.25 Lorentz forces exerted on individual phases and the total Lorentz force under short circuit conditions

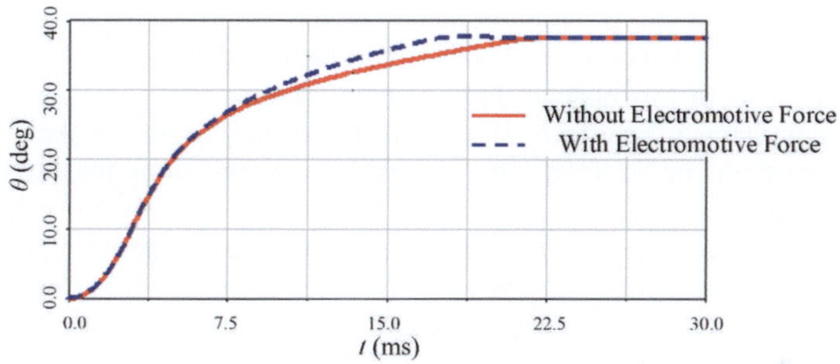

Fig. 5.26 Influence of the Lorentz force on the angular displacement curve of the main shaft of the operating mechanism during the breaking process under short circuit conditions

Figure 5.26 illustrates the influence of the Lorentz force on the angular displacement curve of the main shaft of the operating mechanism during the breaking process. The contact movement is accelerated and the contact separation time is shortened as a result of the repulsive Lorentz forces.

5.6 Summary

The method and procedure to calculate and analyse the Lorentz force in circuit breakers are introduced in this chapter. The Holm force appearing on the physically contacting tips of the electrical contacts is a special type of Lorentz force in nature. The appearance of short-circuit current in high-voltage equipment can induce severe Lorentz force that is able to de-shape the solid structure and adversely affect the performance of the equipment. Circuit breakers as important electrical equipment are used for controlling, protecting and switching purposes with movable components. One of the important factors affecting the opening and closing operations of the circuit breakers is the Lorentz force due to short-circuit current. Its calculation and analysis have practical significance.

The most common and severe types of short-circuit faults in power systems are introduced first together with relevant fundamental knowledge and concepts. Three-phase short-circuit faults are selected as examples to calculate the short-circuit currents.

Secondly, the Lorentz force in low-voltage electrical apparatuses when breaking the short-circuit current is simulated. Common low-voltage circuit breakers are involved in the calculation. The simulation process is divided into two parts, model development and implementation. The mathematic expressions for the short-circuit Lorentz forces are established for low-voltage air circuit breakers, moulded case circuit breakers (MCCB). Approximations and simplifications are introduced to improve the calculation efficiency. The models are then implemented in ANSYS and results obtained are then verify the accuracy and effectiveness of the method and procedure. The influences of different factors on the results are explored to provide theoretical support to the optimisation of circuit breaker design.

In another part of this chapter, high-voltage vacuum circuit breakers are used to derive and analyse the effects of Lorentz forces during the breakers' closing/breaking operations under short-circuit conditions. Expressions for attractive and repulsive Lorentz forces are derived separately. The attractive force is generated by the current component parallel to the contact surface planes in the structure designed to induce an axial magnetic field. The repulsive force is generated by the contraction of the current conducting areas approaching the interface of the contacts. Based on the basic methods of calculation, models for calculating the Lorentz forces operating in the closing and breaking processes of the selected vacuum circuit breaker are established by considering the sequence of pre-breakdown in the closing operation and the sequence of current interruption in the three-phase circuit breaker. The dynamic behaviour of the Lorentz forces and the operating mechanism, in terms of the angular displacement of the main shaft, is then obtained and studied. This allows the effects of synchronous and asynchronous pre-breakdowns of the contact gaps between the three phases to be evaluated during the closing operation. Similarly, the effect of the short circuit currents in a three-phase vacuum circuit breaker is studied through dynamic simulation, clearly demonstrating that the repulsive Lorentz forces can accelerate the contact separation speed during the breaking operation under short circuit conditions.

References

1. Zhou H L. The development opportunity of China electric power industry and the thinking about the applications of power electronics[J]. Power Electronics, 2004(06): 10-13.
2. Chen Y G. Research on short-circuit faults in power system based on wavelet analysis and neural network theory[D]. Changsha: Central South University, 2004: 1-51.
3. Liu S. Electrical engineering basics Vol.2[M]. Beijing: Science Press, 2002, 8: 260–273.

4. Nath P K, Ghosh A. Kineto-elastodynamic analysis of mechanisms by finite element method[J]. Mechanism and Machine Theory, 1980, 15(3): 179-197.
5. Bahgat B M, Willmert K D. Finite element vibrational analysis of planar mechanisms[J]. Mechanism and Machine Theory, 1976, 11(1): 47-71.
6. Jiabo Technology. Finite Element Analysis Software-ANSYS Integration and Penetration[M]. Beijing: China Water Resources and Hydropower Press, 2002: 5-20.
7. Erdman A G, Sandor G N. Kineto-elastodynamics-a review of the state of the art and trends[J]. Mechanism and Machine Theory, 1972, 7(1): 19-33.
8. Sun M L, Hu R X, Cui H R, et al. ANSYS10.0 Electromagnetic finite element analysis example tutorial[M]. Beijing: Machinery Industry Press, 2007: 1–439.
9. Chen S. Electromagnetic design and breaking characteristics optimization of molded case circuit breaker[D]. Xi'an: Xi'an Jiaotong University, 2012.
10. Xie C. Simulation and experimental research on key characteristics of dual-breakpoint modeled case circuit breakers [D]. Xi'an: Xi'an Jiaotong University, 2012.
11. Ma R G. Research on key technologies of large-capacity medium-voltage DC interruption based on air medium arc extinguishing [D]. Xi'an: Xi'an Jiaotong University, 2014.
12. Stammberger H. Force calculations for the movable electrode of circuit breakers. 19th International Conf. on Electrical Electrode phenomena [C]. Nuremberg, Germany, 1998: 369–374.
13. Faltin C. Electrodynamic repulsion between electric electrodes with arbitrary current-density distribution[J]. IEEE transactions on components, hybrids, and manufacturing technology, 1986, 9(2): 188-189.
14. Zhang G S. Theoretical basis of electrical appliances [M]. Beijing: Machinery Industry Press, 1996: 1-323.
15. Kawase Y, Mori H, Ito S. 3-D finite element analysis of electrodynamic repulsion forces in stationary electric electrodes taking into account asymmetric shape[J]. IEEE Transactions on Magnetics,1997, 33(2): 1994-1999.
16. Feng C Z. Electromagnetic field [M]. Second Edition. Beijing: Higher Education Press, 1983: 1-493.
17. Wang J M. Vacuum switch theory and its application [M]. Xi'an: Xi'an Jiaotong University Press, 1986: 1–152.
18. Qiu G Y. Circuit [M]. Fourth Edition. Beijing: Higher Education Press, 1996: 1-501.
19. Cao R J, Gu N H. Operating conditions of high voltage AC circuit breakers [M]. Beijing: Beijing University of Technology Press, 1999: 1-532.
20. Shang Z Q, Guo W Y. High-voltage electrical appliances [M]. Xi'an: Xi'an Jiaotong University Press, 2000: 1–192.
21. Yang W. Study on the dynamic characteristics and on-line detection method of key state parameters of high voltage circuit breaker mechanism [D]. Xi'an: Xi'an Jiaotong University, 2002.
22. Niu C P, Chen D G, Zhang J Z, et al. Research on the breaking characteristics of low-voltage circuit breaker with the effect of electrodynamic repulsion force[J]. Transactions of China Electrotechnical Society, 2005, 7: 34-38.
23. Ito S, Kawase Y, Mori H. 3-D finite element analysis of magnetic blowout forces acting on the arc in molded case circuit breakers[J]. IEEE Transactions on Magnetics, 1997, 33(2): 2053-2056.

Chapter 6
Calculation and Analysis of Performance Parameters of Operating Mechanisms in Switchgear

6.1 Introduction

Operating mechanisms, also commonly referred to as driving mechanisms, are used in switchgear apparatuses to open or close the main and arc contacts during the breaking or closing operations of the equipment. They are a key component of the switchgear and directly affect the reliability and stability of the operation. Common operating mechanisms include spring mechanism, hydraulic mechanism, electromagnetic mechanism, permanent magnet mechanism, and disc spring mechanism. Among them, electromagnetic mechanisms are mostly used in low-voltage apparatuses. Spring mechanisms and permanent magnet mechanisms are generally used for medium voltage (around 10 kV to 70 kV) applications, while hydraulic mechanisms and disc spring mechanisms are mainly used in high voltage field (above 70 kV). These mechanisms can be used alone or in pairs. At the same time, there are applications of new types of mechanisms, such as servo-motor based operating mechanism [1]. For switchgear, the operating mechanisms must be able to start and stop their operation rapidly with fast speed, short stroke of travel, and severe mechanical impact, making the design, manufacturing and fault diagnosis a complicated task.

Traditionally, the design of operating mechanisms follows several steps, including preliminary design, prototyping, field testing, design improvement and mass production. This often requires numerous iterations of the design-test-design cycle to make the product meet the application requirements. The product development model has apparent issues such as high cost, long development cycle, limited amount of tests, and incomplete performance optimisation. This seriously restricts the capacity to improve product quality, reduce development costs, and accelerate market access [2]. The design of operating mechanisms based on computer-aided engineering (CAE) is an inevitable trend of technological development that has made rapid progress in the past decade and is becoming a mainstream design tool for operating mechanisms. Among them, virtual prototyping technology is a CAE technology developed in the 1980s, with the ability to systematically simulate the kinematics and dynamics of

M. Rong et al., *Computing Methodology for Switchgear*,
https://doi.org/10.1007/978-981-95-2840-0_6

devices with multiple components. With it, the operational characteristics of an operating mechanism can be obtained with reasonable accuracy but without performing expensive and labour-intensive tests. The product development cost and time can also be reduced.

The virtual prototyping technology is used not only in the design of operating mechanisms but also in their fault diagnosis. A reliability survey conducted by the China Electric Power Research Institute (CEPRI) revealed that out of 550 faults, developed between 1998 and 2002 in switchgear products across the voltage range of 6 to 500 kV, mechanical failures accounted for 37% of the total faults, one of the main fault types encountered in field operation [3]. The development of a mechanical fault can mean that the switchgear equipment will not open or close in the normal way, seriously endangering the safety of the power system. Therefore, it is of great importance that the mechanical conditions of the operating mechanisms can be detected on-line, and any hidden problems identified in advance. However, there is little information that can be obtained by on-line detection (apart from the stroke of travel of the operating mechanisms), so accurately evaluating the mechanical conditions of the operating mechanisms is a key issue. There are numerous components in an operating mechanism and they link to each other in a complex manner, so it would be difficult to quantitatively relate the detected information to the mechanical state of the components through test. In contrast, virtual prototyping technology has clear advantages. It can assume all types of possible faults and produce the corresponding stroke curves, thus forming a database of stroke curves (Fig. 6.1). It provides the basis for fault diagnosis of the operating mechanism [4–6] in the virtual environment.

Different software packages exist in the field of virtual prototyping technology and ADAMS (Automatic Dynamic Analysis of Mechanical System) is a common

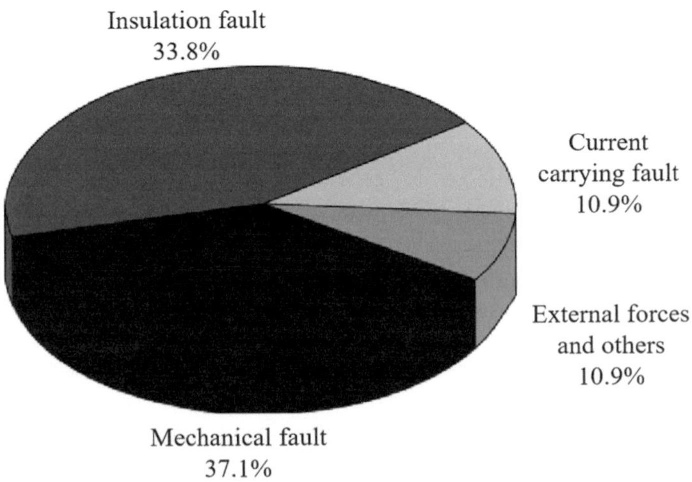

Fig. 6.1 Fault type distribution of high voltage switchgear in China (1998–2002)

choice, developed by MDI. ADAMS is a simulation and analysis software for multi-body systems and the governing equations describing the dynamics and kinematic features of the mechanical system are pre-implemented together with a built-in solver. ADAMS is commonly used for the calculation and analysis of spring operating mechanisms. Linked with the electromagnetic analysis software such as ANSYS, it can also be used to simulate electromagnetic operating mechanisms and permanent magnet mechanisms. For hydraulic operating mechanisms, the most commonly used virtual prototyping software is AMESim that is developed by IMAGINE. AMEsim can be used to simulate systems where fluid dynamics, thermal process and system control are coupled. It can also be linked with ANSYS and other software packages to simulate more complex operating mechanisms [7]. Both ADAMS and AMESim can be used with UG [8], Pro/E [9] and other solid geometry modelling software to first build an accurate 3D geometric model and then import the latter into the virtual prototyping software to add motion-driving sources and constraint conditions. Following this, the dynamic and kinematic behaviours of an operating mechanism can then be obtained through the solution procedure.

The optimal design and fault diagnosis methods for the operating mechanisms of switching apparatuses are introduced in this chapter. The methods are demonstrated by the use of ADAMS and AMESim. Together with ANSYS, Ansoft and another electromagnetic analysis software, a spring mechanism, hydraulic mechanism and permanent magnet mechanism are, respectively, studied. It is expected that these methods will attract more applications following the inevitable trend of technological development [10].

6.2 Calculation and Analysis of Spring Operating Mechanisms

A spring operating mechanism is a mechanical system that uses springs for energy storage and is made of one or more cams, connecting rods and other components. This type of operating mechanism does not need a large capacity power supply and the power rating and cost of the motor are also low, attracting wide applications in switchgear. However, the operating mechanism also has inherent limitations. In design, it is not straightforward to match the output force characteristics of the spring with the load requirement of the switching device. This is because the opening and closing operations of the mechanism need to take place in several tens of milliseconds yet the effects due to mechanical impact or shock between different components have to be controlled to an appropriate level [8].

In terms of fault diagnosis, the stability of the spring mechanism can be an issue and its opening and closing stroke curves will drift over long service, making it difficult to accurately diagnose the conditions of the operating mechanism.

The establishment of simulation model using ADAMS and the approach to achieve mechanical fault diagnosis for a medium-voltage spring mechanism used in switching devices are introduced in this section. Coupled simulation based on ANSYS and ADAMS is also used to obtain the optimal design of the spring operating mechanism used in low voltage switches.

6.2.1 Modelling and Fault Diagnosis of Spring Operating Mechanism in Medium Voltage Switchgear

1. A Brief Introduction to Multi-Body Dynamics

Operating mechanisms in switching devices are characterised by their short response time, rapid operation (fast speed), short contact gap, and strong mechanical impact due to collision. This makes it difficult to establish the non-linear governing equations that describe the dynamics of the mechanisms. An analytic solution to the governing equations requires onerous mathematical operations. In the past decades, a new branch of discipline, multi-body dynamics, has been developed from classical computational mechanics. The research object of multi-body dynamics is a mechanical system composed of multiple rigid bodies or elastic bodies connected mechanically by joints. Its main task is to study how to establish the motion equations of a system automatically by a computer, and how to solve these equations quickly and accurately to execute a kinematic and dynamic simulation of the mechanical system. Multi-body dynamics provides a solution methodology for simulating the operating mechanisms of switching devices.

ADAMS is a multi-body dynamics simulation software package widely used in the fields of automobile manufacturing, aerospace, railway transportation and others. A parameterised solid model (virtual prototyping model) can be conveniently established and solved using ADAMS. Users only need to define the structure, connection, externally applied forces and other relevant parameters appropriate to the prototype in the graphical user interface (GUI). Compared with traditional simulation methods, the use of ADAMS avoids the tedious work of setting up and solving the governing equations, allowing the users to focus on the concerned physical problems thus performing efficient simulation analysis. ADAMS makes use of an interactive graphical environment, component library, constraint library and force library to rapidly build a three-dimensional parametric model of the mechanical system based on the concept of pre-defined building blocks. Using ADAMS, it can evaluate the mechanical system performance and range of movement, implement collision detection, obtain the peak load and produce the input information for finite element computation. A variety of modules are available, including the interactive graphics environment module (ADAMS/View), simulation solver (ADAMS/Solver), finite element interface (ADAMS/FEA), advanced animation generation module (ADAMS/Animation), post-processing module (ADAMS/Postprocessor) and others. Each module is defined independently using specific programming functions.

2. Modelling of Spring Operating Mechanism for Medium Voltage Switchgear

The spring operating mechanism of VS1 vacuum circuit breaker is used as the object of modelling in this section due to its popularity in medium voltage applications. The rated voltage of the circuit breaker is 12 kV, the rated current is 1.25 kA, and the rated short-circuit breaking current is 31.5 kA. As shown in Fig. 6.2, this operating mechanism has a flat design and is placed in the chassis in front of the arc extinguishing chamber. It includes two closing springs (10), three opening springs (35), three contact springs (to generate contacting force, 32) and one oil buffer (45), making up the main power sources for closing and opening operations. The moving part from the spindle to the moving contact is typically a four-bar mechanism, including the spindle drive arm (37), the connecting plate (36), the drive arm (34) and the insulated rod (33), generating the translational movement of the contact from the rotation of the spindle (46).

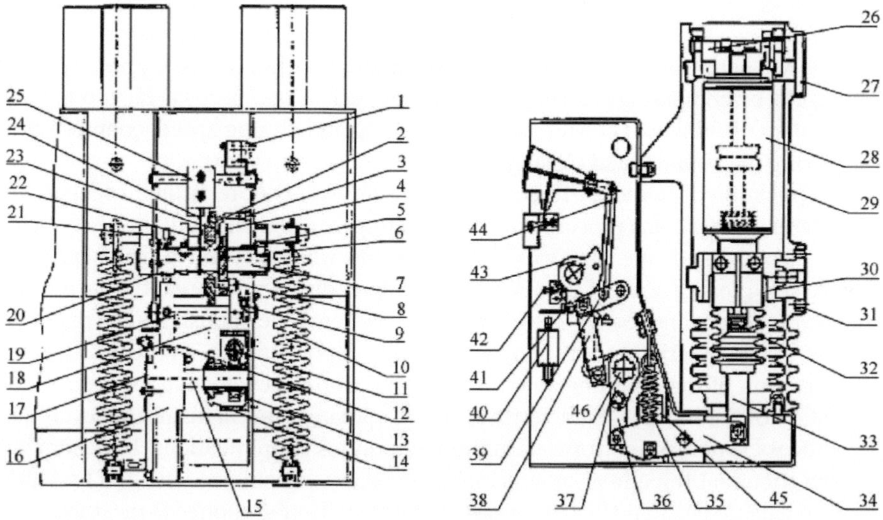

Fig. 6.2 Structural diagram of the operating mechanism (VS1) in a vacuum circuit breaker. 1-Toggle switch when full energy storage is achieved; 2-Pin; 3-Positional limit rod; 4-Slider; 5-connecting lever; 6-energy storage transmission wheel; 7-energy storage shaft; 8-roller; 9-energy storage holding pawl; 10-closing spring; 11-manual energy storage handle; 12-closing electromagnet; 13-manual energy storage drive gear; 14-motor drive sprocket; 15-motor output shaft; 16-energy storage motor; 17-interlocking drive bending plate; 18-drive chain; 19-energy storage holding shaft; 20-locking electromagnet; 21-connecting lever; 22-cam; 23-energy storage drive sprocket; 24-side plate; 25-energy storage sign; 26-upper bracket; 27-upper outlet seat; 28-vacuum interrupter; 29-insulated cylinder; 30-lower bracket; 31-lower outlet seat; 32-contact spring; 33-insulated rod; 34-transmission crutch arm; 35-opening spring; 36-transmission connecting plate; 37-Spindle drive lever; 38-closing holding lever; 39-connecting plate; 40-opening electromagnet; 41-opening half shaft; 42-manual opening ejector rod; 43-cam; 44-on–off sign connecting plate; 45-oil buffer; 46-spindle

Aiming for the above operating mechanism, a multi-body dynamics simulation model is established using ADAMS, and the process generally includes five steps.

(1) Structuring and assembly of mechanism parts: each part is modelled and assembled as an entity according to its physical shape and size. Circuit breaker manufacturers often have 3D model files of the operating mechanism, which can be imported into ADAMS to simplify the definition process of the operating mechanism.
(2) Application of driving power: the driving force for circuit breaker opening and closing changes rapidly with time, and there are many influencing factors. For example, the driving force of the opening process mainly comes from the opening spring and the contact spring, where the compression of the contact spring is difficult to model accurately because in real product the pre-compression of the spring may have deviations due to the assembling process. At the same time, the collision between the connecting rods (a small gap may exist in connection) and the shape of the cam can have a substantial influence on the transfer of force, rendering the driving forces difficult to calculate accurately. Therefore adjustment to model parameters is often necessary based on test results by comparing measurable quantities such as the speeds at the moment of contact opening and closing, average opening and closing speed, contact travel during the opening and closing processes, and so on.
(3) Application of retarding force: there are many kinds of retarding forces in the opening and closing processes of the operating mechanism, including the gravity of the moving contact and moving components when they move against gravity, the self-closing force of the vacuum interrupter due to pressure difference when the breaker opens, the resisting force of the bellows, the opposing force of the oil buffer and the friction between the moving components, etc. Appropriate representation of the retarding force is important and necessary for accurate simulation of the motion characteristics of the operating mechanism. However, this process is not straightforward and often requires simplifications. For example, the frictional force in a vacuum circuit breaker is a complex situation to define and it has to be divided into two parts: rotational friction applied to the rotation of the spindle and sliding friction applied to the connecting rods (for three-phases). Other frictional forces of a minor nature are ignored. Similar to the situation relating to the application of the driving power, adjustment to the relevant model parameters also needs to be considered based on experimental results.
(4) Constraint relationship: the constraint relationship in an operating mechanism includes fixed constraint, rotation constraint, translation constraint, collision, etc. The addition of constraint relationships should be based on full consideration of the possible constraints in the actual operating mechanism; omission or misapplication of a constraint relationship may lead to incorrect results.
(5) Simulation control: the overall calculation time, time step size and accuracy in results should be decided through a balanced consideration of the simulation objectives for design or fault identification. Batch simulation with parameter

changes within defined ranges can be made automatic to facilitate the analysis of the influence of different design parameters.

In the above steps, the implementation of Steps 2–4 has no fixed order and can be arranged according to the actual needs. The multi-body dynamic model of the operating mechanism (VS1) of the vacuum circuit breaker is shown in Fig. 6.3. The size of the geometric model is consistent with that of the circuit breaker but the structure is simplified to eliminate the stationary and auxiliary components that have no effect on the results, such as the upper bracket (26 in Fig. 6.2), energy storage sign (25 in Fig. 6.2) and so on.

Based on the settings for the structural parameters, driving forces, retarding forces and constraints in the model, ADAMS automatically establishes the Lagrangian equations of motion for the mechanical system, where each rigid body contains six Lagrangian equations within a system of generalised coordinates, multipliers and the corresponding constraint equations, as described below:

$$\begin{aligned} &\frac{d}{dt}\left(\frac{\partial K}{\partial q_j}\right) - \frac{\partial K}{\partial q_j} + \sum_{i=1}^{n} \frac{\partial \psi_i}{\partial q_j}\lambda_i = F_j \\ &\psi_i = 0 \quad (i = 1, 2, \ldots m) \end{aligned} \tag{6.1}$$

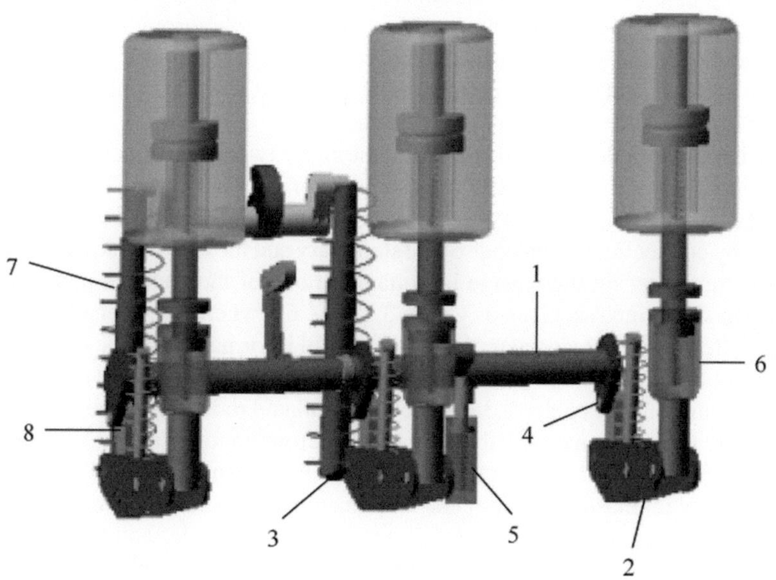

Fig. 6.3 Multi-body Dynamics Model of the VS1 operating mechanism. 1-Spindle; 2-Bottom clevis pin of insulated rod; 3-clevis pin of the middle phase; 4-clevis pin of the right side phase; 5-Oil buffer; 6-Contact spring; 7-Closing spring; 8-Opening spring

where K is the kinetic energy, q_j is the generalised coordinates, ψ_i the constraint equations, F_j the generalised forces in the direction of each generalised coordinate; and λ_i the Lagrangian multiplier matrix of size m × 1.

ADAMS uses the Cartesian coordinates at the centre of mass of rigid body i and the Euler angle (or generalised Euler angle) reflecting the orientation of the rigid body in place of the generalised coordinates, that is:

$$q_i = [x, y, z, \psi, \theta, \varphi]_i^T, q = [q_1^T, ..., q_n^T]^T \tag{6.2}$$

Due to the use of generalised coordinates, the system dynamic equations consist of a large number of closely coupled differential–algebraic equations, which can be solved by a method based on a sparse matrix. Using the modified Newton–Raphson iterative algorithm, ADAMS is able to efficiently solve the non-linear partial differential Eq. (6.1) and obtain the dynamic and kinematic characteristics of the whole mechanical.

To confirm whether the simulation model is consistent with the simulated object, it is necessary to measure the opening and closing stroke curves of the circuit breaker and compare them against the simulation results. The main stroke curves include the angular displacement of the spindle and the linear displacement of the insulated rod, detectable by an angular displacement sensor and a linear displacement sensor, respectively. The installation of the sensor is shown in Fig. 6.4, where the angular displacement sensor is installed at one end of the spindle through a fixed bracket; the linear displacement sensor is installed on the base of the circuit breaker, and the sliding rod is connected with a certain phase insulated pull rod, which is consistent with the linear movement of the insulated pull rod.

Figure 6.5 shows the detection results of the above two displacement sensors in the opening process, and the contact state change signal for analysis. The latter is used to signify the change in the close/open state of the stationary and moving contacts, obtained by connecting the far ends of the stationary and moving contacts to a resistive circuit and monitoring the change in voltage on one of the contacts under no-load conditions. This signal is given as Curve 3 in Fig. 6.5 where a sudden change of the voltage reflects the contact separation time. Useful information can be extracted. For example, compared with the linear displacement curve of the insulated rod (Curve 2), contact separation takes place 2.5 ms after the insulated rod starts to move, indicating that the linear displacement of the contact tip before contact separation is affected by the return stroke of the contact spring. The angular displacement of the spindle is approximately in phase with the linear displacement of the rod (Curve 1).

A comparison between the predicted and measured angular and linear displacements is given in Fig. 6.6a, b under normal conditions, respectively. The two curves in both figures math well.

The performance of the model for situation with a specific fault is also assessed. The opening spring of the mechanism in a phase is removed in the experiment and corresponding changes are made in the model. The comparison presented in Fig. 6.7 suggests that good agreement in the angular displacement stroke curve is achieved, demonstrating the capability and accuracy of the model.

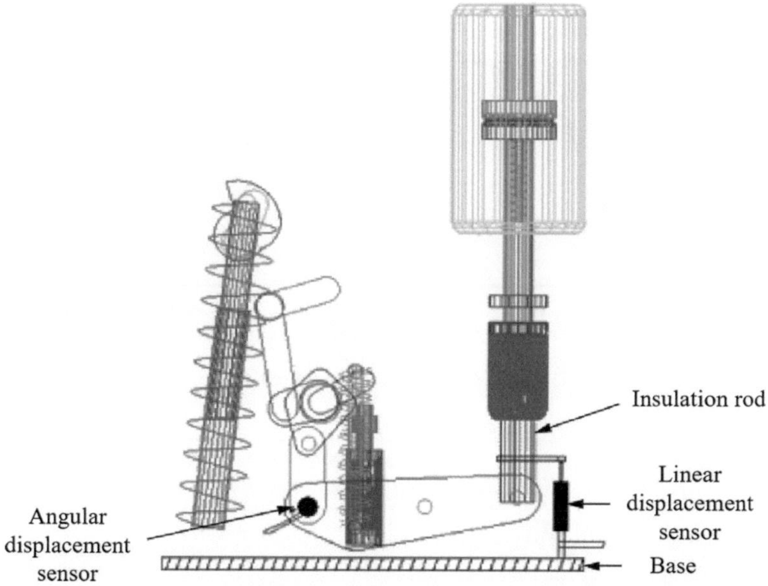

Fig. 6.4 Location of installation of the displacement sensors in the circuit breaker

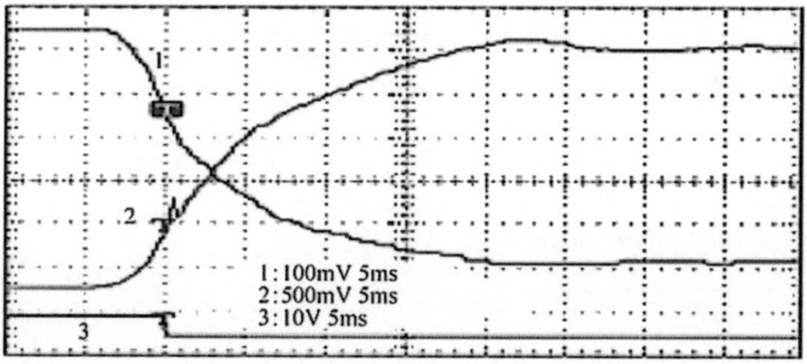

Fig. 6.5 Output of displacement sensors during the circuit breaker opening process. 1-spindle angular displacement curve; 2-insulated rod displacement curve; 3-contact state signal

The simulation model for the closing process of the circuit breaker also needs to be verified. the most uncertainty is with the friction coefficient of the mechanism among other model parameters, which can lead to relatively large errors if not set properly. Again, experimental results are needed to guide the selection of its value to improve the model accuracy. When the simulated opening and closing stroke curves under normal and typical fault conditions of the mechanism match those from the measurement, the model is then considered to be effective.

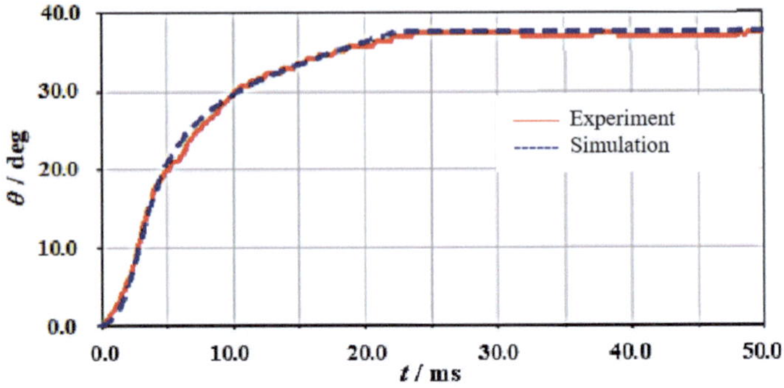

(a) Spindle angular displacement curve

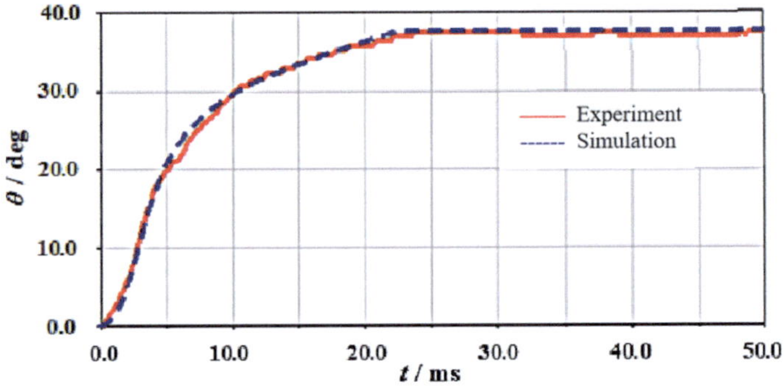

(b) Linear displacement curve of insulated rod

Fig. 6.6 Comparison between simulation and measurement of stroke curves in opening process of the circuit breaker at the rated current

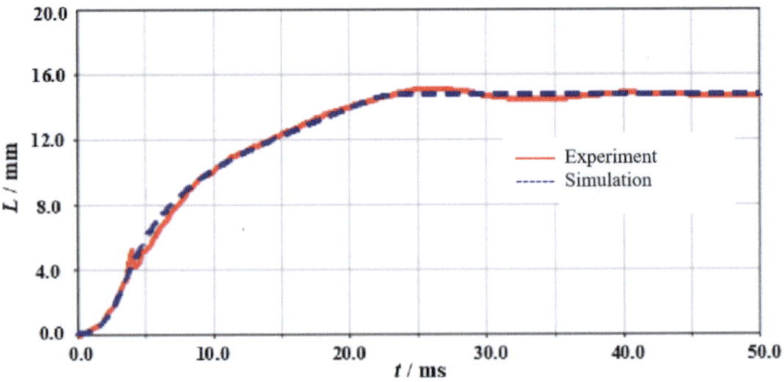

Fig. 6.7 A comparison of simulated and measured spindle angular displacement curve when the opening spring in a phase is removed (equivalent to spring failure)

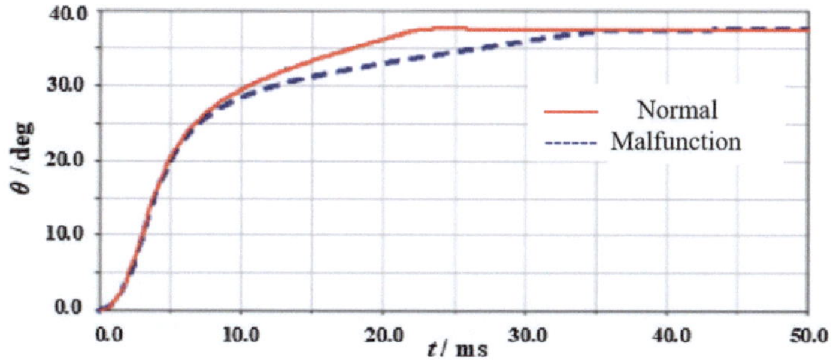

Fig. 6.8 Spindle angular displacement stroke curve in the presence of opening spring failure for one phase

3. Mechanical Fault Diagnosis Based on a Multi-Body Dynamics Model

As mentioned earlier in this section, an important usage of the multi-body dynamics model is to diagnose a potential fault in an operating mechanism. Considering the common fault types of the operating mechanism during its opening process, a fault diagnosis method based on the multi-body dynamics model is introduced below. These fault types include: failure of an opening spring in a phase, shedding of a shaft pin, oil leakage from the buffer and other typical ones. The angular displacement stroke curves under these fault conditions are obtained by simulation and compared with the curves under normal conditions to identify a fault. Furthermore, the stroke curves under different conditions can be parameterised to establish a quantitative criterion for fault diagnosis of the operating mechanisms.

(1) Failure of the opening spring in one phase

When the opening spring for a phase fails, the total driving force decreases and the duration of the opening process is prolonged. The simulation results of the spindle angular displacement stroke curve with the presence of the failure are shown in Fig. 6.8.

(2) Failure due to malfunction of a shaft pin

When a shaft pin falls off from the mechanism, the contacts in the concerned phase become separated from the operating mechanism and the opening spring and contact spring are no longer able to supply a driving force to the moving contact for its opening operation. The duration of the opening process is prolonged because the lost driving force is much greater than the frictional force. The opening process is not affected by the friction in this phase. As can be seen from Fig. 6.9, complete separation of the contacts is delayed from 23 ms under normal conditions to 38 ms under fault conditions.

(3) Oil leakage from buffer

The spindle angular displacement stroke curve of the mechanism with a severe leakage developed in the oil buffer is given in Fig. 6.10 for the opening process.

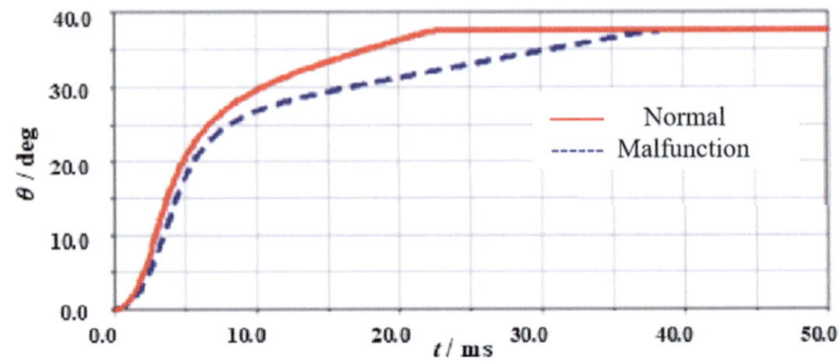

Fig. 6.9 Angular displacement stroke curve of the spindle after simultaneous failure of opening spring and contact spring in a phase

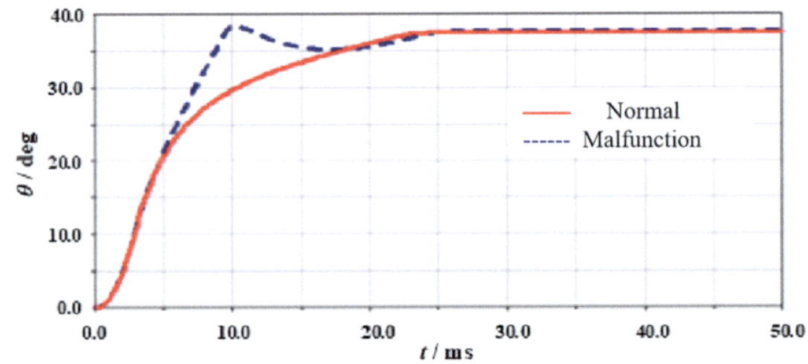

Fig. 6.10 Angular displacement stroke curve of the spindle when the operating mechanism suffers from severe oil leakage

It is clear that due to a decrease in the retarding force from the oil buffer, the contact speed before 10 ms remains high, resulting in stronger bouncing of the contact.

Figures 6.8, 6.9 and 6.10 show a noticeable difference in the spindle angular displacement stroke curve between the normal operation of the mechanism and the faulted mechanism. However, to enable fault diagnosis by a computer or embedded system, it is necessary to parameterise the stroke curve and identify a fault type based on one or more feature parameters. According to the provision of a driving force, source of retarding force and the movement of the constituent components of the operating mechanism, the opening process of the vacuum circuit breaker can be divided into four stages.

- Stage 1: begins with the movement of the spindle until the contacts in all three phases are separated. After the completion of this stage, the contact springs no longer act. During this stage, the moving components of the operating mechanism are mainly affected by the contact spring power, opening spring power and retarding force due to friction.
- Stage 2: starts from the instant of contact separation until the buffer arm on the spindle activates the oil buffer. During this stage, the moving components of

the operating mechanism are mainly affected by the opening spring power and friction.

- Stage 3: starts from the activation of the buffer until the oil buffer is fully compressed and the circuit breaker opening position is locked by a mechanism. During this stage, the oil buffer produces a retarding force to slow down the rotation of the spindle and the moving components are mainly affected by the opening spring power, the regarding force of the buffer and the friction.
- Stage 4: starts from the instant the oil buffer is fully compressed until the opening process of the circuit breaker finishes. In a normal operation of the circuit breaker, the angular velocity of the spindle is negligible unless there is significant oil leakage from the buffer.

In each of the above four stages, the average slope of the segment of the angular displacement stroke curve is taken as a characteristic quantity, leading to four parameters of $K1$, $K2$, $K3$ and $K4$, respectively. They represent the average angular speed of these four stages, as shown in Fig. 6.11. During the opening process of the circuit breaker under different operating conditions, the four characteristics will be different, and the results are shown in Table 6.1.

Table 6.1 indicates that the four characteristic parameters describing the spindle angular displacement of the vacuum circuit breaker follow different patterns of change under different conditions. Therefore, by simulating the opening and closing operations of a circuit breaker under different fault conditions with a verified model, the type of fault can be identified by matching the characteristic quantities derived from the measured angular displacement stroke curve with one of the cases simulated. These characteristic parameters can thus be used to form the baseline data of an expert system to provide quantitative criteria for fault diagnosis.

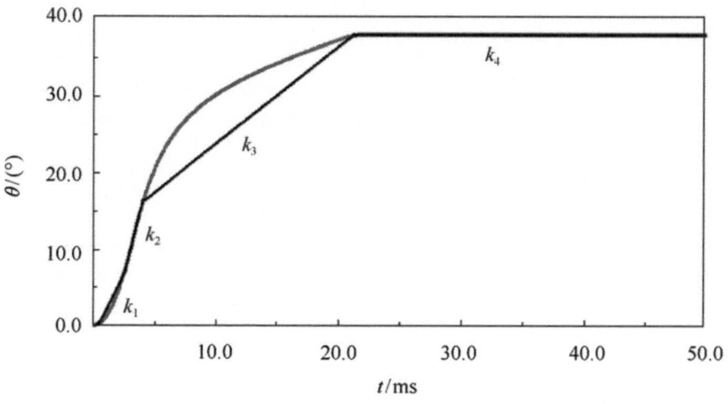

Fig. 6.11 Parameterisation of the angular displacement stroke curve of the spindle

Table 6.1 List of four characteristic parameters of the spindle angular displacement during the opening process of the VS1 circuit breaker under different fault conditions

Fault type	Mechanical state	$K1$ (°/ms)	$K2$ (°/ms)	$K3$ (°/ms)	$K4$ (°/ms)
1	Normal operation	3.80	5.12	1.09	−0.001
2	Simultaneous failure of one-phase openingspring and contact spring	3.10	4.10	0.55	0.0
3	Failure of one-phase opening spring	3.70	4.96	0.63	0.0
4	Failure of one-phase contact spring	3.21	4.25	0.94	0.0
5	Oil buffer failure	3.80	5.12	3.12	−0.047

6.2.2 *Modelling and Optimal Design of Spring Operating Mechanisms in Low Voltage Switchgear*

An air circuit breaker of a certain type with a rated voltage of 660 V, rated current of 2 kA, and rated short circuit breaking current of 40 kA will be used as a representative operating mechanism for low voltage applications. A spring operating mechanism is used and the spring force drives its opening and closing operations. However, the repulsive Lorentz force can become significant when a short-circuit current is dealt with and increases the bouncing of the moving contact in the opening process. This leads to a risk of arc reignition and hence a breaking failure. Therefore, it is necessary to optimise the operating mechanism to reduce the contact bouncing and improve the reliability of the circuit breaker. ANSYS is used in this section to obtain Lorentz force data under short-circuit conditions which are then input into the operating mechanism model based on ADAMS to simulate the dynamic and kinematic characteristics. Based on the output contact bouncing in the circuit breaker can then be minimised.

1. Simulation of motion characteristics under short circuit conditions

Figure 6.12 is a schematic diagram of the operating mechanism in an open position of the DW45 air circuit breaker, consisting of a five-bar mechanism (connecting rods 1–4 and a ground rod). When the mechanism starts to close, connecting rod 4 is held up by the ejector rod and does not move; the energy storage spring (6) in its compressed state pushes the energy storage lever to rotate anti-clockwise; Connecting rod 3 and connecting rod 2 are then forced to shift towards the left-hand side in the diagram; the lower end of the connecting rod 1 is finally driven to move to close the contacts. When connecting point Oc is below the line passing Od and Ob, the mechanism reaches the dead point and the contacts are in a stable closed state.

When the opening process starts, the half shaft (4) is freed by a tripping mechanism; connecting rod 1 starts to rotate anti-clockwise under the action of contacting force, opening spring force and repulsive Lorentz force; the moving contact then separates from the stationary contact. Meanwhile, connecting rod 4 is freed from the

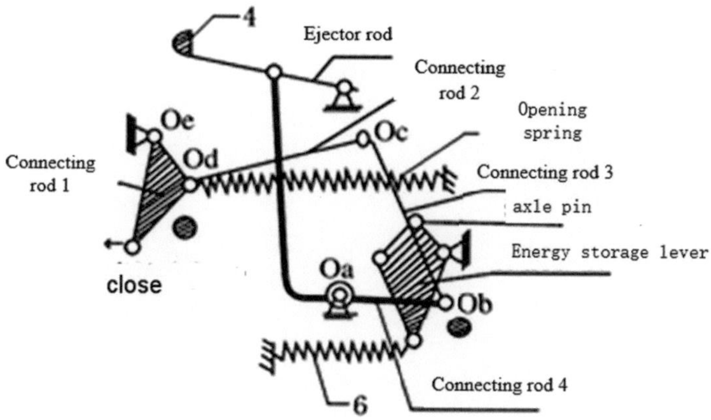

Fig. 6.12 Schematic diagram of the operating mechanism of DW45 circuit breaker in its open position

ejector rod and rotates clockwise around Oa, and the point Ob moves downwards. When point Oc is above the connection line between Od and Ob, the mechanism accelerates to the right-hand side until the circuit breaker is in a stable opened state [11].

The operating mechanism is modelled by ADAMS in a way similar to that introduced in Sect. 2.1. First, the file output by a three-dimensional structure generation package is imported into ADAMS and the structure is simplified, for example, mechanical components that do not contribute to the motion characteristics of the mechanism are removed. Second, constraint relationships between the components are added, including the rotation constraint between a connecting rod and a rotating shaft, the contacting constraint between two components, and the fixed constraint between a rotating shaft and the base. Finally, simulation parameters such as driving power and retarding forces are set. When the Lorentz force due to a short-circuit current needs to be considered, it is necessary to import the Lorentz force data due to short-circuit obtained by ANSYS into ADAMS. The mechanical model of the DW45 air circuit breaker that is used for dynamic simulation is shown in Fig. 6.13.

(1) Calculation and loading of short-circuit Lorentz force

ANSYS is used to calculate the Lorentz force needed for the dynamic simulation of the circuit breaker under short-circuit conditions. The method is introduced in Chap. 5 and only a brief description is given here. An examination of the contact system and arc interrupter of the circuit breaker, as shown in Fig. 6.14, is needed first. The non-ferromagnetic components in the structure have little influence on the Lorentz force and can be removed before being imported into ANSYS to form the model for calculating the Lorentz force. As shown in Fig. 6.15, the model mainly consists of: the moving conductive bar, moving contact, stationary conductive bar, stationary contact, arc chute, and the arc slider.

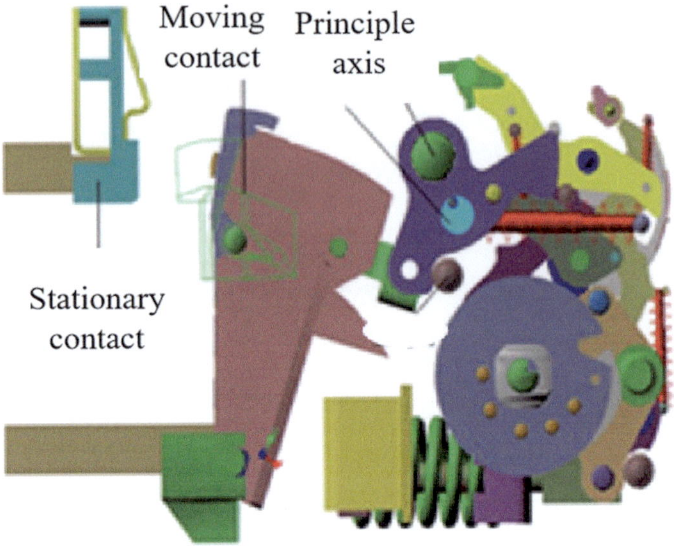

Fig. 6.13 Structure of the DW45 air circuit breaker after simplification for simulating the dynamics of the contact movement

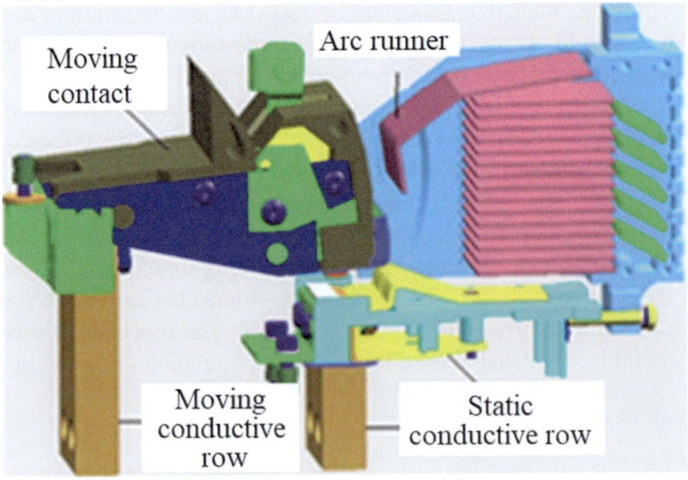

Fig. 6.14 Structure of the contact system and interrupter of the air circuit breaker under study

The calculation of the Lorentz force involves two stages: the closed stage of the contacts when the gap length is reducing to zero and the Holm force dominates, and the arcing stage after contact separation when the volumetric Lorentz force prevails. The key to the calculation of the latter is to establish equivalent geometry models of the arc column. Since the arc characteristics change with the contact gap

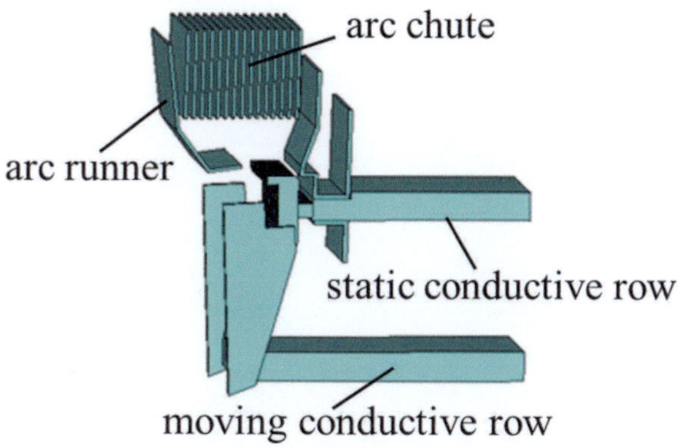

Fig. 6.15 Simplified geometry model for calculating the Lorentz force in the air circuit breaker

length, it is therefore necessary to establish a series of equivalent models of the arc column, corresponding to different instants in the arcing process, to obtain the discrete data of the Lorentz force over the whole opening process. The short-circuit current data comes from experimental records or prescribed. Figure 6.16 shows the experimentally recorded current and arc voltage waveform at 40 kA where the short-circuit current lasts for about 18 ms before contact separation and arc initiation. The arc then lasts for about 12 ms before extinction.

The calculated Lorentz force in a discrete form is then imported into the multi-body dynamics model implemented in ADAMS to simulate the dynamic characteristics of the whole mechanism. Importing is carried out by a three-dimensional spline function (Spline in ADAMS) with the independent variables x, y and z, representing the current, Lorentz force and the contact gap, respectively.

(2) Simulation results under short-circuit conditions.

The multi-body dynamics simulation model considering the short-circuit current-induced Lorentz force produces data that describe the variation of the spindle angular displacement, contact gap length and Lorentz force in the opening process as a

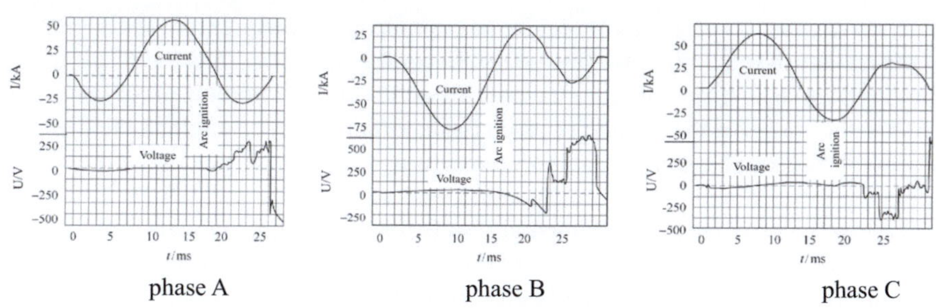

Fig. 6.16 Experimentally recorded voltage and current waveform for three phases of the air circuit breaker with a short-circuit current of 40 kA. The vertical line indicates the instant of arc initiation

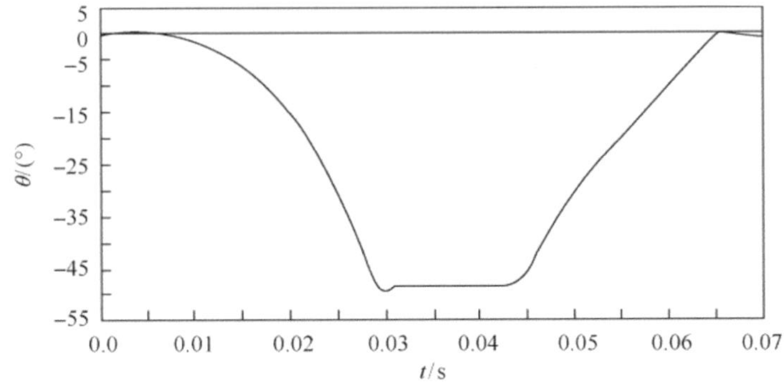

(a). Predicted closing -> opening sequence under short-circuit conditions for the air circuit breaker under study.

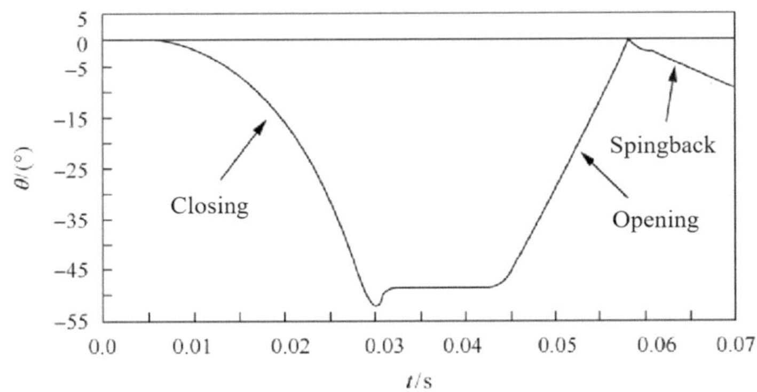

(b) Predicted closing -> opening sequence under no-load condition for the air circuit breaker under study.

Fig. 6.17 Angular displacement stroke curve of the spindle of the air circuit breaker under study in the sequence of closing - > opening under short circuit and no-load conditions

function of time. As given in Fig. 6.17a, the spindle takes a while from the beginning to the instant of maximum rotation angle. In contrast, Fig. 6.17b is the result of the spindle angular displacement under no-load condition (no Lorentz force) where it takes longer time for the spindle to reach its maximum rotation angle in the opening process. It is clear that the Lorentz force speeds up the opening process of the spring operating mechanism.

A representative set of experimental results for the spindle angular displacement is shown in Fig. 6.18 when the air circuit breaker interrupts a 40 kA short-circuit current. The time taken from the beginning of the spindle rotation (20 ms) to the instant of maximum rotation angle (36 ms) is between 15.8 and 16.2 ms for multiple opening tests, which is close to the simulation result of 16.3 ms. The severe bouncing

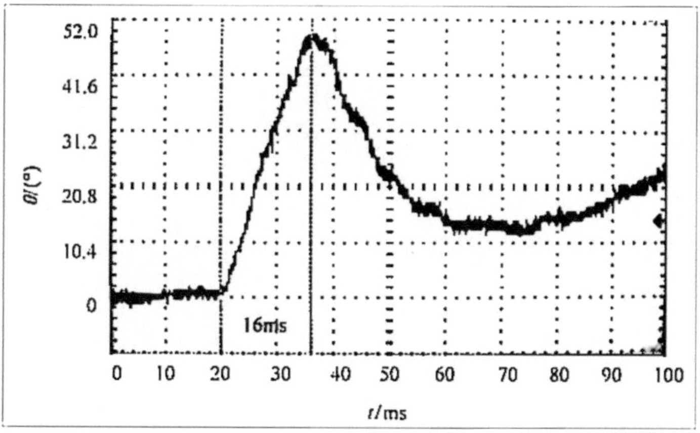

Fig. 6.18 Experimental record of the spindle angular displacement curve during the opening process of a 40 kA short-circuit current

implies that there is a possibility of arc reignition due to the contacts approaching each other again and it is necessary to optimise the anti-bouncing design.

2. Anti-Bouncing Design Of Low Voltage Spring Operating Mechanism

In order to eliminate any design failure of the anti-bouncing mechanism for the moving contact, a simulation based on ADAMS and ANSYS is performed to test the proposed design changes. With the DW45 air circuit breaker shown in Fig. 6.13, it opens rapidly under the combined action of the contact spring, opening spring and Lorentz force in response to a short-circuit current. When the mechanism comes in contact with the stop that is fixed on the supporting frame, the contact separation reaches its maximum and then the moving contact starts to bounce back, reducing the contact gap length. An effective anti-bouncing mechanism should hold contact when the contact gap length reaches its maximum and minimise the bouncing.

Figure 6.19 is a structural diagram of the anti-bouncing mechanism. The stop and the side plate are fixed on the frame of the circuit breaker and the transmission component is welded onto the main shaft and they rotate together. The swinging component is connected to the transmission component through the shaft pin and can rotate around the shaft. The positional limiting rod is fixed on the swinging component and extends into the limiting hole to limit the rotation of the swinging component. And the reset spring holds the positional limiting rod onto the upper end of the limiting hole simultaneously. The anti-bouncing mechanism makes use of the principle of inertia. Once a short-circuit fault occurs in the circuit, the transmission component drive the swinging component to swing downwards. When the contact gap reaches its maximum, the swinging component tends to continue its swinging due to inertia and this counteracts the bounding of the contact, allowing the swinging component to be held on the stop to forcibly prevent rebound of the moving electrodes.

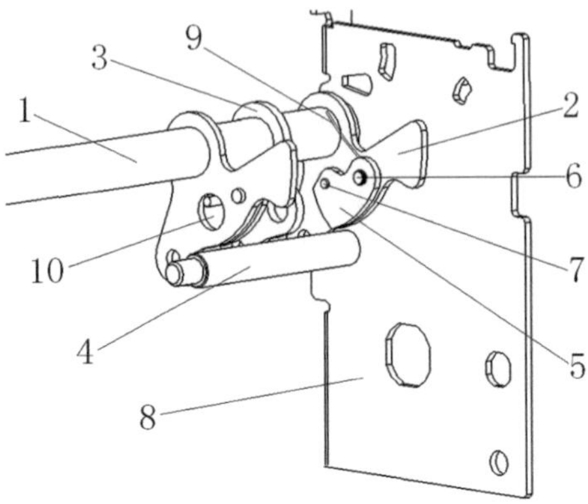

Fig. 6.19 Structural diagram of the anti-bouncing mechanism. 1-spindle; 2-transmission component; 3-connecting lever; 4-stop rod; 5-swing component; 6-shaft pin; 7-shaft pin; 8-side plate; 9-reset spring; 10-limit hole

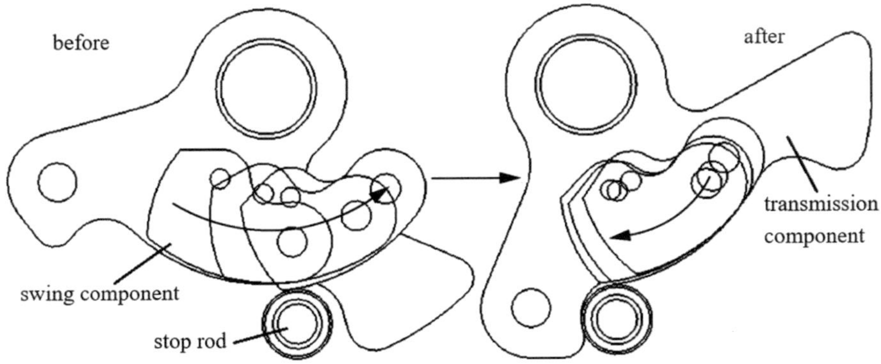

Fig. 6.20 Diagram showing potential failure of the anti-bouncing mechanism of the air circuit breaker. The arrows mean the movement of the transmission component

Simulation results show that the short-circuit Lorentz force may cause failure in stopping the swinging component, rendering the anti-bouncing mechanism ineffective. Figure 6.20 shows the mechanical relationship in the case of an anti-bouncing failure. When the circuit breaker is being opened, the swinging component moves towards the right-hand side, driven by the transmission component. During this process, it swings slightly downwards. When it touches the stop rod, it bounces back and swings upwards slightly. Thus, when the transmission component collides with the stop rod and tends to bounce back, the swing component fails to hold it onto the stop rod and bouncing then takes place.

In order to improve the reliability of the anti-bouncing mechanism, an optimal design is proposed, as given in Fig. 6.21, and explained below.

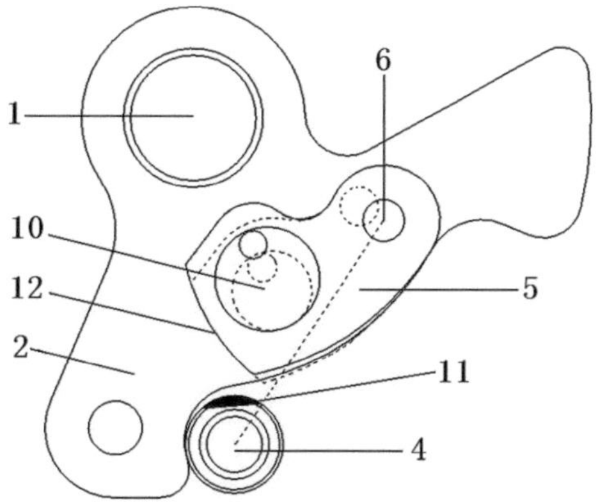

Fig. 6.21 Schematic diagram of improved deign of the anti-bouncing mechanism. 1-spindle; 2-transmission component; 4-stop rod; 5-swing component; 6-shaft pin; 10-limit hole; 11-portion of stop rod marked black for reference; 12-left edge of swing component

- Satisfying the mechanical strength requirement, the position of the shaft pin is moved to the upper left part of the transmission component (Fig. 6.21) to increase the inertial effect of the downward-moving swinging component;
- Reducing the size of the limit hole so that the upper edge of the swing component shifts downwards to allow the centre of mass of the swing component to shift downwards to increase the inertial effect.
- The contact point between the stop rod and the swing component must be on or very close to the line passing the shaft pin and the centre of the stop rod (dotted straight line in Fig. 6.21) when the swing component work properly to limit the movement of the transmission component. The small portion of the stop rod, as marked black in Fig. 6.21, can be removed so it no longer causes any blockage to the downward swing of the swing component.
- After the black portion of the stop rod is removed, the lower edge of the swinging component can shift downwards, helping to lower the centre of mass of the swing component and increase its inertial effect to weaken its upward bouncing when it collides with the stop rod.
- When the swing component is held up by the stop rod, there may be a possibility that it slips upwards to misplace. To avoid this, the centre point of the left edge of the swing component can be shifted from the central position of the shaft pin slightly towards the lower-right direction.

The design of the anti-bouncing mechanism is improved according to the above discussion and also in consideration of the practicality. Simulation analysis is then performed with different parameter values. Results clearly show that the modified mechanism exhibits reliable anti-bouncing performance for short-circuit currents at 30 kA and above. The predicted spindle angular displacement stroke curve of the modified mechanism during an opening process is shown in Fig. 6.22 where the bouncing is reduced to about 5°.

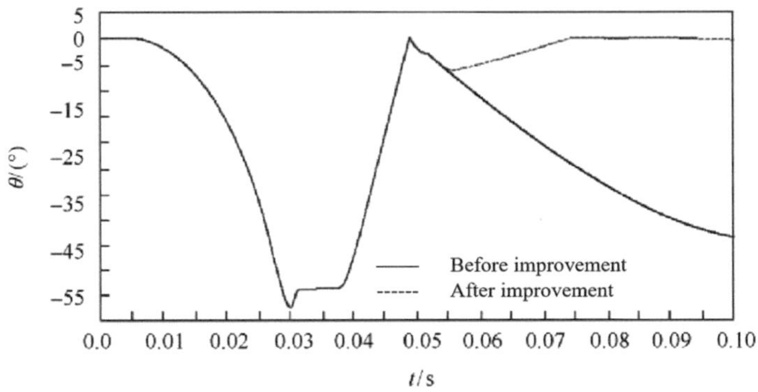

Fig. 6.22 A comparison of the spindle angular displacement curve before and after design changes, predicted by the dynamic model

6.3 Calculation and Analysis of Hydraulic Operating Mechanisms

6.3.1 Model for Hydraulic Operating Mechanisms

The very first circuit breaker in the world using a hydraulic operating mechanism was produced by Dalle in France in 1970s, but the hydraulic mechanism did not exist as an operating mechanism, instead, it was only used to transmit signals. Around the 1950s, ETNA, also located in France, began to use the hydraulic mechanism as an operating mechanism in the circuit breakers produced by Dalle. In the 1970s, Mitsubishi in Japan developed a hydraulic mechanism suitable for single-pressure SF6 arc interrupter and a series of high-voltage circuit breakers, which greatly increased the breaking current level and promoted the application of the hydraulic operating mechanisms [12]. Hydraulic technology for use in circuit breaker operating mechanisms started to be used in China also in the 1970s and significant technological development has since been made [13]. For example, XD Group and Pinggao Group developed their own hydraulic operating mechanisms and have since become leading suppliers in the Chinese domestic market. Hydraulic operating mechanisms are almost the only type of operating mechanism used in ultra-high voltage (UHV) circuit breakers.

The operation of a hydraulic operating mechanism relies on the oil stored in a high-pressure tank. Hydraulic oil is continuously pumped into the accumulator (high-pressure tank) of the mechanism by a motor to maintain a high pressure in the accumulator. Oil is used as the medium to transmit power and force. The hydraulic mechanism can provide a large amount of energy within a short time [14, 15] during the breaking and closing processes of a circuit breaker. To provide enough power, a hydraulic mechanism normally amplifies the energy flow in stages, such as using a two-stage or three-stage hydraulic valve configuration to amplify the power. UHV circuit breakers generally use a three-stage hydraulic valve design to obtain the required power to drive the contact.

The main advantages of the hydraulic operating mechanisms are: (1) it can provide large output power; (2) the action is fast; (3) the noise is relatively small; (4) the speed of shaft is controlled by the oil flow rate, which is relatively simple to implement; (5) the maintenance is simple and the service life is long. At the same time, it also has some issues that cannot be ignored, such as oil leakage. In addition, due to the complex internal flow environment of the hydraulic mechanisms, pipeline controlled by shift valve and so on is also difficult [16].

The principle of the hydraulic operating mechanism of a 550 kV GIS(Gas Insulated Switchgear) is introduced in this section and its operation is simulated and analysed based on AMESim and Ansoft [7]. The internal structure of the buffer is an important aspect that affects the operational stability and mechanical lifetime of the operating mechanism. For this, the influence of important parameters, such as the size of the buffer steps and orifice area, on the cushioning process will be analysed and an optimisation scheme proposed to improve the smoothness of the operation of the system and prolong its mechanical lifetime.

6.3.2 Modelling of Hydraulic Operating Mechanisms

1. Operation principle of the hydraulic operating mechanism of a 550 kV tank circuit breaker

The hydraulic operating mechanism includes three main parts: hydraulic control valves, accumulator and hydraulic cylinder. Each part will be explained below and the associated pressure loss discussed.

(1) Multistage Hydraulic Control Valves

The hydraulic valves form the central control of the operating mechanism. Their operational performance and reliability directly determine the performance and reliability of the operating mechanism [17]. The large component weights and high opening and closing speeds required for UHV circuit breakers mean that the operating mechanism must be able to act quickly on a heavy load. The hydraulic control valves therefore have to accommodate short action time and high oil flow rate. Multi-stage amplification is commonly adopted. The operating mechanism under study uses a three-stage spool control, including one set of pilot valves for closing and two sets of pilot valves for opening, one set of amplifying valves for closing and one set of amplifying valves for opening, and one set of two-position three-way main valve [18]. The pilot valves are each operated by an electromagnet and they in turn control the corresponding amplifying valves. It is the two amplifying valves that control the main valve to switch between its closing and opening positions. Cascading the controls, the valves work together to generate the power needed to meet the requirement for

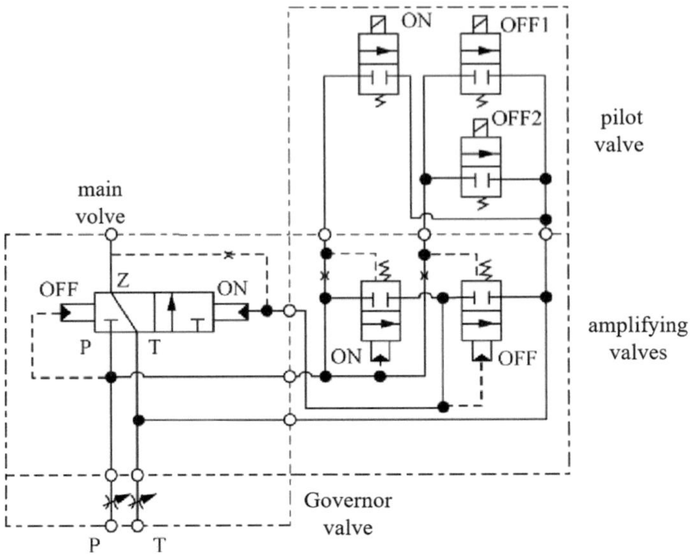

Fig. 6.23 Diagram showing the cascade structure of the hydraulic control valves used in the hydraulic operating system of the 550 kV tank circuit breaker

both response time and driving force during the closing and opening processes of the circuit breaker. The principle of the hydraulic control valves is shown in Fig. 6.23.

During the closing process, the closing electromagnet is first energised by a current from the central control, producing a magnetic force on the iron core to open the pilot valve for circuit breaker closing (stage 1 spool). The opened pilot valve clears the path for high-pressure oil present in the port of the amplifying valve (stage 2 spool) to return to the low-pressure oil tank. The sudden decrease in pressure creates a force imbalance on the spool to drive it to the open position, which clears the path for oil in the main valve to flow. With the opening of the amplifying valve, one port in the main valve is pressurised by the high-pressure oil which then drives the spool in the main valve to move to the position for circuit breaker closing. When this happens, the cylinder of the hydraulic operating mechanism is connected to the accumulator and the sudden increase in pressure in the cylinder creates a strong force and pushes the piston rod and the moving components of the circuit breaker that are mechanically connected to the rod to develop a high speed to execute the closing operation.

During the opening process, the opening electromagnet is first energised by a current from the central control, producing a magnetic force on the iron core to open the pilot valve for circuit breaker opening (stage 1 spool). The opened pilot valve clears the path for high-pressure oil present in the port of the amplifying valve (stage 2 spool) to return to the low-pressure oil tank. The sudden decrease in pressure creates a force imbalance on the spool to drive the amplifying valve to the open position, which clears the path for oil in the main valve to flow. With the opening of the amplifying valve, another port in the main valve is connected to the low-pressure oil tank and

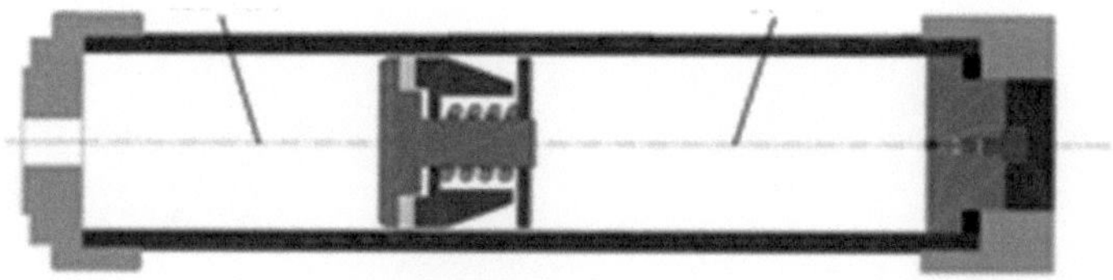

Fig. 6.24 Structure of the accumulator in a hydraulic operating mechanism where high pressure of oil is maintained

the pressure difference drives the spool in the main valve to move to the position for circuit breaker opening. When this happens, the cylinder of the hydraulic operating mechanism is connected to the low-pressure oil tank and the pressure imbalance drives the piston and rod to develop a high speed to execute the opening operation.

(2) Accumulator

The load of a UHV circuit breaker should be sufficiently high to achieve higher operating speed, but the operation frequency is low. This means the hydraulic operating mechanism needs to be able to pass a high flow rate of oil, but does not need a stable high pressure oil source [19]. The cylinder-piston type accumulator is usually adopted as it has the advantages of simple structure, easy maintenance, and the ability to provide, for a short duration, a large flow rate of oil at high pressure. The down side is that it does not absorb well the pressure fluctuation in the system. The structure of the accumulator is shown in Fig. 6.24.

(3) Operating cylinder

The operating cylinder, shown in Fig. 6.25, is a single-piston hydraulic cylinder. A step-style buffer mechanism is fixed on the end of the piston to provide a cushioning effect when the piston approaches its maximum stroke. The opening buffer has a seven-step structure while the closing buffer is five-step. The cylinder is connected to the accumulator through a connector located on the right-hand side of the cylinder. The left port of the cylinder is connected to the opening and closing ports of the main valve by a manifold located on the bottom of the cylinder. The pressure in the left port only temporarily becomes high.

During the opening process of the circuit breaker, the opening electromagnet is energised to initiate the necessary valve operations, leading to the rod-free port on the left-hand side connected to the low-pressure oil tank. The pressure in the rod-free port drops immediately and the pressure difference on both sides of the piston exerts a force to move the piston and rod towards the left direction (Fig. 6.25) at high speed to carry out the opening operation of the circuit breaker. During this process, the accumulator supplies high-pressure oil to the cylinder.

Similarly, during the closing process, the closing electromagnet is energised to activate the cascaded valve controls. As a result, the main cylinder port around the rod is connected to the accumulator and the rodless cavity is also connected to the

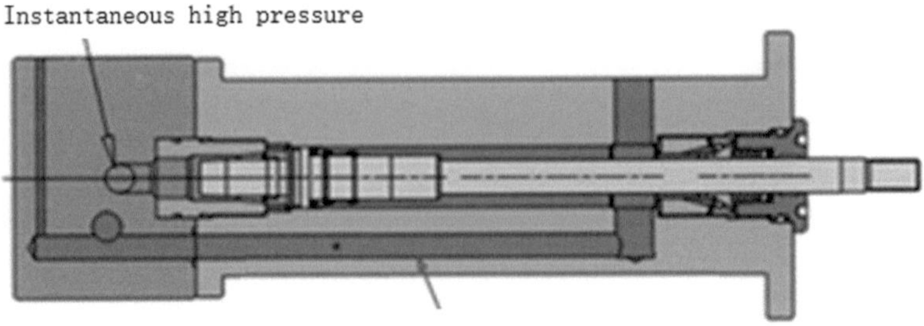

Fig. 6.25 Structure of the operating cylinder of the hydraulic operating system of a 550 kV circuit breaker

accumulator through a manifold at the bottom of the cylinder. The oil pressure in the rod-free port suddenly increases and becomes equal to that in the main cylinder port. It is this pressure rise that creates a pressure force on the piston that is larger than the force on the piston from the main cylinder due to the difference in the oil-contacting areas on both sides. The net force towards the right side in Fig. 6.25 thus pushes the piston and rod to move at high speed to complete the closing process. During the process, high-pressure oil in the main cylinder port returns to the accumulator.

(4) System pressure loss

Flow of oil in the valves, pipes and cylinder results in pressure losses which will each have a impact on the operational characteristics of the operating mechanism. Pressure losses will need to be considered when establishing the model. The pressure loss in the main valve is:

$$\Delta P_m = \frac{Q_{c0}^2 \rho}{2C_q^2 A_m^2} \tag{6.3}$$

where ΔPm is the pressure difference between the inlet and outlet of the main valve; ρ is the density of the oil; Am is the cross-sectional area of the oil inlet and outlet; and Cq is a coefficient associated with the oil flow rate; Q_{c0} is buffer chamber flow. The pressure loss between the accumulator and the operating cylinder is:

$$\Delta P_p = \lambda \frac{l}{d} \frac{\rho v^2}{2} + \sum \xi \frac{\rho v^2}{2} \tag{6.4}$$

where λ is the pressure loss coefficient along the pipes; v is the speed of the oil in the pipe, l is the length of the pipe; ξ represents a local pressure loss coefficient associated with the pipes; and d is the inner diameter of the pipes.

2. Simulation Method Of Hydraulic Operating Mechanism

Simulation of the hydraulic operating mechanism requires a complex model that considers fluid flow, electromagnetic field and forces, and mechanical dynamics. It

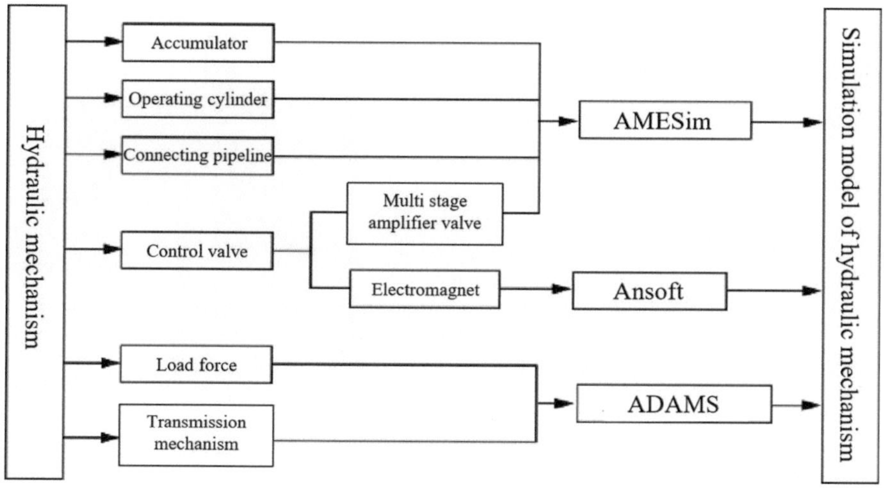

Fig. 6.26 Diagram illustrating the constituent models and overall structure of the simulation model for the operation of the hydraulic system

needs to be most probably carried out through co-simulation of the different processes using individual models implemented in different software packages. For the operating system under study, the model to simulate the fluid system, including the accumulator, operating cylinder, pipelines and multi-stage valve system is implemented in AMESim; the electromagnetism model is implemented in Ansoft; and the model to simulate the mechanical load and transmission mechanism is implemented in ADAMS. Outputs from Ansoft and ADAMS are fed into AMESim to establish a complete model for the operation of the hydraulic mechanism. The construction of the simulation model is illustrated in Fig. 6.26.

(1) Simulation model of multi-stage hydraulic control valves

The hydraulic control valves consist of the electromagnet-operated pilot valve, the amplifying valve and the main valve. Simulations based on Ansoft are carried out to obtain the relationships between the magnetic force, the current and the air gap. Data defining the relationships are then imported into AMESim software to represent the operation of the electromagnet.

The electromagnet is a solenoid type and its iron core, driven by a magnetic force, will move the spool of the pilot valve to open its port. The structure of the electromagnet is shown in Fig. 6.27. It has an axisymmetric structure and a model for only half of its structure is needed. The simulation domain for the electromagnet and the computed magnetic lines of force are shown in Fig. 6.28.

The electromagnetic attraction on the ion core is induced by the coil and it is the driving force in the pilot valve. The static attraction is often taken as the basic characteristic of the electromagnet. It is related to the air gap and coil current. Figure 6.29 displays the relationship between the static electromagnetic attraction and the air

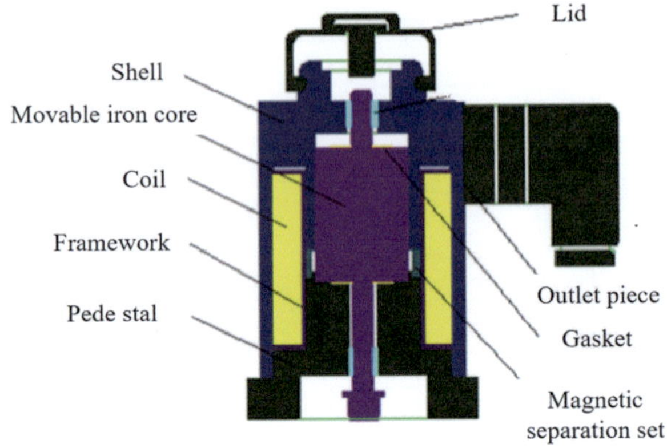

Fig. 6.27 Structure of the electromagnet in the pilot valve

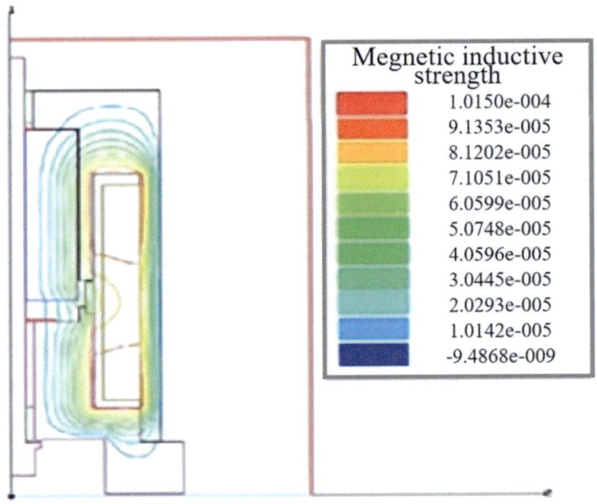

Fig. 6.28 Simulation domain and magnetic lines of force line

gap of the electromagnet at the rated coil current. The air gap refers to the distance between the movable iron core and the yoke. By changing the air gap length and current within a certain range, a three-dimensional data space of the electromagnetic attraction can be obtained, as shown in Fig. 6.30. The obtained three-dimensional data set will be used as input for the simulation of the operation of the pilot valve based on AMESim.

Based on the configuration of the valve control system, four sub-models are needed for the components, including the electromagnet, pilot valve, magnifying valve and the main valve. The basic structure of hydraulic valves is shown in Fig. 6.31. The four sub-models created in AMESim are displayed in Fig. 6.32. It is worth mentioning that a simulation model can be set up in AMESim using a graphical user interface (GUI) and also displayed in an intuitive manner based on standard ISO icons and

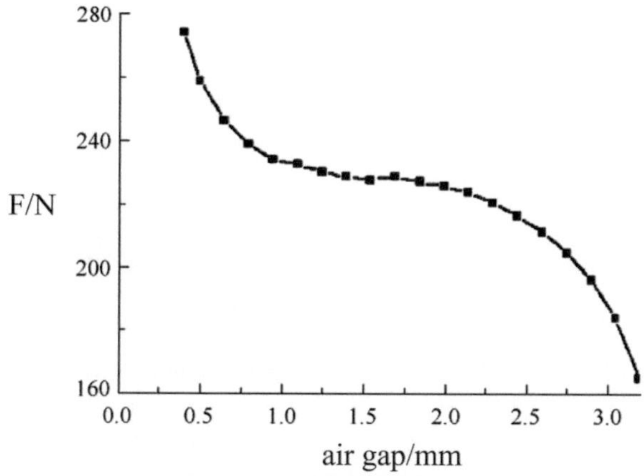

Fig. 6.29 Relationship between the static electromagnetic attraction and the air gap size of the electromagnet used in the hydraulic operating mechanism under study

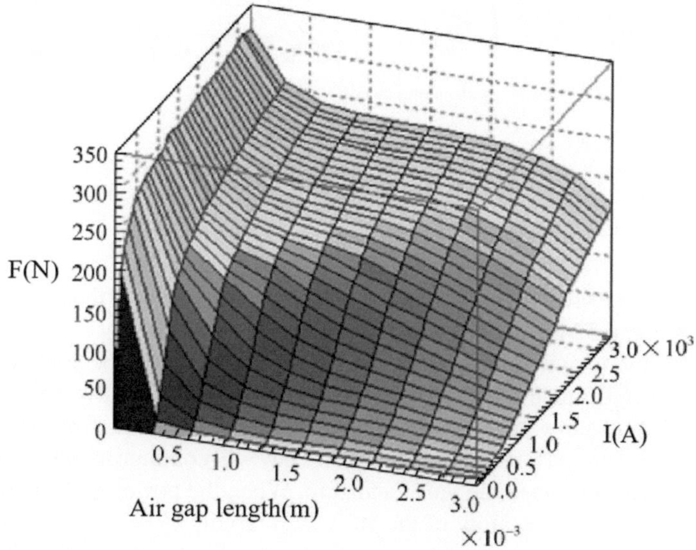

Fig. 6.30 Three-dimensional visualisation of the computed dependence of the static electromagnetic attraction on the air gap length and coil current in the pilot valve under study

multi-port block diagrams, thus offering great convenience to users in setting up the simulation model for a complex system.

An integrated electromagnet module, EMLT40, is available in AMESim to enable the coupling of magnetic circuits, mechanical and hydraulic systems. The three-dimensional data set for the electromagnetic attraction obtained from Ansoft is

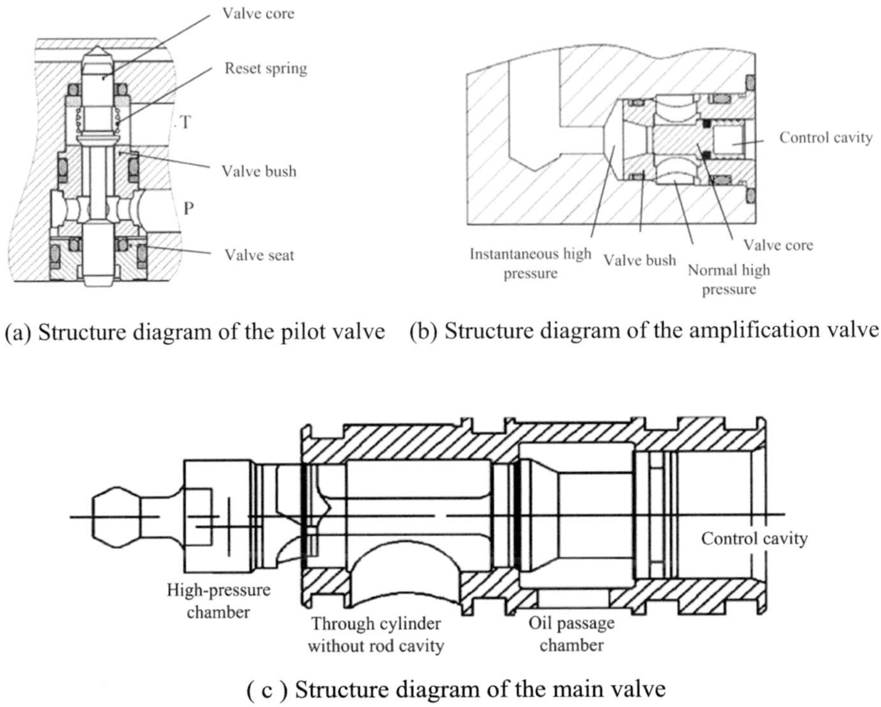

Fig. 6.31 Structure of the hydraulic control valves used in the hydraulic operating system under study

imported into the EMLT40 module to define the characteristics of the electromagnet, as shown in Fig. 6.32a.

Driven by the magnetic force induced by the current coil, the iron core of the electromagnet will move together with the spool of the pilot valve for 1.5 mm to open the valve and allow the high-pressure oil at the bottom of the spool to discharge into the low-pressure oil tank through the port. They will continue to move for another 1.5 mm to reach fully open the valve and stop. When the power to the coil is cut, a reset spring fixed on the bottom of the valve will rapidly close the valve due to the diminishment of the magnetic force. The pilot valve is a typical two-position, two-way cone valve. Based on the types of the valve port and spool movement pattern, a model for the pilot valve can be built in the HCD (hydraulic component design) module of AMESim, as shown in Fig. 6.32b.

The magnifying valve is a two-position, two-way hydraulic control valve, and its port is connected to the pilot valve. While the pilot valve is opened to the low-pressure tank, the oil in the magnifying valve originally at high pressure is depressurised and flows towards the tank. The pressure difference on both sides of the spool drives it to move quickly to open the amplifying valve. A model for the amplifying valve can

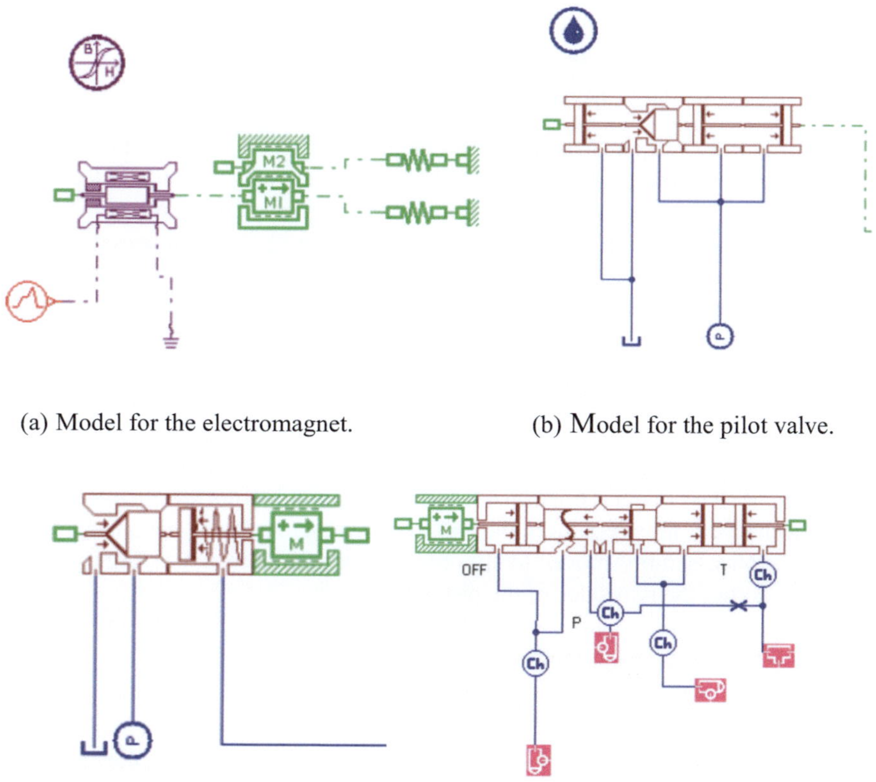

(a) Model for the electromagnet. (b) Model for the pilot valve.

(c) Model for the magnifying valve. (d) Model for the main valve.

Fig. 6.32 Sub-models created by AMESim to simulate the operation of the hydraulic control valve system in the hydraulic operating system in the 550 kV circuit breaker

also be built in the HCD module in a way similar to the pilot valve, as shown in Fig. 6.32c.

The main valve is a two-position three-way directional valve. The port of the valve is controlled by a cone type valve with the spool sliding to open or close in a sealed space. The oil pressure in the oil control chamber is regulated by the respective opening or closing magnifying valve. When the main valve closes, the oil in the control chamber is pressurised (through the amplifying valve) and the spool moves to the closing position under a force difference. The closing valve port is then opened while the opening valve port is closed. When the main valve opens, the oil pressure in the control chamber of the main valve is depressurised through the amplifying valve and the spool moves to the opening position under pressure difference. The opening valve port is then opened while the closing valve port is closed. The model of the main valve is shown in Fig. 6.32d that is built in a way similar to those of the pilot valve and the magnifying valve.

(2) Simulation model of the accumulator and operating cylinder

As shown in Fig. 6.25, the accumulator is the high-pressure oil source of the whole hydraulic mechanism. It is of a piston-cylinder structure with one side of the piston filled with nitrogen and the other side filled with oil. For the hydraulic operating mechanism under consideration in this section, the pressure of the compressed nitrogen in the accumulator is 18 MPa.

The operating cylinder, as shown in Fig. 6.25, serves as the actuator of the hydraulic operating mechanism. It is a single-piston hydraulic cylinder with the main rod-containing chamber connected to the accumulator via pipe and the other rod-free chamber connected to the main control valve where a temporary high pressure can be created. In the closing operation of the mechanism, the oil in the rod-free chamber suddenly changes to high pressure and the piston-rod assembly moves to the closing position under the action of a hydraulic force. During the opening process of the mechanism, the pressure in the rod-free chamber suddenly becomes low and the piston-rod assembly moves to the opening position under the large pressure difference across the two sides of the piston.

Both the accumulator and the operating cylinder are simulated by models set up in the hydraulic component module in AMESim (Fig. 6.33).

(3) Simulation of transmission mechanism and load force

For the 550 kV tank circuit breaker used as an example in this section, the piston-rod assembly of the operating mechanism does not directly drive the moving contact in the arcing chamber. It does so through a transmission mechanism. The driving force and it direction on the moving contact is transmitted from the piston through connecting rods, bell cranks with fixed and moving pivots, and levers, as shown in Fig. 6.34. It can be seen that there is a clear difference in the motion characteristics between the piston-rod assembly in the operating cylinder and the insulating rod. The displacement and speed of the insulating rod represent the output characteristics of the whole hydraulic

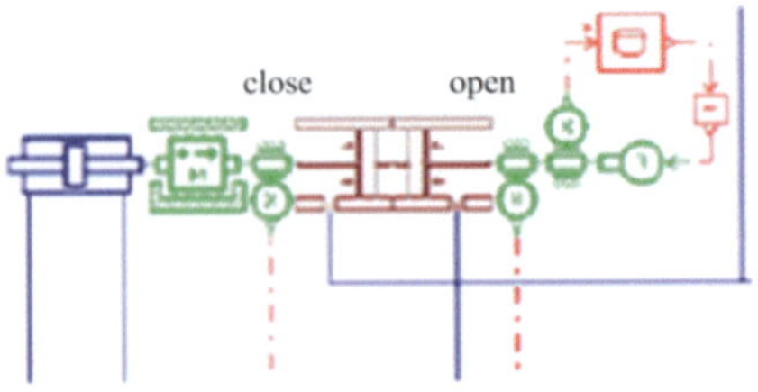

(a) Model for the accumulator.

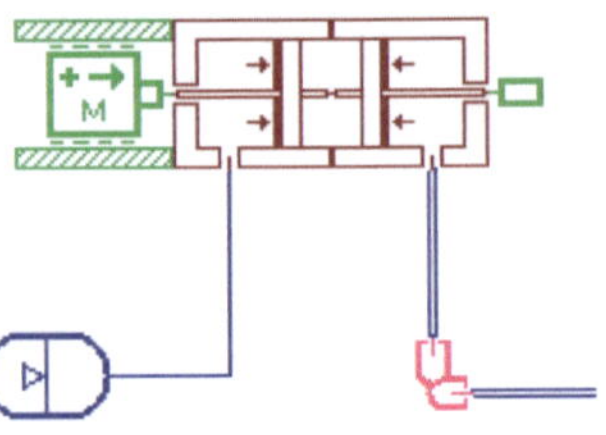

(b) Model for the operating cylinder.

Fig. 6.33 Simulation model for the accumulator and operating cylinder, generated in the HCD hydraulic component module in AMESim

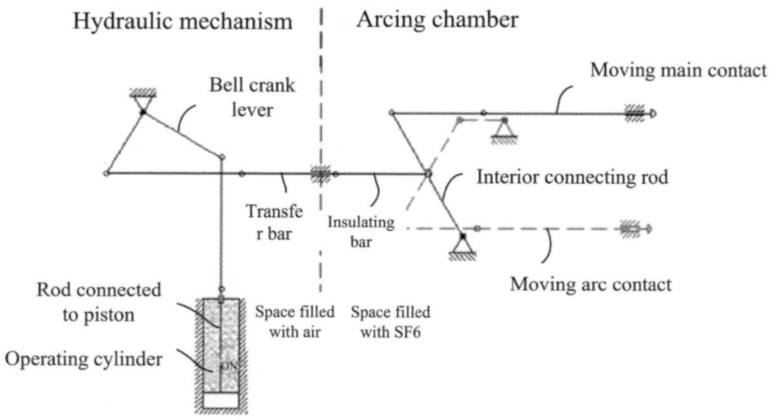

Fig. 6.34 Schematic diagram of the mechanical transmission system of the 550 kV tank circuit breaker under study

operating mechanism. In the actual design, this output characteristics needs to match the motion characteristics of the load (the moving contact in the arcing chamber), which in turn determines the motion characteristics of the piston-rod assembly.

The different motion characteristics of the piston-rod assembly and the insulating rod can be quantitatively obtained through multi-body dynamics simulation based on ADAMS or other code without much complexity. Defining the displacement ratio of the insulating rod to that of the piston as the output displacement amplification ratio, and the velocity ratio of the two as the output speed amplification ratio, the relationship is shown in Fig. 6.35. Inputting the characteristic curves into the dynamic simulation model (implemented in AMESim), the motion characteristics of the whole hydraulic mechanism is then obtained.

Apart from the requirement on the kinematic characteristics, the dynamic characteristics of the load is also important. The load force is often difficult to measure by experiment, it is thus necessary to obtain the load force curve generated in the arcing chamber during the opening and closing processes based on simulation using ADAMS. The reaction force curve consisting of load force is then loaded into the AMESim model.

To sum up, the hydraulic operating mechanism can be simulated and analysed by using a dynamic model, such as that implemented in AMESim. The calculation of the Lorentz force of the electromagnet and the kinematic and dynamic characteristics of the load is still not perfect within the model in AMESim, so it is often necessary to use electromagnetic simulation software such as Ansys, Ansoft and multi-body dynamics simulation software such as ADAMS for auxiliary calculation, and the results of calculations are taken as the input parameters of the AMESim model to perform the systematic simulation of the hydraulic operating mechanism. The model also needs to be calibrated and corrected by experiments. It is commonly compared with the stroke curve measured in experiment and the model parameters are adjusted

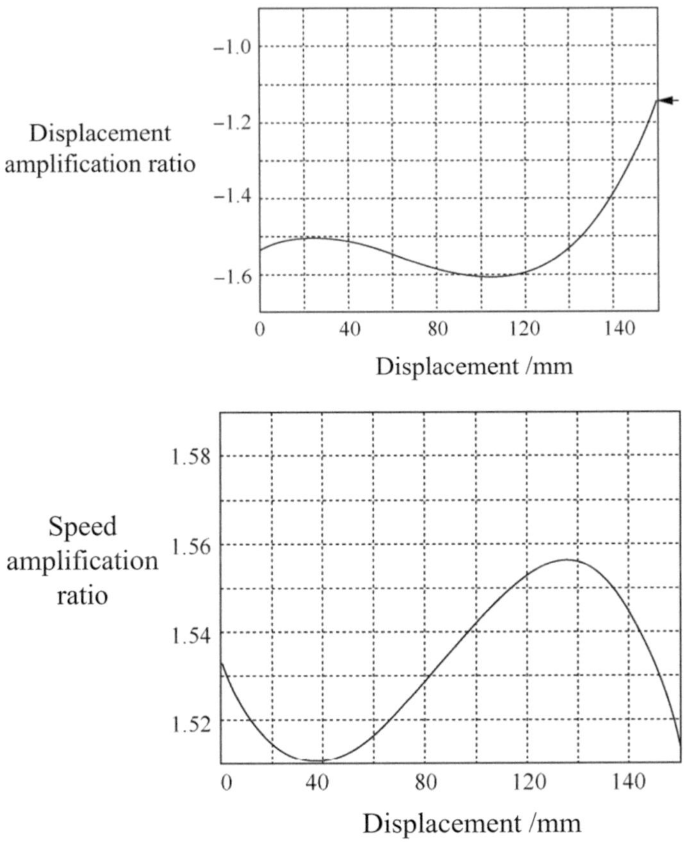

Fig. 6.35 Amplification ratios for displacement and speed of the insulation rod over the piston in the operating mechanism of the 550 kV circuit breaker under study

to make the simulation results close to reality. The corrected model can then be used for optimal design and fault diagnosis of the hydraulic operating mechanism.

6.3.3 Optimal Design of Buffer Structure in Hydraulic Operating Mechanisms

The hydraulic operating mechanism of the 550 kV tank circuit breaker features a large operating power and high speed and needs an appropriately designed buffer to reduce the impact on the mechanism during braking of the movement to improve the mechanical life and stability of the mechanism. For the opening operation, the buffering effect depends on the speed of the contact during the sudden deceleration of the movement. The smaller the speed is, the lower the impact force is, and the better the buffering effect is. In addition, the buffering effect also needs to consider the influence of the buffering force on the internal components of the hydraulic operating cylinder. A lower peak pressure in the buffer chamber imposes a smaller impact on the internal components of the operating cylinders, resulting in a better buffering effect. Therefore, the opening speed and maximum pressure in the hydraulic buffer chamber are taken as the evaluation indicators of the cushioning effect.

Hydraulic buffers are divided into two types: pressure relief type and throttle type. The hydraulic operating mechanism discussed in this section uses a throttle buffer with the advantages of simple structure, low cost, stable performance, small size and good cushioning effect [20]. The working principle of the throttle buffer is as follows: a cushioning structure is fitted inside the cylinder and it quickly reduces the oil discharge cross-sectional area from the respective chamber (buffer chamber) when the movement of the piston requires braking. During this process, the oil pressure in that buffer chamber rises quickly due to increased compression of oil and this results in a counter force to decelerate the movement of the piston.

The hydraulic operating mechanism discussed in this section adopts a multi-step throttle buffer structure, where there are seven steps in the opening buffer structure and five steps in the closing buffer structure. The annular steps of different diameters are attached to the piston. Its structure is shown in Fig. 6.36. It is found that the main parameters affecting the cushioning effect include the gap width between the moving steps and the buffer sleeve, andthrottle hole area in the opening plunger. In view of the opening process, the AMESim simulation model is used to optimise the design of these two parameters in order to improve the cushioning performance of the hydraulic operating mechanism.

To further optimise the design, the seven steps seen in Fig. 6.36 are also examined. They have initial values as shown in Table 6.2. According to the sequence in which the steps enter the buffer sleeve in the opening process, they are respectively expressed as g1–g7. The flow area of the throttle hole in the opening plunger is 20 cm^2, which is expressed in Table 6.2 as *s*.

For this case, curves showing the variation of the opening speed and the pressure in the open buffer chamber with displacement are shown in Fig. 6.37. The termination speed of the opening process is 2.3 m/s, and the peak pressure in the open buffer chamber is about 800 bar. These two parameters are on the high side. The gap width (g1-g7) between the buffer step and the buffer sleeve and the flow area of the long

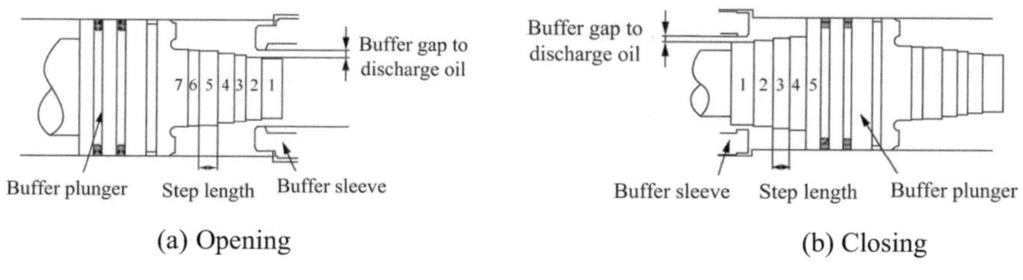

Fig. 6.36 Schematic diagram of the buffer structure in the process of opening and closing

Table 6.2 Initial values used in design optimisation of the gaps between the buffe step and buffer sleeve, and flow area of the throttle hole. g1 means gap 1 and s means area of the throttle hole

g1 (mm)	g2 (mm)	g3 (mm)	g4 (mm)	g5 (mm)	g6 (mm)	g7 (mm)	s (mm^2)
0.5	0.45	0.4	0.3	0.225	0.15	0.09	20

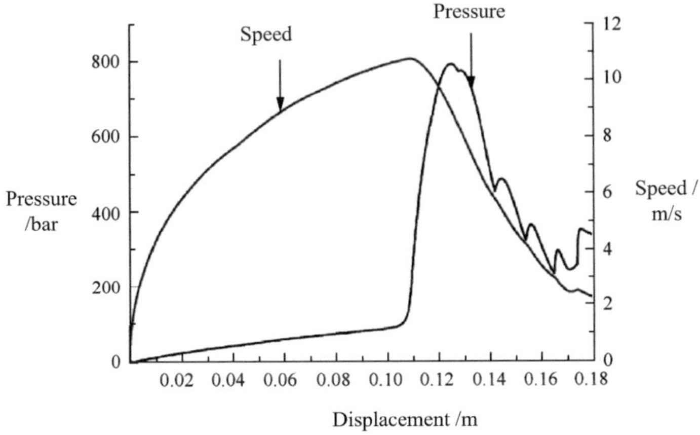

Fig. 6.37 The variation curve of velocity and internal pressure of buffer cavity with displacement in the process of opening

throttle hole is altered within a reasonable range, and a multi-parameter optimisation process is carried out by a simulation model established in AMESim, and the best effect is shown in Fig. 6.38.

As can be seen from Fig. 6.38, the termination speed in the opening process is reduced by 0.15 m/s after optimisation, and the peak pressure inside the buffer chamber is reduced by 120 bar. The main parameters before and after optimisation are shown in Table 6.3.

6.4 Calculation and Analysis of Permanent Magnet Operating Mechanisms

The first model circuit breaker using a permanent magnet mechanism was built at the University of Manchester in the late 1980s. Since then, this technology has been widely used in medium- and low-voltage circuit breakers. The permanent magnet operating mechanisms take the magnetic force produced by a current coil as the driving force and the force produced by the permanent magnet as the holding force, which avoids the complex tripping and locking systems used by the spring mechanism. They have the following advantages:

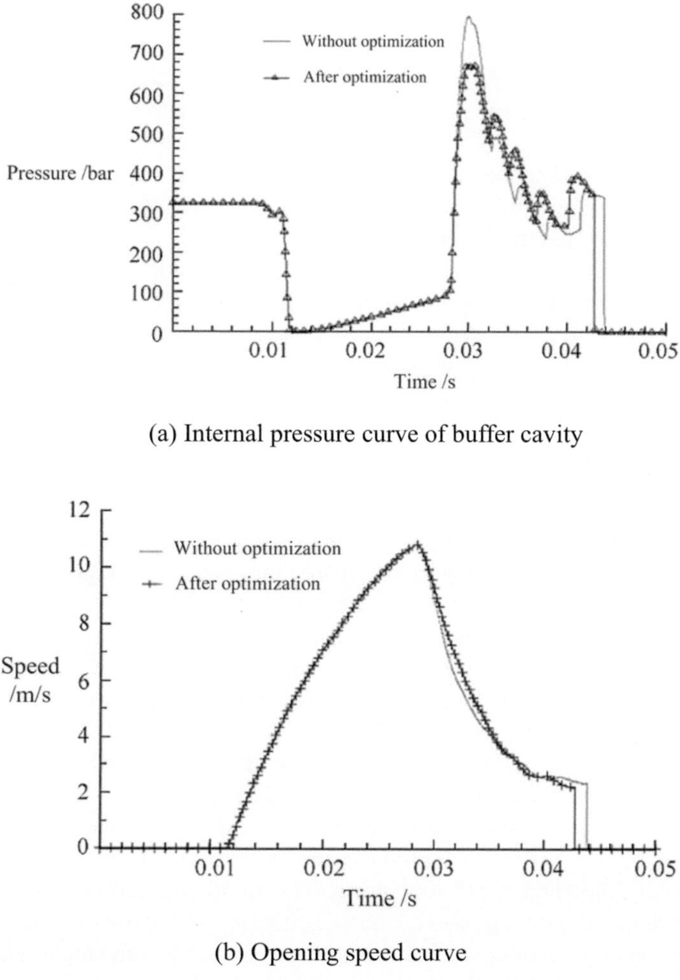

Fig. 6.38 Comparison of the cushioning effects in the operating mechanism before and after design optimisation

Table 6.3 Comparison of main parameters before and after design optimisation of the buffer mechanism

	g1 (mm)	g2 (mm)	g3 (mm)	g4 (mm)	g5 (mm)	g6 (mm)	g7 (mm)	s (cm^2)
Before optimisation	0.5	0.45	0.4	0.3	0.225	0.15	0.09	20
After optimisation	0.56	0.49	0.43	0.3	0.2	0.11	0.05	25

- Fewer moving components. The number is about 30–40% of that of the spring mechanism [21], and there is no need for mechanical tripping and locking devices. Generally speaking, the failure rate of circuit breakers increases with the number of components, and fewer potential sources of fault help improve the reliability [22].
- Long service life. The mechanical life of the permanent magnet operating mechanism is more than 100,000 operations which is obviously higher than that of the ordinary spring operating mechanism.
- Accurate opening and closing control. The phase angle of the AC current for opening and closing can be controlled accurately, for example, opening at the zero crossing of the AC current or closing near the zero crossing of the voltage. This can effectively reduce the overvoltage or inrush current to improve the safety of the power systems and the lifetime of the equipment [23]. The scattering in the opening and closing times of the permanent magnet operating mechanisms is small, which allows the possible realisation of phase-controlled open and close of circuit breakers.

The permanent magnet mechanisms also have limitations. For example, the control circuit is vulnerable to electromagnetic interference, leading to logic errors; the current coil is prone to overheating and burn-out.

There are mainly three types of permanent magnet operating systems: double coil permanent magnet mechanism, single coil permanent magnet mechanism, and multi-magnetic circuit permanent magnet mechanism. The double coil permanent magnet mechanism uses two sets of coils to control the opening and closing operations separately, such as that adopted by the VM1 circuit breaker (ABB). The single-coil permanent magnet mechanism uses the same coil to control the opening and closing operations, commonly used in conjunction with a spring mechanism. The closing power is provided by the magnetic force of the excitation coil and the opening power is provided by the force from the spring. The multi-magnetic circuit permanent magnet mechanism normally uses a single excitation coil but uses separate magnetic circuits for closing, opening and state-holding. For clarity, the single- and double-coil permanent magnet operating mechanisms both use the permanent magnet to hold the open and close positions while in the multi-magnetic circuit permanent magnet mechanisms, the permanent magnet is only used to hold the close position.

In this section, ADAMS and Ansoft are used to implement the models and analyse the switching operations of a medium-voltage permanent magnet mechanism for fault diagnosis. ANSYS (finite element solver), MATLAB and ADAMS (mechanic dynamics) are used to perform the simulation and analysis. Fault identification methods are proposed for different fault types occurring in the operating mechanisms of modern medium-voltage switchgear.

6.4.1 Modelling and Fault Diagnosis of Permanent Magnet-Based Operating Mechanisms in Medium-Voltage Switchgear

The permanent magnet operating mechanism of a 12 kV vacuum circuit breaker is chosen in this section. Its opening and closing coils share the same iron core. The closing operation is driven by the magnetic force generated by the closing coil, which at the same time, also energises the opening spring. The opening operation is co-driven by the force from the opening spring and the magnetic force from the opening coil.

In this section, the magnetic force is simulated by Ansoft and the multi-body dynamics model for the whole mechanism is implemented in ADAMS. Results from magnetic force calculation are coupled to the multi-body dynamics model. Opening and closing stroke curves of the moving contact are obtained through simulation of the normal and typical fault conditions, which provides baseline data for fault diagnosis.

The multi-body dynamics model of the whole operating mechanism is shown in Fig. 6.39, considering the permanent magnet, moving iron core, spindle, four-bar linkage, and opening spring as the main components. A number of stationary components, such as the excitation coil surround the moving core for opening and closing. This mechanism is similar to that shown in Fig. 6.3. In fact, the permanent magnet operating mechanism described in this section is an improved version the VS1 spring operating mechanism. The main difference is that the closing spring and its auxiliary components are replaced by a permanent magnet and excitation coil, meaning that there are fewer components in the mechanism which is easier to control. In this section, the full details of the mechanism and its multi-body dynamics model will not be repeated and the focus will be on the simulation of the magnetic force and its coupling with the multi-body dynamics model.

1. Simulation and Analysis of Magnetic Force and its Coupling with Multi-Body dynamics Model

A cross-sectional view of the permanent magnet and excitation coil is shown in Fig. 6.40a where a stationary iron core, a magnetic short circuit ring, a moving iron core and a lower magnetic yoke form a closed magnetic circuit. The opening and closing coils and the permanent magnet are surrounded by the magnetic circuit. The magnetic field, as shown in Fig. 6.40b for a typical distribution, can be simulated and analysed using Ansoft to obtain the magnetic force acted on the moving iron core which is then taken as the input parameter for the multi-body dynamics model.

The magnetic force induced acting on the moving iron core can be obtained through the magnetic calculation model with the current in the opening and closing coils and the distance between the moving and stationary iron cores as the input parameters. The multi-dimensional array of magnetic force obtained for different coil currents and the position of the moving iron core are then imported into the multi-body dynamics model implemented in ADAMS, and the parameter surface

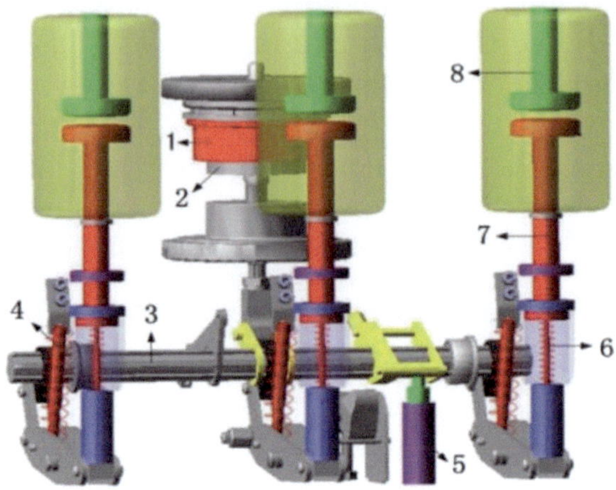

1-permanent magnet; 2-moving core; 3-spindle; 4-opening spring; 5-oil buffer;
6-contact spring; 7-moving contact; 8-stationary contact

Fig. 6.39 Components included in the simulation model of the permanent magnet operating mechanism used in the 12 kV vacuum circuit breaker

can be obtained with the electromagnetic force as the dependent variable and the distance between the opening and closing current and the dynamic and static core as the independent variables. This means a three-dimensional surface plot can be built up where the magnetic force is the dependent variable and the coil current and the gap length between the stationary and moving iron cores are independent variables. Furthermore, the measured current waveforms for the opening and closing operations can be curve-fitted using an interpolation function such as the AKISPL and input into ADAMS. This allows accurate force changes during the opening and closing processes be loaded into the multi-body dynamics model.

To verify the simulation model, a linear displacement sensor can be installed on the insulated rod to obtain the stroke curves in the opening and closing processes. The experimental data for the stroke curves are compared with the simulation results, as shown in Fig. 6.41, clearly showing that the simulation closely follows the experimental results with a maximum error of 5%. The accuracy of the model is thus verified.

2. Fault Diagnosis of Permanent Magnet Operating Mechanism Based on Multi-Body Dynamics Simulation

During field operation, the common fault types of permanent magnet-based vacuum circuit breakers include faulted coil circuit of the permanent magnet mechanism, opening spring fault, and oil buffer failure. In this section, the coil current waveform under fault conditions and the angular displacement stroke curve of the spindle are simulated for the opening and closing processes, and the results are compared

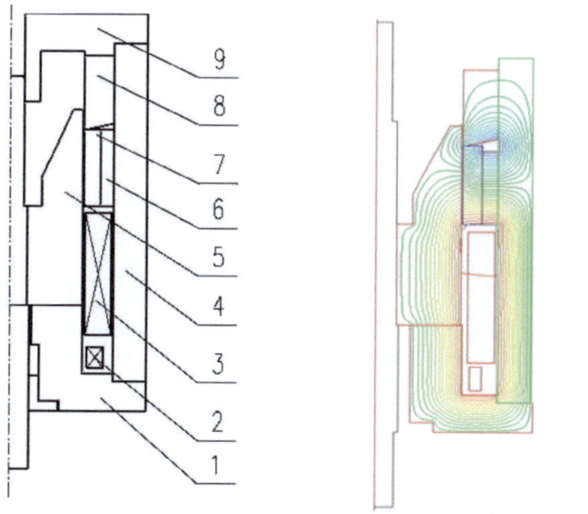

(a) Sectional view. (b) Representative magnetic field distribution.

Fig. 6.40 A cross-sectional view of the structure of the permanent magnet and excitation coil and the magnetic field distribution in close position. 1-lower yoke; 2-opening coil; 3-closing coil; 4-stationary core; 5-moving core; 6-permanent magnet; 7-internal yoke; 8-magnetic short circuit ring; 9-non-conductive end cover

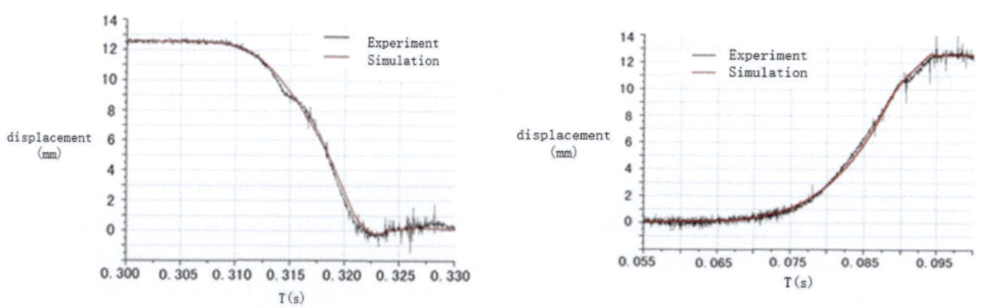

(a) Opening. (b) Closing.

Fig. 6.41 Comparison of the simulation results against measured displacement of the moving contact during the opening and closing operations of the vacuum circuit breaker

with those under normal conditions to provide key information for mechanical fault diagnosis.

A coil circuit fault of the permanent magnet mechanism can be caused by many factors, including voltage fluctuation, poor electrical connection in the circuit, and short-circuits between the coil turns [24]. The direct symptom is always with the current, in that the current is either too small or too large, and the duration of the current is too short or too long. Too high a coil current or too long a duration of the

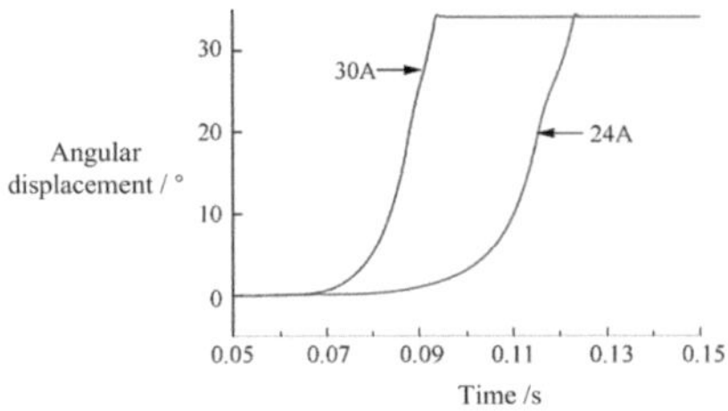

(a) Closing.

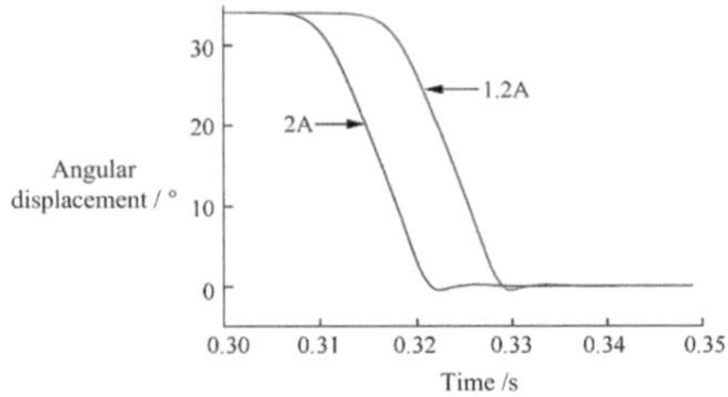

(b) Opening.

Fig. 6.42 A comparison of the angular displacement stroke curves at normal and threshold current of the coils in the permanent magnet operating mechanism of the vacuum circuit breaker

current can lead to coil burnout, so circuit breaker products often come with an upper limit of the coil current. The current limit is obtained from the test and can be used for reliable detection of the coil fault. On the other hand, when the coil current is too small or the duration of the current is too short, there is a general lack of quantitative indicators and threshold values to judge the fault. In such a situation, the operating mechanism in reality may not be able to complete the opening or closing operation. However, simulation can be used to obtain the lower limits for the coil current and duration of current corresponding to the opening and closing processes, and the low threshold values can then be used for identification of the coil circuit fault.

For the permanent magnet mechanism circuit breaker described in this section, the standard current of the closing coil is 30 A, and the standard current of the opening coil is 2 A. The simulation results show that the circuit breaker cannot complete the closing operation when the closing coil current is lower than 80% of its standard value (i.e. 24 A), and the circuit breaker cannot complete the opening operation when the opening coil current is lower than 60% of its standard value (i.e. 1.2 A). Considering the variation of the frictional coefficient of the interrupter in practical applications (such as those arising from the differences between individual products and the increase in friction after prolonged non-action of the circuit breaker), the lower threshold of the opening and closing coil current should be set higher to retain sufficient margin. A comparison of the angular displacement stroke curves between the standard current and the threshold current is shown in Fig. 6.42.

The driving force in the closing process is mainly the moving core magnetic force, while the driving force in the opening process includes the forces from the moving core and the opening spring. Simulation results show that the maximum magnetic force (Lorentz force) is 4700 N and the maximum force from the opening spring is 3000 N. With the opening spring, the lowest current of the opening coil is higher than that of the closing coil.

Fault in the opening spring is also a common fault type, which mainly manifests as the change in the spring stiffness coefficient and the shedding of the single-phase spring. The standard value of the stiffness coefficient of the opening spring is $K = 34$ N/mm. A simulation shows that when the stiffness coefficient increases by more than 20% ($K = 40.8$ N/mm), the mechanism cannot be reliably closed, while if the stiffness

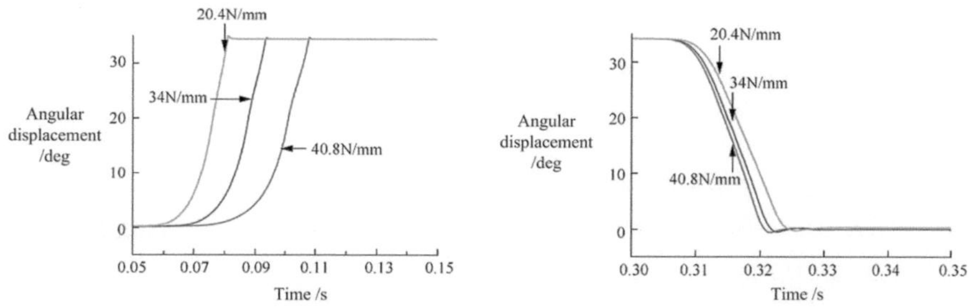

(a) Closing. (b) Opening.

Fig. 6.43 The influence of the spring stiffness coefficient on the angular displacement stroke curve of the spindle of the operating mechanism

coefficient is reduced by more than 40% (K = 20.4 N/mm) the mechanism cannot open reliably. A comparison of the angular displacement curves of the closing and opening spindles are shown in Fig. 6.43 where the stiffness coefficient of the opening spring is set to 20.4 N/mm, 34 N/mm and 40.8 N/mm. It is clear that the angular displacement stroke curve is sensitive to the change in the stiffness coefficient.

The shedding of a single-phase opening spring reduces the driving force for opening and the retarding force for closing. The influence on the stroke curve is similar to a decrease in the stiffness coefficient of the opening spring.

Severe oil leakage from the oil buffer can lead to weakened cushioning effect, resulting in significantly increased bouncing of the contact that is classified as a failure of the oil buffer. Figure 6.44 shows a comparison of the angular angle displacement stroke curves between an oil leakage failure and the normal operation where the bouncing in the opening process is clearly seen.

It can be seen from Figs. 6.42, 6.43 and 6.44 that the spindle angular displacement stroke curves are obviously different between normal and faulted conditions, and they are also different for different types of fault. Similar to Sect. 2.1, the angular displacement stroke curves can be parameterised and an operational state can be characterised by a unique set of parameters, which enables an intelligent online

Fig. 6.44 A comparison of the spindle angular displacement stroke curves with severe oil leakage and normal operation

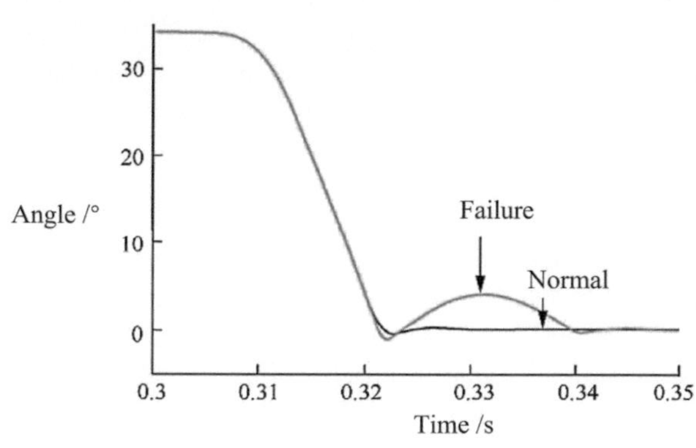

monitoring system to reliably differentiate a fault state from the normal state to realise the identification of typical types of fault.

6.4.2 *Modelling and Optimal Design of Permanent Magnet-Based Operating Mechanism in Low-Voltage Switchgear*

A contactor of unclassified type, equipped with a permanent magnet operating mechanism, is to be analysed through simulation in this section on its performance to optimise its functional design [25]. Compared with the traditional magnetic mechanism-based contactor, the contactor under study has a permanent magnet placed at the central position of the bottom end of the stationary iron core, thus its performance, features and mode of control are all altered.

When the contactor is in the close position, the magnetic force produced by the permanent magnet is greater than the counter force from the spring and the close state can be maintained by the magnet without energising the coil. It offers some advantages. First, it is energy-saving, particularly suitable for applications with low frequency of operation. Second, in its closed state, no noise is produced, so it is suitable for applications where a low noise level is required.

However, this type of permanent magnet mechanism contactor also has some shortcomings. On the one hand, a contactor must offer voltage drop protection as forced in the standard, that is, it needs to open automatically when the system voltage is lower than a threshold as specified in the standard. Therefore, the permanent magnet mechanism contactor should be equipped with a voltage drop protection device. However, the voltage drop protection device is powered by a capacitor which can discharge its stored energy in case of prolonged power loss. This means that, when power loss and capacitor discharge lasts for a considerable length of time, the contactor may not automatically open if it is forced to close by human action or for any other reason. On the other hand, a permanent magnet contactor either uses a spring or a permanent magnet to open, both leading to certain issues. When a spring is used to drive the contactor to open, it offers poor controllability and the opening process cannot be regulated in real time. When magnetic suction is used to open the contactor, the force at the initial stage of the opening process is smaller than expected and larger towards the end of the process. It does not match the ideal characteristics of the opening process.

A number of aspects are covered in this section. First, the magnetic field of the permanent magnet contactor is calculated in ANSYS and the characteristics of the static magnetic force of the operating mechanism is obtained to develop the criteria that can be used to judge whether the moving and stationary contacts can be reliably held together or expelled apart reliably. Second, the governing differential equations describing the dynamic process of the contactor are solved in MATLAB by coupling the calculation in ANSYS into the solution procedure to obtain the relationship

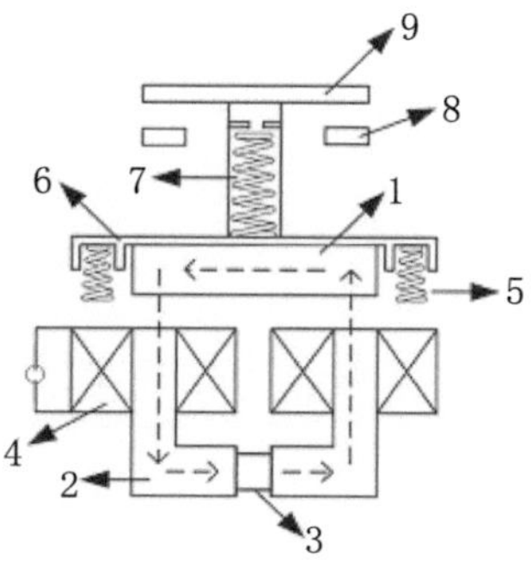

Fig. 6.45 Structure of the contactor equipped with a single coil permanent magnet operating mechanism with monostable state. 1-moving iron core; 2-stationary iron core; 3-permanent magnet; 4-coil; 5-counter spring; 6-plastic shell; 7-contact spring; 8-stationary contact; 9-moving contact

between the force and the displacement of the moving contact of the operating mechanism. Third, the data describing this relationship are then imported into ADAMS to supplement the multi-body dynamics model for the contactor. By simulating the dynamic process and analysing the solutions, the motion characteristics of the iron core and contact can be obtained together with their changes under the influence of different physical parameters. Finally, a parallel magnetic circuit based permanent magnet operating mechanism is proposed and optimised to improve the voltage drop protection performance and the contact separating speed of the contactor.

1. Operating Principle and Calculation of Static Magnetic Force in Permanent Magnet Mechanism Contactors

The main structure of the contactor is shown in Fig. 6.45 [26]. It is a modification of a magnetic mechanism-based contactor employing a single coil. The main improvement is the placement of a permanent magnet in the central hollow-out of the bottom end of the stationary iron core. The contactor uses the magnetic circuit and coil for both opening and closing. The operating mechanism has a simple structure and the contactor can be held in the close state by the permanent magnet. It is thus a single coil permanent magnet operating mechanism with a mono-stable state.

Figure 6.46 shows the operating principle of the permanent magnet mechanism contactor which is also briefly described below:

When the contactor is in the open position (Fig. 6.46a), the gap length between the moving and stationary iron cores reaches the maximum and the permanent magnetic suction on the moving iron core is the smallest. There is no current in the coil and no magnetic suction from the coil is produced. The force from the spring is greater than the magnetic suction, holding the contactor in its open state.

When the contactor receives the closing signal (Fig. 6.46b), a DC current flows in the designated positive direction of the coil and induces a magnetic field in the same direction as that of the permanent magnet. The enhancement in the magnetic field increases the magnetic suction between the moving and stationary iron cores. The downward electromagnetic attraction is greater than the counter force from the spring and the moving iron core begins to move downward until its hold position. In the closing process, due to the continuous reduction in the air gap length between the moving and stationary iron cores, the electromagnetic attraction on the moving iron

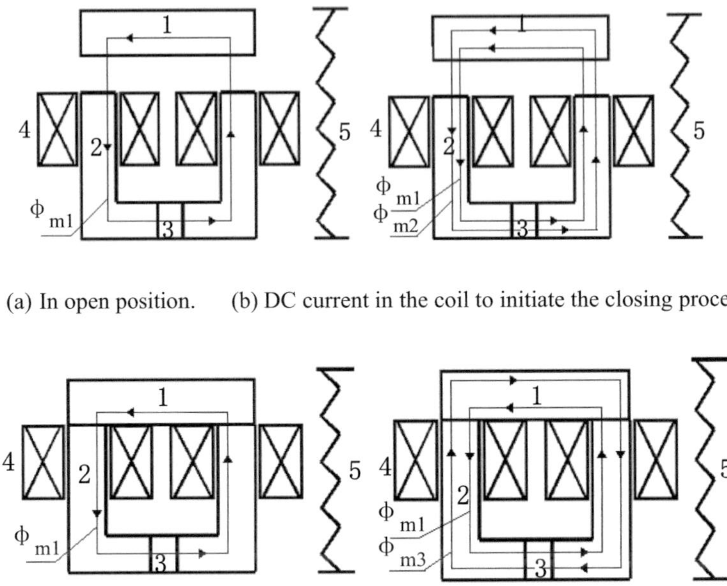

(a) In open position. (b) DC current in the coil to initiate the closing process.

(c) In close position. (d) Reversed DC current in the coil to initiate the opening process

Fig. 6.46 Diagrams illustrating the operating principle of the contactor equipped with a single coil permanent magnet mechanism. 1-moving iron core; 2-stationary iron core; 3-permanent magnet; 4-coil; 5-spring; φ m1-permanent magnet flux; φ m2-coil closing flux; φ m3-coil opening flux

core continues to increase. Despite the increase in the spring force, a faster increase in the electromagnetic attraction accelerates the closing process of the contactor.

When the contactor reaches its hold position (Fig. 6.46c), the electromagnetic attraction produced by the permanent magnet is about twice that of the spring force and this allows the cut-off of the coil current while holding the contactor in its close position.

When the contactor receives the opening signal (Fig. 6.46d), a DC current flows in the negative direction of the coil, producing a magnetic field that cancels that of the permanent magnet. As a result, the downward electromagnetic attraction is less than the spring force and the moving iron core begins to move upwards until reaches its open position.

The static characteristic of the permanent magnet mechanism contactor refers to the change of the magnetic force on the moving iron core as a function of the displacement of the moving core at a constant coil current. The effect of the electromotive force induced in the coil during the movement of the iron core is not considered, thus the term static. Generally, the static magnetic force can be used to judge whether the moving and stationary contacts can be reliably held together or detached. Calculation of the static magnetic force in the permanent magnet mechanism follows the steady-state magnetic field equations by ignoring the time-varying terms. The force can be calculated by solving the equations using a finite element solver such as ANSYS.

The basic equation for calculating the static magnetic field in the current-free region is given in Eq. (6.5). The equation based on vector magnetic potential is shown in Eq. (6.6). The equation for calculating the magnetic suction in Maxwell's stress tensor form is shown in Eqs. (6.7) and (6.8).

$$\nabla^2 \varphi_m = 0 \tag{6.5}$$

$$\nabla^2 A = -\mu J \tag{6.6}$$

$$F = \int [Tm]dS \tag{6.7}$$

$$T_m = H@B - \frac{H.B}{2}[l] \tag{6.8}$$

where, φ_{m} is the scalar magnetic potential/A; A is the vector magnetic potential/Wb·m^{-1}; μ is the magnetic permeability/H-m-1; J is the current density/A·m^{-2}; F is the electromagnetic attraction on the object/N; $[T_{\mathrm{m}}]$ is the magnetic stress tensor/N·m^{-2}; S is the surface of the object/m^2; H is the magnetic field strength/A·m^{-1}; B is the magnetic flux density/T; @ is the meridional vector operator; and $[l]$ is the 2nd order unit tensor.

The excitation source of the static magnetic force of the permanent magnet mechanism is the permanent magnet and a DC current. Its field distribution is in a steady state. A set of equations for the calculation of the magnetic field can be derived from the relevant Maxwell's electromagnetic field equations, as shown in Eq. (6.9), which can be solved in ANSYS based on the finite element method. The symmetric structure of the permanent magnet mechanism allows only a quarter of its space modelled. Parallel magnetic lines as a boundary condition is imposed on the symmetrical boundary of the selected domain.

For the other boundary conditions of an asymmetric nature, there are normally two methods to construct the boundary conditions. The first one is to use the far-field boundary condition, i.e. using the far-field element INFIN111 to consider the far-field dissipation of the magnetic field and mark the exterior surface of the far-field boundary as INF (infinite surface). The second option is to set a fixed value on the main degrees of freedom of nodes, for example, setting the value of magnetic vector as zero in some cases. Theoretically, the far-field boundary condition is more accurate, however, since discrepancy always exists between the simulation results and reality, a choice on the type of boundary conditions will need to be based on a comparison with the experimental results.

$$\left.\begin{aligned}\frac{\partial^2}{\partial x^2}A_x + \frac{\partial^2}{\partial y^2}A_x + \frac{\partial^2}{\partial z^2}A_x = -J_x\mu \\ \frac{\partial^2}{\partial x^2}A_y + \frac{\partial^2}{\partial y^2}A_y + \frac{\partial^2}{\partial z^2}A_y = -J_y\mu \\ \frac{\partial^2}{\partial x^2}A_z + \frac{\partial^2}{\partial y^2}A_z + \frac{\partial^2}{\partial z^2}A_z = -J_z\mu\end{aligned}\right\} \tag{6.9}$$

Two models implemented in ANSYS can be used for the calculation of the static magnetic force. The first one is based on the scalar magnetic potential to obtain the static force on the moving iron core. The second is based on the vector magnetic potential to obtain the magnetic linkage flux in the coil. Results for the static magnetic force on the moving iron core suction and the magnetic flux linkage of the coil are the basis of subsequent dynamic magnetic force simulation. The main degree of freedom in the scalar magnetic potential model is the scalar magnetic potential, and its advantage is that loading of the current can be done conveniently using the pseudo-current finite element in ANSYS through a macro type command. There is no need to include the space where current flow in the computational domain, which simplifies the process of model set-up and meshing. On the other hand, the main degree of freedom in the vector magnetic potential model is the vector magnetic potential which requires the inclusion of the current-carrying region in the computational domain. The finite element model is then set up according to the geometric model (all components in the domain). While the second approach complicates the model set-up and meshing processes, the advantage is however that the loaded current can be displayed conveniently and intuitively, and more information can be extracted from processing the results.

Figure 6.47 shows the magnetic flux density distribution obtained by the vector magnetic potential model (without using the far-field boundary condition), where the gap lengths between the moving and stationary iron cores are 13 mm and 1 mm, respectively. A-A' and B-B' are symmetrical planes. Because of the symmetry in structure, it only shows the magnetic flux density distribution in a quarter of the main magnetic circuit. As can be seen from the figure, when the distance between the moving and stationary iron cores is 1 mm, the flux is almost completely concentrated in the air gap and there is little magnetic flux leakage.

The scalar and vector magnetic potential models are tested for their performance in simulating the magnetic circuit using both the far-field boundary conditions and non-far-field boundary conditions. Four sets of results on the magnetic force and displacement are obtained. To verify the simulation models, a single coil permanent magnet-activated contactor with a rated current of 250 A was built and experiments were designed to use a force sensor and a displacement sensor to measure the magnetic force on the moving iron core at different gap lengths. The gap length between the iron cores can be adjusted through a screw thread mechanism. The experimental set-up is shown in Fig. 6.48a, and the force measurement is compared with the simulation results in Fig. 6.48b. It can be seen from the figure that the results of several calculation methods are similar to the experimental results, and

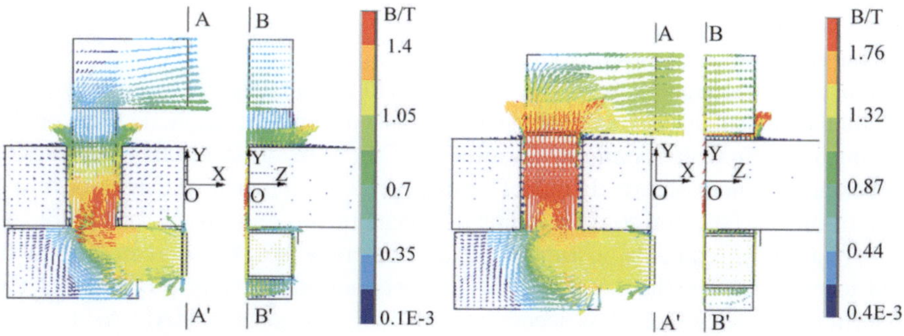

(a) Core spacing of 13 mm.

(b) Core spacing of 1 mm.

Fig. 6.47 Magnetic flux density distribution obtained from the vector magnetic potential model with two gap lengths of 13 mm and 1 m respectively

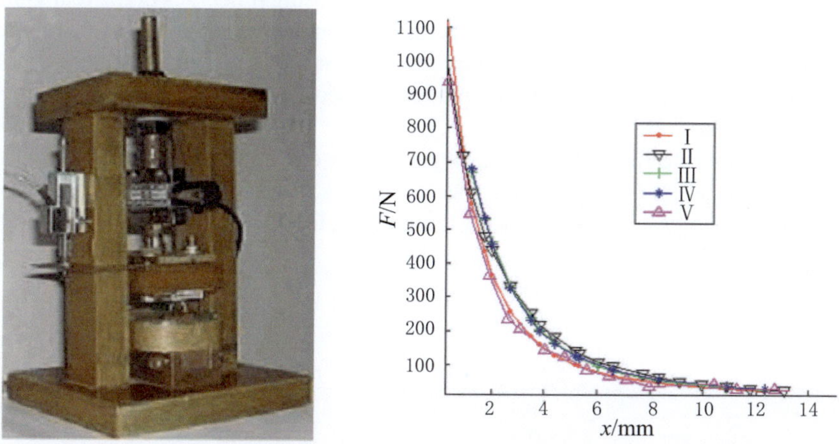

(a) Magnetic force measurement device.

(b) Force-displacement characteristics of the moving iron core.

Fig. 6.48 Magnetic force measurement set-up and comparison of measured and simulated magnetic force. I-vector magnetic potential model without using the far-field element; II-scalar magnetic potential model without using the far-field element; III-scalar magnetic potential model using far-field element; IV-vector magnetic potential model using far-field element; V-measurement

for the specific experimental configuration used in this section, the vector magnetic potential model without using far-field boundary conditions produces the closest agreement with measurement.

2. Calculation of dynamic magnetic force in permanent magnet mechanism contactor

In the process of opening or closing the contactor, the magnetic flux, through the moving and stationary iron cores and the coil, changes with the air gap length between the moving and stationary iron cores. This leads to the induction of an electromotive potential in the coil as well as a change in the coil current. The magnetic force on the moving iron core is also continuously affected. Therefore, the static characteristics of the contactor cannot accurately reflect the changes in the magnetic parameters during the movement of the iron core, including the magnetic force. As discussed before, the static relationship can only be used to judge whether the two contacts can be reliably held together or detached. To describe the opening and closing processes of the contactor more accurately, it is necessary to use software such as MATLAB to solve the governing differential equations describing the dynamic process of the contactor on the basis of the static finite element simulation in ANSYS [26]. The simulation of the dynamic characteristics of the contactor takes into account the influence of the magnetic parameters on the electromagnetic attraction on the moving iron core as well as the combined effect of the electromagnetic attraction and the opposite spring force, so it is expected to reflect the opening and closing processes of the real system.

For dynamic simulation of the operating system, the equation of Kirchhoff's voltage law is applied on the coil circuit consisting of the voltage source, resistor and the inductor, the D'Alembert equation is used for the kinematics of the mechanism and Maxwell's equations are followed for the magnetic field. These equations describe the closely coupled physical process involved in the opening and closing processes of the contactor. The governing differential equations describing the dynamic process of the contactor are given below:

$$\frac{d\psi}{dt} = u - i(x, \psi) * R \tag{6.10}$$

$$\frac{dv}{dt} = \frac{[F_m(x, \psi) - F_f(x)]}{m} \tag{6.11}$$

$$\frac{dx}{dt} = v \tag{6.12}$$

where Ψ is the coil magnetic chain/Wb·n; t is the time/s; u is the DC coil voltage/V; R is the DC coil resistance/Ω; i is the coil current/A; v is the moving iron core movement speed/m·s^{-1}; x is the moving iron core displacement /m; F_{m} is the electromagnetic attraction /N; F_{f} is the spring reaction force/N; M is the mass of the moving part/kg.

Equations (6.10)–(6.12) are solved using the fourth-order Runge–Kutta method. Part of the data needed in the solution process are provided by the static magnetic model decried in Sect. 6.1) above. Using a discretised time step, t_i, the algebraic equations used in the iterative solution process based on the fourth-order Runge–Kutta method are given in Eqs. (6.13)–(6.16), producing an approximate solution to the governing differential Eqs. (6.10)–(6.12):

$$\psi_{i+1} = \psi_i + \frac{\Delta t}{6}(k_{i1} + 2k_{i2} + 2k_{i3} + k_{i4}) \tag{6.13}$$

$$v_{i+1} = v_i + \frac{\Delta t}{6}(N_{i1} + 2N_{i2} + 2N_{i3} + N_{i4}) \tag{6.14}$$

$$x_{i+1} = x_i + \frac{\Delta t}{6}(M_{i1} + 2M_{i2} + 2M_{i3} + M_{i4}) \tag{6.15}$$

$$\Delta t = t_{i+1} - t_i \tag{6.16}$$

where Δt is the xxxx, and K_{ij}, N_{ij} and M_{ij} (j = 1, 2, 3, 4) are, respectively, the rate of change of Ψ, *v*, and *x*, i.e. *dΨ/dt*, *dv/dt* and *dx/dt,* at four different instants within the time interval of Δt. They can be expressed by Eqs. (6.17)–(6.19):

$$k_{ij} = u - i(\psi_{ij}, x_{ij})R(j = 1, 2, 3, 4) \tag{6.17}$$

$$N_{ij} = [F(\psi_{ij}, x_{ij}) - F_f(x_{ij})]/m(j = 1, 2, 3, 4) \tag{6.18}$$

$$\begin{aligned} M_{ij} &= v_i + p_{ij}N_{j-1} \\ p_{i1} &= 0; p_{i2} = p_{i3} = \Delta t/2; p_{i4} = \Delta t \end{aligned} \tag{6.19}$$

An examination of Eqs. (6.17)–(6.19) reveals that the key to the solution process is to obtain the values of the current $i(x,\psi)$ and the electromagnetic attraction of the moving iron core $F_m(x,\psi)$. Using the static simulation model, only the values for $F(x, i)$ and $\psi(x, i)$ can be obtained. Thus, an important step is to derive the values for $F(x, i)$ and $\psi(x, i)$ using the static model with the given current and position of the moving iron core. Then the current $i(x,\psi)$ and the electromagnetic attraction $F_m(x,\psi)$ can be obtained through curve fitting. Taking the current as an example, the values for $\psi(x, i)$ are calculated using the static magnetic model as given in Table 6.4. Through interpolation using third-order polynomials, the current in the dynamic model $i(x,\psi)$ is obtained using MATLAB, as given in Table 6.4. Similarly, the electromagnetic attraction $F_m(x,\psi)$ can be obtained in a similar way.

The curve family for $F(x, i)$ and $\psi(x, i)$ is obtained by simulation using the vector magnetic potential-based three-dimensional finite element model of the permanent magnet operating mechanism with discretised coil current values, *i,* and discretised displacement values of the moving iron core, *x.* When using discretised values, the

Table 6.4 $\psi(x, i)$ obtained by static simulation

	…	i_{h-1}	i_h	i_{h+1}	…
…					
x_{k-1}		$\Psi_{k-1,h-1}$	$\Psi_{k-1,h}$	$\Psi_{k-1,h+1}$	
x_k		$\Psi_{k,h-1}$	$\Psi_{k,h}$	$\Psi_{k,h+1}$	
x_{k+1}		$\Psi_{k+1,h-1}$	$\Psi_{k+1,h}$	$\Psi_{k+1,h+1}$	
…					

Table 6.5 $i(x, \psi)$ for dynamic simulation

	...	Ψ_{h-1}	Ψ_h	Ψ_{h+1}	...
...					
x_{k-1}		$i_{k-1,h-1}$	$i_{k-1,h}$	$i_{k-1,h+1}$	
x_k		$i_{k,h-1}$	$i_{k,h}$	$i_{k,h+1}$	
x_{k+1}		$i_{k+1,h-1}$	$i_{k+1,h}$	$i_{k+1,h+1}$	
...					

number of data points for current and displacement is limited. More data points (less interval between two points) will lead to higher accuracy but the computational cost will also increase (Table 6.5).

Based on the static characteristics of the magnetic force obtained from ANSYS, the governing differential equations describing the dynamic process are solved in MATLAB to obtain the dynamic characteristics of the permanent magnet mechanism contactor. As examples, the relationships between the static magnetic suction $F(x, i)$ on the moving iron core and the coil current, and between the static coil flux $\psi(x, i)$ and the coil current, are shown in Fig. 6.49 with the displacement of the moving iron core as a parameter. The dynamic electromagnetic attraction on the moving iron as a function of the displacement of the moving iron core is shown in Fig. 6.50. It is to be noted that the gap length between the moving and stationary iron cores is equal in magnitude to the displacement of the moving iron core.

3. Dynamic simulation of the permanent magnet mechanism contactor

The structure of the permanent magnet mechanism considered in the multi-body dynamics model, established in ADAMS, is shown in Fig. 6.51. It includes the structural components of the operating mechanism and the contact system, such as the stationary iron core, moving iron core, moving contact, stationary contact, contact spring, counter-force spring, and the connecting components. The structure

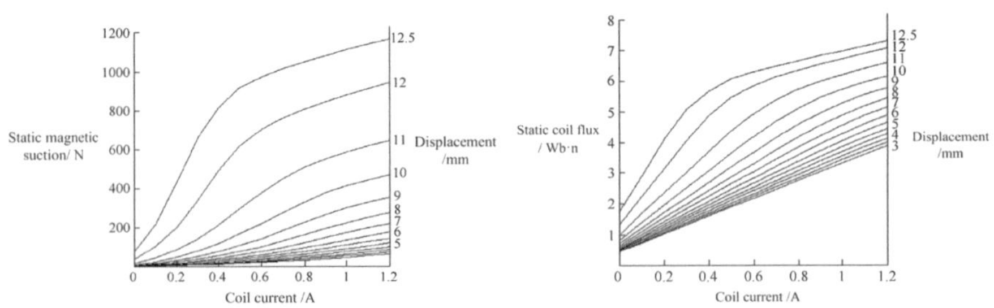

(a) $F(x, i)$ (b) $\psi(x, i)$

Fig. 6.49 The dependence on coil current of **a** the static magnetic suction $F(x, i)$ acted on the moving iron core and **b** the static flux linkage $\psi(x, i)$, with the displacement of the moving iron core (x) and the current (i) as parameters

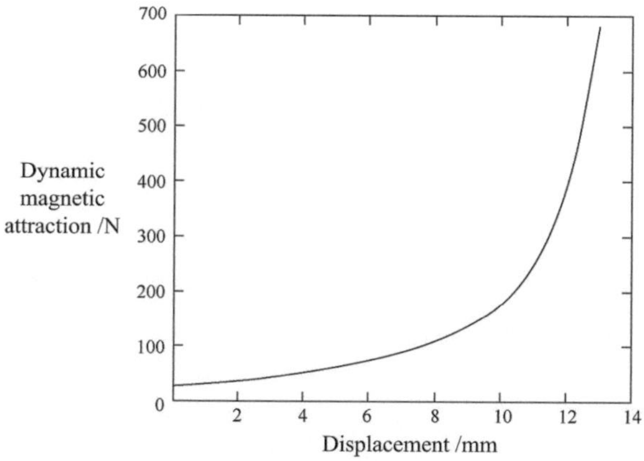

Fig. 6.50 Relationship between the dynamic magnetic attraction and the displacement of the moving iron core

of the permanent magnet mechanism contactor is simplified in the simulation model without affecting its motion characteristics. The relational data for the force and displacement, as obtained from the magnetic dynamic simulation in ANSYS, are fed into the mechanical dynamic simulation model for the operating mechanism defined in ADAMS. Details of the simulation model based on ADAMS have been discussed in Sect. 2.1.

Figure 6.52a is the simulation of the closing stroke curve of the moving electrode head, and the comparison between the curve and the experimental results. The experimental test device is shown in Fig. 6.52b, wherein the linear displacement sensor is fixed at the upper end of the test device and the electrode is fixed at the bottom of the test device. The linear displacement sensor and the moving electrode are connected by the connecting rod, so that the motion of the moving electrode drives the movement of the displacement sensor, and the displacement curve of the moving electrode with time can be obtained. It can be seen from Fig. 6.52a that the simulation results of the displacement–time characteristics of the moving electrode are close to the experimental results, particularly the time when the moving electrode ceases movement is consistent. This verifies the correctness of the ADAMS multi-body dynamics model. At the same time, because the data of the ADAMS model comes from the dynamic electromagnetic force simulation model of ANSYS, it also indirectly verifies the correctness of the dynamic electromagnetic force simulation model.

4. Contactors Based on Permanent Magnet Mechanism with Parallel Magnetic Circuits

The force generated by the permanent magnet in the close state of the contactor is greater than the counterforce from the spring. So it can hold the contactor in a closed state without the need to pass a current in the coil, making the design advantageous in terms of energy-saving and low noise. As mentioned above, the disadvantage is

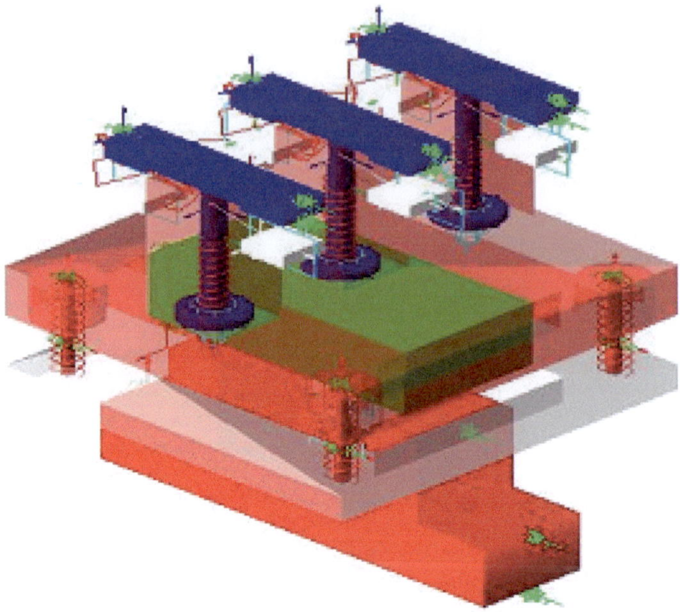

Fig. 6.51 Structure and components of the permanent magnet mechanism contactor considered in the multi-body dynamic model

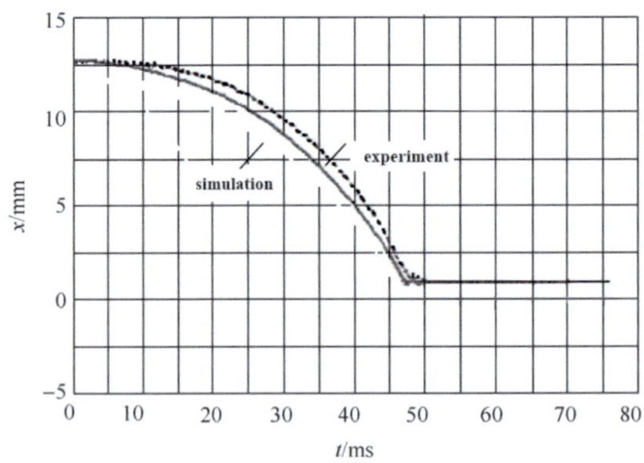

(a) Stroke curve

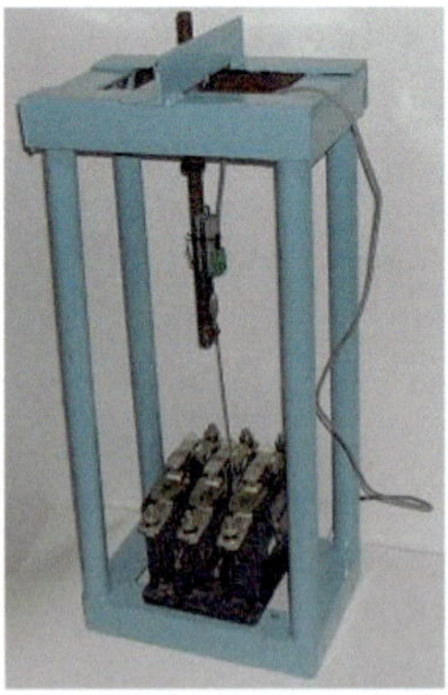

(b) Experimental set-up for stroke curve measurement.

Fig. 6.52 Experimental set-up for contact travel measurement and comparison between simulation prediction and measurement

that it cannot protect the mechanism against voltage drop. While at the moment all permanent magnet contactors are required to equipped with a voltage drop protection device, it nevertheless cannot provide 100% reliability for voltage drop protection. This is because most of protection devices use a capacitor energy storage as the energy source and there may be not sufficient stored energy after prolonged power loss in the network. Therefore, when the contactor is accidentally closed due to human or other reasons, the contactor will not separate.

On the other hand, the driving force for opening in the permanent magnet contactor adopts one of two ways: the first is to use the spring force to open the contactor, which has poor controllability and cannot effectively regulate the opening process; the other way is to use the magnetic force to perform the opening. However for a bi-stable permanent magnet mechanism the electromagnetic attraction is weak at the beginning of opening process and much stronger at the end of the opening process, which is contrary to what is expected for an ideal opening operation. This restricts the opening performance of the contactor. Thus a problem for research is to understand and propose a methods to gain better controllability of the opening process and increase the speed of the moving contact when it just separates from the stationary contact. For this purpose, a permanent magnet mechanism based on parallel magnetic circuits is proposed and the design parameters are optimised using simulation to achieve the following objectives:

- Removal of the need for voltage drop protection;
- Efficient energy saving;
- Increased average speed of contact separation and also speed when the contacts just separate.

The basic structure of the contactor based on a permanent magnet mechanism with parallel magnetic circuits is shown in Fig. 6.53. The magnetic circuit of the coil is made up by the stationary iron core, the coil and the moving iron core. The magnetic circuit of the permanent magnet is made up by the permanent magnet, the moving iron core and a specific stationary iron core for the permanent magnet. The moving iron core is shared by the parallel magnetic circuits.

The copper plate (No. 2 in Fig. 6.53) with low permeability is used to hold the permanent magnet in position as well as to isolate the two magnetic circuits. A special feature of this design is that the fluxes induced by the coil and by the permanent

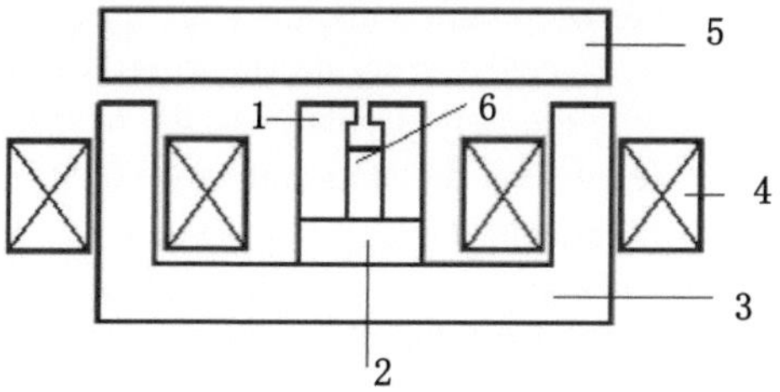

Fig. 6.53 Structural diagram of the proposed permanent magnet mechanism contactor with parallel magnetic circuits. 1-stationary iron core surrounding the permanent magnet; 2-copper plate; 3-stationary iron core; 4-coil; 5-moving iron core; 6-permanent magnet

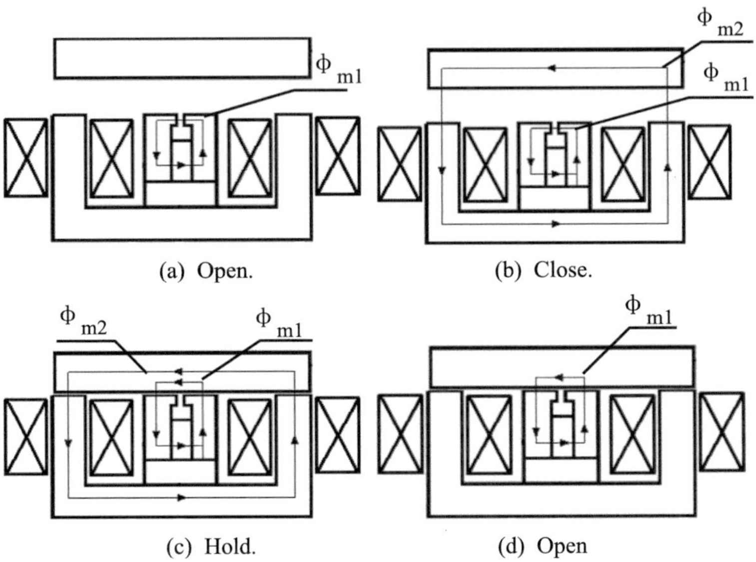

Fig. 6.54 Diagrams illustrating the operating principle of the proposed permanent magnet operating mechanism with parallel magnetic circuits. The arrows indicate the direction of the magnetic flux

magnet are confined in their respective magnetic circuits and the two circuits do not interfere with each other. In the close position, half of the suction force is provided by the permanent magnet and half from the coil. The total suction is comfortably larger than the spring force keeping the contactor in its close state. With a power loss to the coil due to voltage drop, the magnetic force produced by the permanent magnet becomes smaller than the spring counter force and the contactor will open naturally to offer voltage drop protection to the contactor.

Figure 6.54 shows schematic diagrams of the operating principle of the contactor based on the proposed new permanent magnet mechanism, as explained below.

When the contactor is fully opened, as shown in Fig. 6.54a, the air gap between the moving and stationary iron cores is the largest and there is no current flowing through the coil. The magnetic suction produced by the permanent magnet is weak and the spring force is able to keep the contactor in its open position.

When the contactor receives a closing signal, a current flows through the coil and the magnetic flux produced by the coil circulates in the magnetic circuit composed of the moving iron core, the air gap, the stationary iron core and the coil as shown in Fig. 6.54b. A downward magnetic force is exerted on the moving iron core. The total magnetic suction overtakes the spring force and the contactor starts to close. Over most of the closing process, the magnetic suction from the permanent magnet is relatively small and the movement of the iron core is driven by the magnetic suction from the coil.

After the contactor reaches its hold position (fully opened) in Fig. 6.54c, the magnetic force generated by the permanent magnet equalises that from the coil to

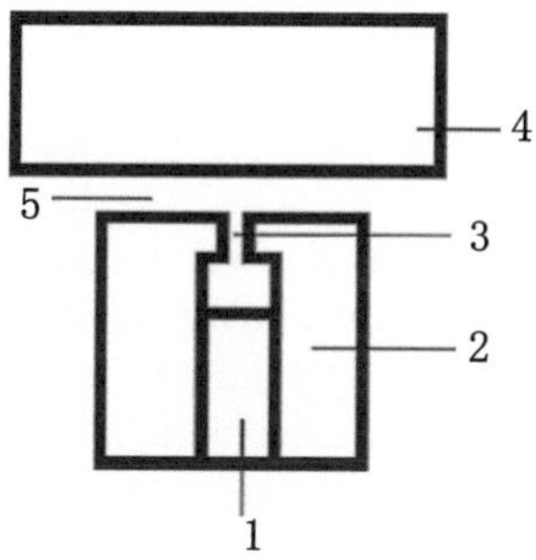

Fig. 6.55 Structure of the shunt magnetic circuit of the permanent magnetic circuit. 1-permanent magnet; 2-stationary iron core surrounding the permanent magnet; 3-regulating air gap; 4-moving iron core; 5-air gap

hold the contactor in its close state as a result of the minimised magnetic reluctance in the magnetic circuit of the coil. The fluxes induced by the coil and by the permanent magnet flow in their respective magnetic circuits do not interfere with each other.

As shown in Fig. 6.54d, when an opening signal is received by the contactor, the coil current is cut off, and the magnetic force generated by the permanent magnet only is smaller than the spring force, the contactor starts to open until it is fully opened.

The opening process of the contactor is driven by the spring force while the magnetic suction from the permanent magnet opposes the movement of the contact. In order to minimise the adverse influence of the permanent magnet, a shunt magnetic circuit is added to the design in Fig. 6.54. The characteristics of the permanent magnet suction is thus altered by the magnetic shunt to increase the magnet suction in the close position. As the length of the air gap increases during opening, the permanent magnet suction decreases rapidly. This design modification not only saves operating energy but also minimises the reduction in the speed of the contact. Figure 6.55 shows the shunt magnetic circuit of the permanent magnet. Figure 6.56 illustrates the principle of adjustment to the permanent magnetic force by the shunt circuit.

As can be seen from Figs. 6.53 and 6.54, the permanent magnet, the air gap, the moving iron core and the stationary iron core make up the main circuit of the permanent magnet, while the permanent magnet, the additional flux regulating air gap

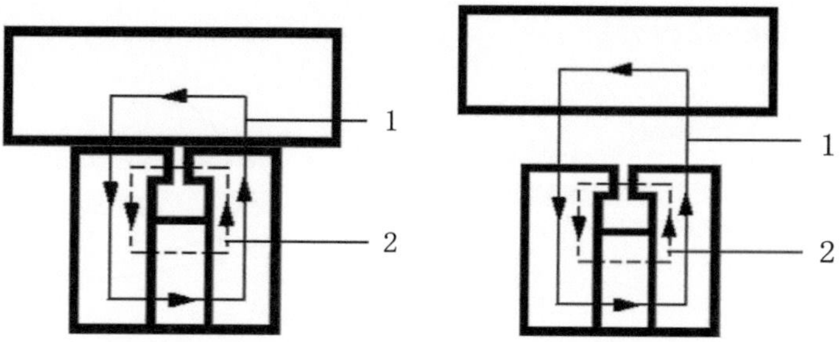

Fig. 6.56 Principle of adjustment to the magnetic force from the permanent magnetic with the presence of the shunt magnetic circuit. 1-Main magnetic circuit. 2-Regulating shunt magnetic circuit

(auxiliary air gap) and the immediate stationary iron core form the regulating shunt magnetic circuit of the permanent magnet. In the close position of the contactor, the main air gap is very small while the auxiliary air gap is significantly larger, so most of the flux produced by the permanent magnet circulates in the main magnetic circuit due to much lower magnetic reluctance. The magnetic suction onto the moving iron core reaches its maximum. Once the moving and stationary iron cores separate, the main air gap quickly increases and becomes larger than the auxiliary air gap and most of the magnetic flux produced by the permanent magnet thus starts to circulate in the shunt magnetic circuit. The magnetic force on the moving iron core rapidly diminishes, minimising the influence of the permanent magnet on the opening speed of the contactor. In addition, the parallel magnetic circuit arrangement also allows for a lighter moving core that also helps improve the opening speed of the contactor.

The key to the effectiveness of the above design based on parallel magnetic circuits is to optimise the parameters of the contactor. Optimisation through simulation greatly reduces the manpower, material usage and time taken. Following the simulation procedure introduced above, the characteristic relationship for the static magnetic force is first obtained in ANSYS; Based on this the relationship for the dynamic magnetic force is then obtained with MATLAB; Thirdly, the relevant data are imported into the multi-body dynamics simulation model in ADAMS to obtain the displacement curves of the contactor under different values of the design parameters. Finally, the parameters are optimised to arrive at a contactor design with parallel magnetic circuits.

During the optimisation process of the contactor design, the auxiliary air gap length as an important parameter is first optimised within ANSYS. The suction force from the permanent magnet is calculated under different gap sizes, with typical results given Fig. 6.57. It can be seen from the figure that when the adjusting air gap is reduced from 2 to 1 mm, the permanent magnet suction decreases noticeably.

The multi-body dynamics simulation model of the contactor with parallel magnetic circuits for the permanent magnet is shown in Fig. 6.58, through which the movement of the iron core and contact during the opening and closing processes of the contactor can be obtained. A comparison between the simulation and measurement of the opening process is shown in Fig. 6.59 where good agreement is achieved

Fig. 6.57 Magnetic suction force as a function of the displacement of the moving iron core from the permanent magnet under different auxiliary gap lengths. Curve 1-air gap length of 2 mm; 2-air gap length of 1 mm

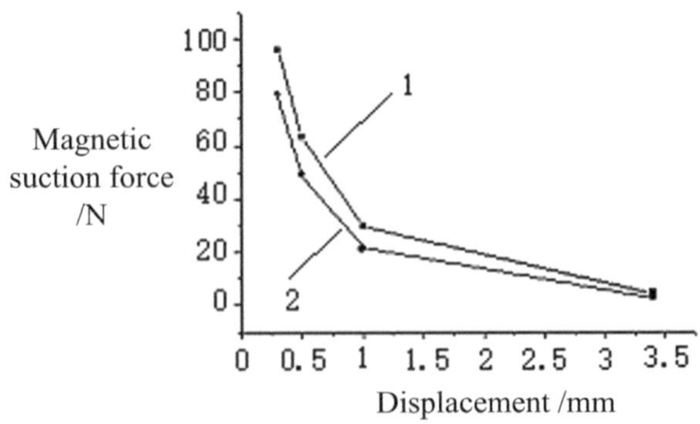

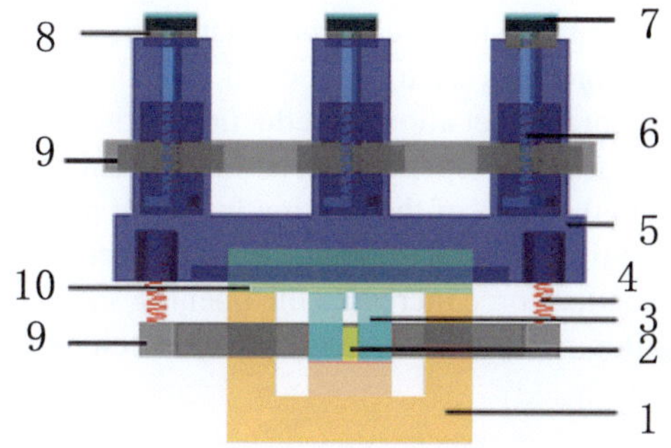

Fig. 6.58 Multi-body dynamics simulation model for the permanent magnet mechanism-based contactor with parallel magnetic circuits. 1-stationary iron core; 2-permanent magnet; 3-stationary iron core surrounding the permanent magnet; 4-spring; 5-moving plastic shell; 6-contact spring; 7-moving contact; 8-stationary contact; 9- housing; 10-stationary iron core

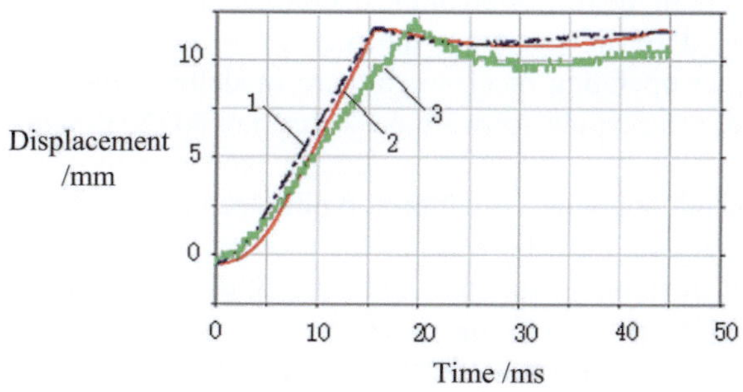

Fig. 6.59 A comparison of the displacement stroke curve of the moving iron core during the opening process of the contactor. 1-Measurement after design optimisation; 2-Simulation for the optimised design; 3-Measurement on the original design before optimisation

for the duration taken by the moving contact to reach its maximum travel at 16 ms, providing strong evidence to support the use of the simulation model for effectively design optimisation of permanent magnet mechanism based contactors. Compared with the original design of the AC contactor, the new design with a parallel magnetic circuit has a shorter opening time and faster averaging speed. From the measurement data in Curves 1 and 3 in Fig. 6.59, it can be found that the initial separating speed of the contactor with the parallel magnetic circuits is 0.71 m/s, increased by 0.11 m/s from 0.6 m/s of the original design (without the parallel magnetic circuits). The proposed modification to the AC contactor with the addition of the shunt magnetic circuit satisfies the design objectives.

In general, the magnet suction force induced by a permanent magnet in AC contactors exhibits a retarding effect on the opening operation by reducing the contact speed. A contactor designed with parallel magnetic circuits for the permanent magnet part is able to achieve a higher opening speed for two reasons: First, the magnet suction force is high only in the close state with a very small air gap between the stationary and moving iron cores. It starts to reduce immediately in significant terms once the stationary and moving iron cores separate. Second, the adoption of parallel magnetic circuits for the permanent magnet allows a reduction in the mass of the moving iron core and thus an increase in the opening speed of the contactor. Among these two factors, the reduction in the mass of the moving iron core plays a leading role in the improvement of the opening speed of the contactor, it further counteracts the retarding effect of the magnet suction force increases the opening speed of the contactor, The time to reach the maximum travel of the iron core has been reduced by a remarkable 4 ms.

6.5 Summary

The focus of this chapter is on the use of virtual prototyping technology for design optimisation and fault identification of the operating mechanisms in switching apparatuses. A selected spring operating mechanism, a hydraulic mechanism, and a permanent magnet operating mechanism were modelled, simulated and analysed using two virtual prototyping software packages, i.e. ADAMS and AMESim, with support from ANSYS, Ansoft and other electromagnetic analysis software.

On the one hand, applying virtual prototyping technology to the design process of switching apparatuses greatly reduces the product development effort, shortens the development time and lowers the development cost. By optimising key design parameters, it also helps enhance the reliability and stability of the products. On the other hand, the virtual prototyping technology can be used to assess the operational state of the switching apparatuses and establish quantitative links between the monitoring parameters and the state variables to provide reference information for fault diagnosis of the operating mechanism of the switching apparatuses.

Virtual prototyping technology has been widely used in the switchgear industry for the design, optimisation and fault identification of the operating mechanisms of switching apparatuses. However, the calculation accuracy and the user-friendliness in setting up the simulation models require further improvement to fully achieve its application potential. There is still significant space for research and technological development.

References

1. Li HM, Lin X, Xu JY, et al. Design and Dynamic Performance Analysis of Permanent Magnet Swing Angle Motor Operating Mechanism for 126 kV Vacuum Circuit Breaker[J]. Power Grid Technology, 2014, 38 (6): 1664-1669.
2. Cui YB, Gao Z, Cao YP, et al. Dynamic Simulation of operating Mechanism of High Voltage Circuit Breaker based on ADAMS[J]. Mechanical design and manufacture, 2006, 4: 66-67.
3. Yuan DL. Review on the Operation status of High Voltage Switchgear in National Power system in the past Ten years[J]. High Voltage switch Industry Newsletter, 1999, (5): 47-53.

4. Rong MZ, Jia SL, Wang XH.Condition detection of electrical equipment[M]. 2007.
5. Rong MZ, Wang XH, Yang W and Jia SL. Mechanical condition recognition of medium-voltage vacuum circuit breaker based on mechanism dynamic features simulation and ANN[J]. IEEE Transactions on Power Delivery, 2005, 20(3): 1904-1909.
6. Wang XH, Rong MZ, Qiu J, et al. Research on mechanical fault prediction algorithm for circuit breakers based on sliding time window and ANN[J]. IEICE Transactions on Electronics, 2008, E91-C(8): 1299–1305.
7. Liu S. Simulation Analysis and optimal Design of the operating Mechanisms in High Voltage Circuit Breakers[D]. 2013.
8. Li SY, Tang FY. Dynamic Simulation and Optimization of the Spring Mechanisms in High Voltage switchgear based on UG and ADAMS[J]. High voltage apparatus, 2008, 44(2): 110-113.
9. Tu K, Zhou Y, Zhang L. Simulation Analysis of High Voltage Circuit Breaker Mechanissm based on Pro/E and ADAMS[J]. Mechanical research and application, 2009, 3: 47-48.
10. Rong MZ, Wang XH. Discussion on the Design method of Intelligent switching Apparatuses [J]. High voltage apparatus, 2010, 46(9): 1-3.
11. Hu ZY. Motion characteristic Simulation and performance improvement of low-Voltage air Circuit Breakers[D]. 2012.
12. Fan S, Hydraulic operating mechanisms in high voltage circuit breakera[M]. 2000.
13. Ding F, Lu YX. Research on the hydraulic operating Mechanism of a New 500kV EHV Circuit Breaker[J]. Chinese Mechanical Engineering, 1998(6): 50-52.
14. Briggs E E, Hambrick R D, Umphrey D M. Hydraulic operating mechanisms for high-capacity circuit breakers[J]. Power Apparatus and Systems, Part III. Transactions of the American Institute of Electrical Engineers, 1953, 72(2): 874–881.
15. Peek H L and Dyer M G. Introduction of hydraulic operation to low-voltage air circuit breakers[J]. Power Apparatus and Systems, Part III. Transactions of the American Institute of Electrical Engineers, 1956, 75(3): 226–232.
16. Lou JF, Design of operating mechanisms in High Voltage Switchgear[M]. 1981.
17. Xu F. Maintenance and improvement of the CY hydraulic operating Mechanism[J]. Zhejiang Electric Power, 2002(5): 68-70.
18. Zhao HF. Modeling and Simulation of hydraulic operating Mechanisms and Arc Chamber characteristics of a 550kV single-break Tank Circuit Breaker[D]. 2011.
19. Zhao SC. Causes and solution methods for frequent pressing of the CY5 hydraulic operating Mechanism[J]. High voltage apparatus, 1994(2): 40-42.
20. Wei ZY, Zhao HF, Liu W, et al. Simulation and Test of cushioning process in the cylinder of a hydraulic operating Mechanism for High Voltage Circuit Breakers[J]. Journal of agricultural machinery, 2010(6): 216-221.
21. Wang JM, Yang R. Survey of permanent Magnet operating system for 126kV vacuum Circuit Breaker[J]. Huatong technology, 2003(1): 25-27.
22. Lin X, Permanent Magnet Mechanism and Vacuum Circuit Breaker[M]. 2002.
23. Fang SH, Lin HY, Yang CF, et al. Magnetic field analysis and control strategy of permanent magnet operating mechanism for low voltage vacuum circuit breakers[C]. 7th International Conference on Power Electronics and Drive Systems, Hong Kong Polytechnic University, 2007: 540–543.
24. Cui YB, Lun ZD. Design of Cam profile of vacuum Circuit Breakers based on ADAMS[J]. 2011, 38(4): 34–36.
25. Lou JY, Rong MZ. Simulation Analysis of dynamic characteristics of single Coil Monostable permanent Magnet Contactor Mechanism[J]. Chinese Journal of Electrical Engineering, 2004, 24(4):120-124.
26. Rong MZ, Lou J. Simulation analysis and experimental study on moving contact characteristics of permanent magnet contactor[J]. Chinese Journal of Electrical Engineering, 2005, 25(1): 109-113.

Chapter 7
Simulation Examples for Switching Arcs

7.1 Introduction

After the relevant basic concepts and fundamentals on the modelling and simulation of switching arcs were introduced in Chap. 1, a number of simulation examples will be provided in this chapter to allow the readers to gain useful experience in setting up and executing switching arc simulation. An air miniature circuit breaker and a high voltage SF_6 circuit breaker are selected for case studies, including specifying the simulation-related conditions. Important issues such as geometry simplification, definition of simulation model, implementation of material properties, and prescribing the boundary conditions are discussed. Typical simulation results are then presented for readers' reference.

7.2 Simulation of Arcs in Miniature Circuit Breakers

7.2.1 *Miniature Circuit Breakers*

Miniature Circuit Breakers (MCB), also known as small-sized circuit breakers in certain countries, are mainly used for power distribution as well as the control and protection of circuits and electrical equipment. They are terminal protection devices widely used in the distribution system [1]. According to statistics, the production scale of MCBs reached 345 million in 2008 in China alone. An MCB is mainly composed of the operating mechanism, trip system, contact system, and arc quenching system. The arc quenching system is the core component of the breaker. For large short-circuit currents, a well-designed arc quenching system is essential to limit the short-circuit current, reduce the arc erosion of the contacts, and quickly interrupt the short-circuit current to protect other electrical equipment in the circuit.

M. Rong et al., *Computing Methodology for Switchgear*,
https://doi.org/10.1007/978-981-95-2840-0_7

The arc quenching system of MCBs is generally composed of the moving contacts, stationary contacts, current carrying conductors, and arc splitters. The main material of the arc splitters is hot-rolled steel (a ferromagnetic material). In order to reduce the difficulties for the arc to enter, arc splitters are commonly designed in certain shapes, such as an inverted V shape. Multiple arc splitter plates are stacked in the arc chamber with the air gap between them providing insulation. Meanwhile, ferromagnetic materials are also added near the moving and stationary contacts to alter the magnetic field in the arc chamber. This helps reduce the dwelling time of the arc roots, thereby easing the erosion of contacts. When a large short-circuit current passes an MCB, the trip mechanism is activated instantaneously and the operating mechanism drives the moving contact to move against the magnetic repulsive force. An arc is drawn between the contacts and stretches with the contact movement. Under the influence of Lorentz force and the aerodynamic effect of the gas, the arc moves quickly towards and into the arc chute and divided into multiple segments of short lengths. The arc is cooled down and quenched in the arc chute eventually.

During the arc quenching process, on one hand, the arc erodes the contacts and the surrounding chamber walls, leading to a reduced lifespan of the breaker. On the other hand, under the influence of the magnetic and gas pressure fields, the arc column is blown into the arc chute and cut. Due to the voltage drops in the electrode layers, the arc voltage increases and current decreases, leading to the interruption of the current.

The quality of arc chamber design has long been the research focus of academics and product development engineers because it directly affects the switching performance of the MCBs. Experimental study of the dynamic behaviour of the arc requires the use and replacement of prototype breakers, which is time-consuming and costly. Additionally, due to the arc's high temperature (tens of thousands of Kelvin), high speed (hundreds of m/s), and short duration (several milliseconds), acquiring experimental data on physical quantities is challenging. Due to this, simulation tools, such as those based on empirical formulas or MHD equations, are also used to study the dynamic characteristics of the arc in the arc chamber.

However, simulation covering the entire arcing process, from arc ignition to its interruption in the arc chute, is quite rare for actual low-voltage circuit breaker products [2–5]. To deal with practical applications, a three-dimensional arc model for a specific MCB product is established based on the theoretical knowledge presented in previous chapters to simulate the dynamic features of the arc. The calculation results have revealed the movement patterns of an arc in the arc chamber and offer information on the transfer of the arc root from the moving contact to the arc runner, on the arc's moving towards the splitter plates and its segmentation process by the arc splitter plates made of ferromagnetic materials. The simulation also clearly shows multiple gas breakdown in the region left behind the arc column during the above arc movement process. The arc also becomes immobile for a considerable length of time at the inflection point of the arc runner, causing intense localised erosion of the contacts. Figure 7.1 shows the structure of common miniature circuit breakers.

As shown in Fig. 7.1, component 1 is the actuator lever, which is used to manually close or open the circuit breaker. The lever also shows the status of the circuit breaker (open or closed). Component 2 is a linkage mechanism used to transmit the

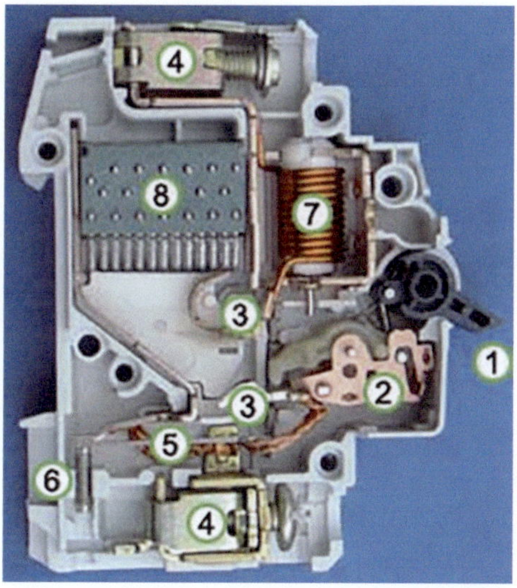

Fig. 7.1 The internal structure of common miniature circuit breakers

mechanical forces used in closing or opening. Components 3 indicate the contacts. Components 4 are the terminals for external connection. Component 5 is a bimetallic strip used for overload protection of the circuit. Component 6 is a screw nut to adjust the position of the metal component connected to it. Component 7 is a solenoid that drives the iron core inside it to move when a short-circuit current flows through the coil. It acts as driving mechanism to open the MCB. Component 8 shows the arc chute where the arc is segmented into multiple short lengths and subsequently extinguished.

7.2.2 Simplifications to the Miniature Circuit Breaker Geometry

A mathematical model derived from the MHD theory is used to simulate and analyse the chosen MCB. In the first step, the geometry involved in the simulation and the computational grid system of the MCB must be established. A four-step procedure is followed for the grid system, as shown in Fig. 7.2.

In the first step, the structure and size of the MCB in the design document is analysed to understand the functionality of the components and their influences on the performance of the MCB. Second, approximations and simplifications are made to the structure for convenience of simulation, making reference to the product prototype if necessary. In the next step, a reasonable mesh (or grid system) is generated, normally using a pre-processing software package. Finally, the generated mesh is verified for its quality. An MCB with a rated current of 63 A is chosen for the simulation work presented in this section. Figure 7.3a, shows the design of the conductive circuit,

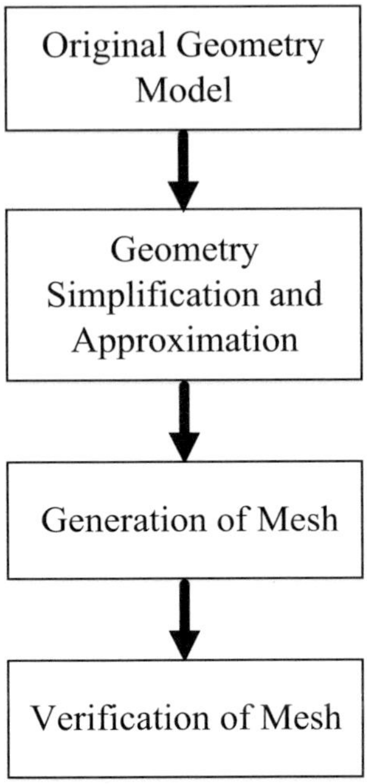

Fig. 7.2 Four steps for generation and verification of the mesh system for MCB simulation

contact system, arc quenching system and operation mechanism. Because it contains details of the components and assembly information, the complexity associated with the structure needs to be reduced by removing unimportant components or details, such as screws, washers, through holes, brackets, etc., before the mesh system is generated. In doing so, the flow, temperature, electric and magnetic fields should not be affected.

Figure 7.3b shows the two-dimensional symmetric geometry after simplification, including the moving and stationary contacts, arc runner, conductive rods, arc splitter plates, gas outlets, etc. In Fig. 7.3b, component 1 is the conductive rod of the moving contact, component 2 is the horizontal conductor to the stationary contact, component 3 is the moving contact, component 4 is the extended conductive rod of the arc runner, component 5 is the arc runner, component 6 is the stationary contact, component 7 are the arc splitter plates, position 8 indicates the right gas outlet, position 9 labels the gas outlet at the back of the splitter plates, and position 10 shows the left gas outlet. The mesh generated for the simplified geometry, using the meshing function of the pre-processor is shown in Fig. 7.4.

Half of the simplified structure divided by the symmetry plane in x-y space is used in the simulation. The calculation domain for thermal and flow phenomena (model set up in Fluent) includes the arc chamber and the gas region outside the outlets, while the domain for electric and magnetic fields (model set up in Ansys) is extended into a space that is a few times larger than the arc chamber. Mesh generation is a crucial step in numerical calculation, involving the discretization of the computational domain.

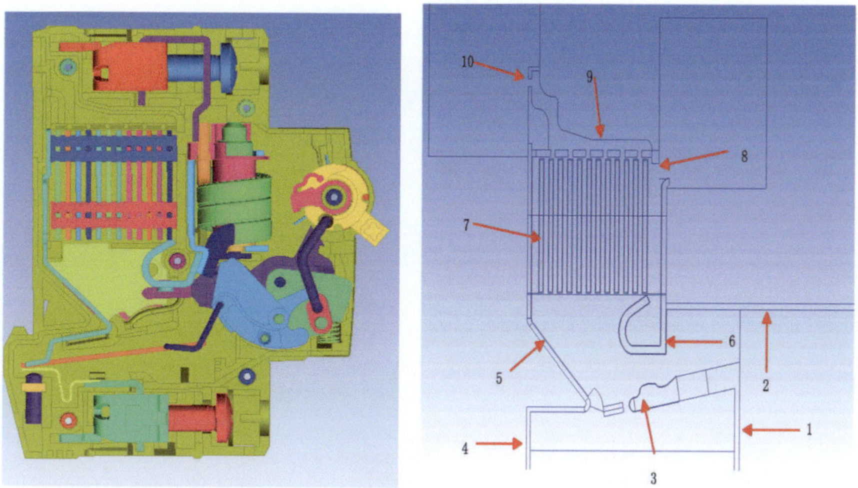

(a) Miniature circuit breaker structure. (b) Miniature circuit breaker simplified geometry.

Fig. 7.3 The original and simplified structure of the miniature circuit breaker under simulation

While not a complex task in theory, the quality of the mesh can have serious impact on the convergence of the solution process and the accuracy of the simulation results. If the mesh is too coarse, the solution may deviate significantly from reality although the solution cycle runs faster. If the mesh is too fine, much more time is needed per computational cycle of the solution process and much more memory is also needed although the accuracy of the solution can be much improved. It is thus essential to strike a balance between the accuracy of the solution and the speed of the computation by using an optimised mesh system. A mesh is regarded as being sufficiently dense when a significant reduction in the mesh size results in only a minor change in the results.

The mesh does not need to be uniform across the computational domain. In fact, due to the complicated three-dimensional structure of the MCB. Fine mesh is needed near the interface between the metallic components, such as the splitter plates and the contact, and air, to spatially resolve the near electrode region. The resulting mesh count can often exceed one million based on a simple meshing algorithm, which is computationally infeasible with available resources. Therefore, it is necessary to divide the computational domain into sub-domains based on knowledge and understanding of the physical phenomena and set an appropriate mesh density in each of the strategic regions. This helps reduce the overall number of elements of the mesh system without compromising the computational accuracy. A few examples are given below.

- When simulating the chosen MCB in this section, the mesh density in the air region is highest while that in the far-field for solving the magnetic field is relatively low.

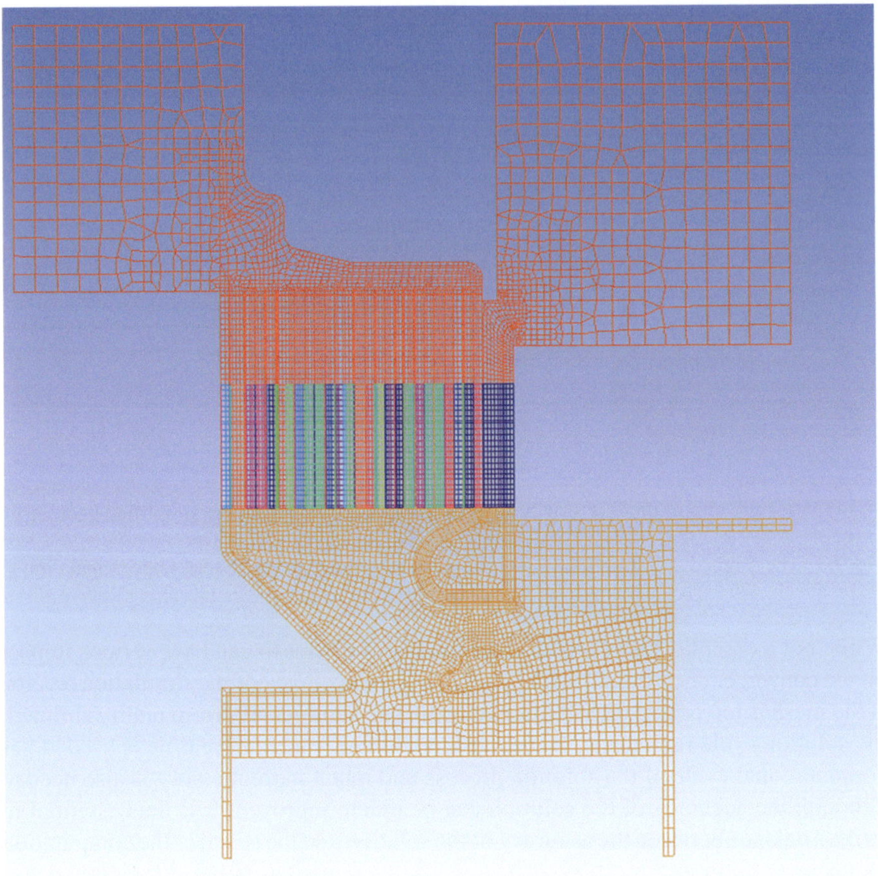

Fig. 7.4 The mesh system generated for the simplified structure of the miniature circuit breaker

- At least four layers of grids are used across the thickness direction of the contacts and splitter plates.
- The mesh in the sheath layer of a metallic surface has a single layer of 0.1 mm in thickness.
- Without counting the mesh representing the sheath layer, at least four layers of grids are generated in the air gap between two splitter plates or between a splitter plate and a contact.
- The local mesh density in the air region surrounding the arc splitter plates is significantly increased because of its heavy influence on arc extinction. Conversely, in other air regions the mesh can be relatively sparse matching the spatial distribution of the physical quantities. Variable mesh size is used for the transition between these two regions with significant difference in mesh density.

After specific settings are made to the mesh system according to the roles of each sub-region, the simulation model set up in Fluent for the MCB has a mesh count of 143,000 and that in Ansys (for magnetic field calculation) is 286,000.

In this example, arc simulation starts with the MCB in the open position, i.e. contact movement is not considered. This means meshing is generated on the simplified MCB structure with the contacts in their open position. Finally, the mesh is improved based on trial simulation and examination of the results before its quality is confirmed to be sufficient.

7.2.3 Material Properties and Their Implementation

The simulation model for this example consists of different types of materials. It is necessary to assign the corresponding material properties to each region. Table 7.1 lists the types of materials corresponding to the numbers in Fig. 7.3b.

In the finite element solution process, copper metal is considered as a solid region with a certain conductivity, and material properties will be supplied to all regions designated as the same material during the solution process. The procedures for defining other materials are similar and will not be repeated here.

In low-voltage air circuit breakers, the most important and complex aspect is the arcing process. However, the physical properties (material properties) of the arcing gas exhibit strong nonlinearity with its temperature and pressure. To accurately model the arc plasma, it is essential to have knowledge of the physical properties of air within a large range of temperature, such as the mass density, specific heat capacities, and other thermophysical parameters, as well as the transport coefficients like viscosity, thermal conductivity, and electrical conductivity [5]. The calculation of the physical

Table 7.1 Materials for different components shown in Fig. 7.3b

Number	Material	Component name
1	Copper	Conductive rod to moving contact
2	Copper	Horizontal conductive rod to stationary contact
3	Copper	Moving contact
4	Copper	Extended conductive rod for arc runner
5	Copper	Arc runner
6	Copper	Stationary contact
7	Lower part is cold rolled steel; upper part is glassfiber reinforced polyester board	Arc splitters
Blank region	Air	Air

properties is only briefly outlined here, and for detailed information please refer to Sect. 7.1 of Chap. 3. Most of the physical properties of the plasma are obtained by calculating its chemical composition. For a plasma in local thermal equilibrium (LTE), the chemical composition depends only on the temperature, pressure, and concentration of chemical elements. So in this example the minimum Gibbs free energy method based on the law of mass action is used to calculate the species concentration. For the air arc plasma, its chemical composition contains multiple species such as O_2, N_2, O^-, N^-, N_2^+, O^+ and so on [6].

7.2.4 Specification of Initial and Boundary Conditions in Miniature Circuit Breakers

In this simulation example, eddy current effects are not considered since the frequency of the current is not high. The boundary conditions to be set for the MCB model are associated with temperature (energy), thermal radiation (energy), pressure outlet (flow), electric and magnetic fields (Ohmic heating and Lorentz force).

1. Temperature boundary

One-dimensional thermal conduction is assumed on the chamber wall to implement the temperature boundary condition:

$$q = q_{rad} - \frac{\lambda_w (T - T_0)}{d} \tag{7.1}$$

where q is the heat flux, q_{rad} the thermal radiation energy flux, T_0 the ambient temperature just outside the wall with a default value of 300 K, T the temperature on the wall of the chamber, λ_w the thermal conductivity of the chamber wall material, and d the thickness of the chamber wall. A flux entering the domain is defined positive.

2. Pressure outlet

The exits of the chamber connect the interior of the arc chamber with the external air. A fluid domain is also established outside the outlet of the arc extinguishing chamber. The pressure outlet is set on the outside of the fluid domain. The gas pressure at the outlet is set to one standard atmospheric pressure.

3. Electric and magnetic fields boundary

In the existing work on arc simulation, the boundary condition for electric field is defined through the current density distribution on the contact-arc interface (arc root attachment). However, it is difficult to measure the current density distribution on the contact surface during an actual arcing process in the arc chamber. Therefore, in existing simulation work, an approximate current density distribution is often used, such as assuming an average current density at the contact-arc interface, or that the current density decreases exponentially with distance from the arc root centre, or that the current density is related to the contact surface temperature and the electrical conductivity of the adjacent arcing gas. Due to the uncertainty in the current density distribution within the arc attachment on the contact surface, it poses a major limitation when the boundary condition for the electric field is set on the contact-arc interface.

In this example, the electric field boundary is set at the surfaces of the moving and stationary contacts, and these surfaces are located at the boundaries of the fluid domain. The conductive rod to the stationary contact is set as the cathode with a potential of zero, and the conductive rod to the moving contact is set as the anode. Set a certain current value on the input surface of the anode. Since the current input surface of conductive rod is sufficiently away from the arc region, the current density on the anode input surface can be assumed to be uniformly distributed. Thus, for a given current I,

$$I = \sum_{i=1}^{N} j_i \cdot S_i \tag{7.2}$$

where j_i is the current density on an element face in the current input plane, S_i the area of the element face, and N is the total number of element faces falling within the anode current input plane.

In this section, the method based on vector magnetic potential is adopted to solve for the magnetic field. The boundary of a magnetic field is the parallel boundary of magnetic field lines. For the symmetry plane of the PCB as part of the boundary, the following relationships hold:

$$\vec{n} \times \vec{A} = 0 \tag{7.3}$$

$$\upsilon \nabla \cdot \vec{A} = 0 \tag{7.4}$$

where υ is the reluctivity (reciprocal of permeability) of the medium in the boundary elements. For boundary elements sufficiently away from the arc column, the vector magnetic potential A is set to zero.

4. Calculation of current through coupling with external circuit

The input current into the arc in this example is not a constant, but related to the circuit that is used for the short-circuit breaking test. The circuit equations need to be solved in the simulation to obtain the current. Simplifying the test circuit into one consisting of a total capacitance C, total inductance L, line internal resistance R, and the arc as a resistor, it is shown in Fig. 7.5.

In Fig. 7.5, the power source is provided by the charged capacitors, CB is a contactor controlling the close/open state of the entire circuit. The contactor in the circuit is primarily used for making connections and isolating the circuit after disconnection, while the breaking operation is performed by the circuit breaker. The MCB is the test object. The arc voltage is measured using a high-voltage probe and the

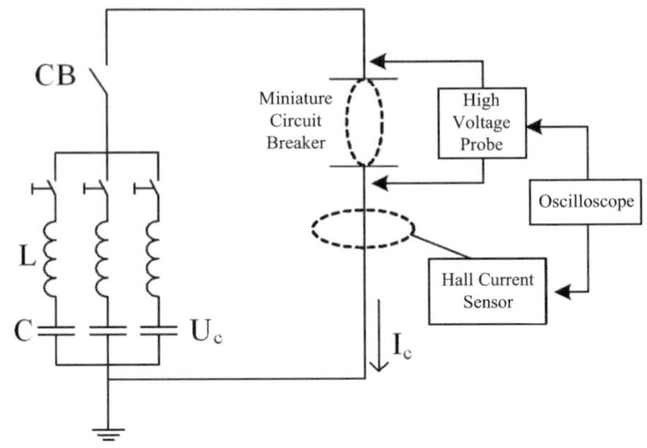

Fig. 7.5 Equivalent test circuit to calculate the arc current when coupled to the arc model

current measured using a Hall sensor. Both voltage and current measurements are fed into an oscilloscope where the waveforms are displayed and recorded.

7.2.5 Calculation Results

1. Results from the arcing process

Based on the pre-calculated material properties of the air arc plasma, the magneto-hydrodynamic (MHD) model is implemented in the commercial computational fluid dynamics (CFD) packages Fluent and ANSYS. MPCCI (Multiphysics Coupling Interface) is used to couple the solution in the flow domain and in the magnetic field domain. The circuit equations are also coupled in the solution process. User defined codes are necessary to implement the simulation model.

As shown in Fig. 7.3, the model is 3D. In order to present the calculation results inside the arc chamber intuitively and concisely, the results on the symmetry plane of the model were selected for analysis in this section. Figures 7.6, 7.7, 7.8, 7.9, 7.10, 7.11 and 7.12 show, respectively, the arc voltage and current, temperature contours, pressure, speed, velocity vector, current density, and the potential contours, respectively in the arc chamber. In the simulation, $C = 35{,}000\ \mu\text{F}$ and $L = 190\ \mu\text{H}$. The peak current is 4kA. The sequences of the colour contours illustrate the evolution of the physical quantities of the arc and reflect the interactions between various factors in the arc chamber.

2. Analysis of arc dynamics

Figure 7.13 shows the temperature distribution in the arc chamber at t = 0.195 ms. At this time, the arc has left the region where the arc is initiated between the moving and stationary contacts. The arc root on the moving contact (not moving in the simulation) has crossed the gap and jumped from the moving contact to the arc

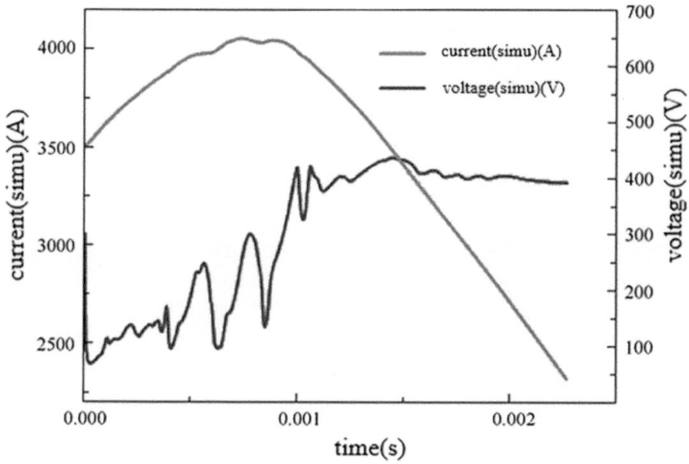

Fig. 7.6 Arc current and voltage waveforms used in the simulation of the chosen MCB

runner. The temperature in the main part of the arc column is high, approximately at 20,000 K. The gas in this region is heated up by strong Ohmic heating and intense thermal ionisation makes it a good conductor.

The current density distribution within the arc chamber at T = 0.195 ms is presented in Fig. 7.14. It is clear that the current primarily flows through the high-temperature arc region. The current in the conductors induces a magnetic field which in turn generated a Lorentz force on the arcing gas and drives it to move towards the arc chute.

Figure 7.15 shows the gas pressure distribution in the arc chamber at t = 0.195 ms. At this time, high pressure has reached the arc chute as a result of pressure wave propagation. A large area of low pressure appears behind the arc and around the moving contact. Back-breakdown phenomenon is observed in this region. Under the combined influence of Lorentz force and pressure gradient, the arc moves upwards.

Figure 7.16 is the velocity distribution in the arc chamber at t = 0.195 ms. A close examination shows that at this moment the maximum speed in the arc chamber appears at the place marked as 4 in the diagram. This region is above the hot arc column and the maximum speed is about 3,000 m/s. The velocity in the main arc column is approximately 1,500 m/s. It is worth noting that in the region labelled as 1, the velocity of the gas reaches about 900 m/s, which is much higher than the gas velocity at the designed outlet as labelled as 2. The results imply that with such a geometric design of the MCB gas is primarily released into the environment through outlet channel 1. In addition, the presence of vortices in the region labelled as 3 suggests that gas exhaustion on the left side of the chamber is less effective compared to that on the right side. The gas flow on the left side of the chamber reverses direction and creates vortices. This contributes to the differences in the movement speed of the arc roots on the left and right sides of the chamber.

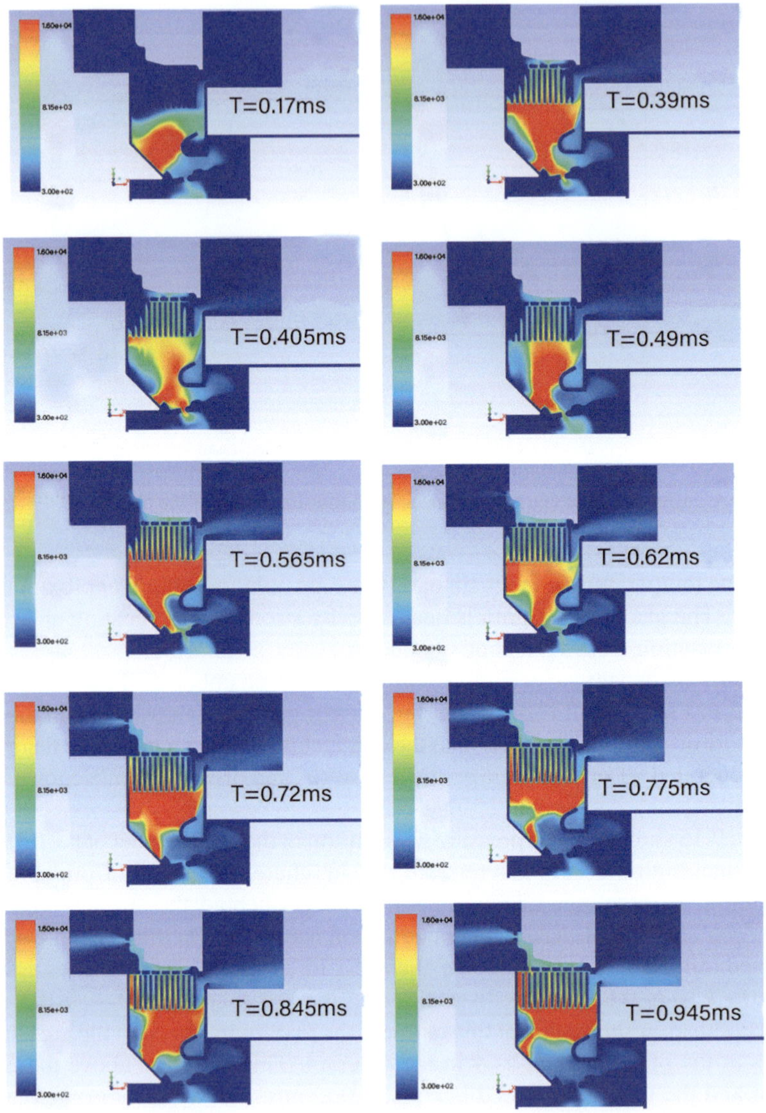

Fig. 7.7 Temperature contours from the simulation of the chosen MCB

The potential distribution at t = 0.195 ms, calculated based on the conductance of the arcing gas, is shown in Fig. 7.17. At this time, the arc voltage has reached 150 V. Compared with the arc voltage at the initial stage when the arc is just formed, the arc voltage has increased significantly due to the elongation of the arc column and the cooling effect of the gas flow.

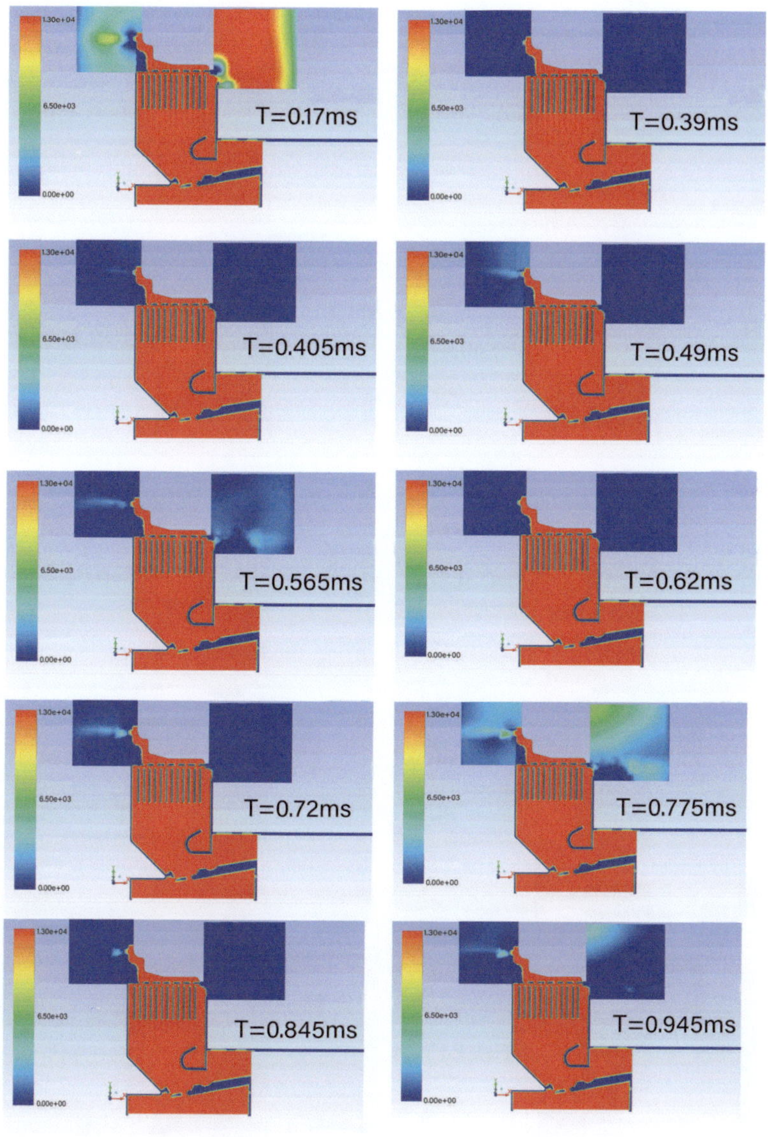

Fig. 7.8 Pressure contours from the simulation of the chosen MCB

The arc column is then driven upwards by the aerodynamic effect and magnetic force until the upper part of the arc reaches the arc chute. Figure 7.18 shows the movement of the arc during the period from the arc reaching the bottom of the arc chute at t = 0.325 ms to the time t = 0.410 ms. At t = 0.325 ms, the arc root on the left side started to move downwards compared to the time of 0.195 ms, the arc root on the right side is almost stagnant, and the upper part of the arc column approaches

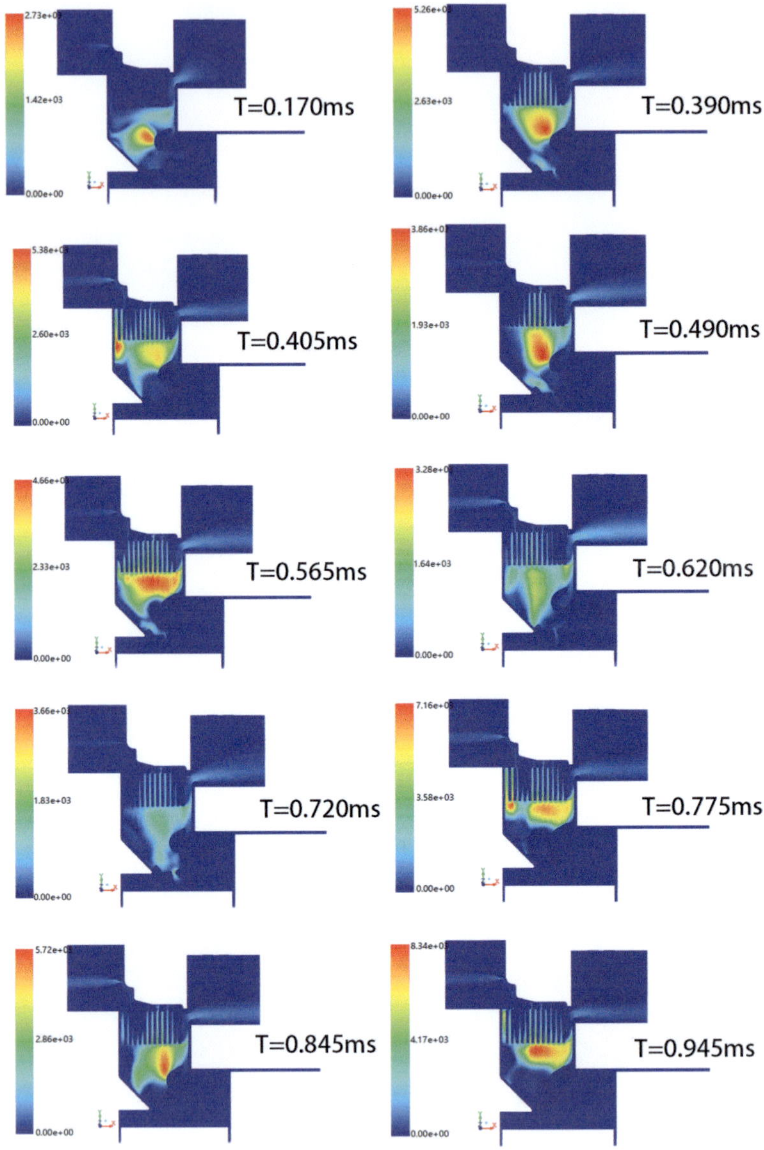

Fig. 7.9 Gas flow speed from the simulation of the chosen MCB

the bottom of the arc chute. At t = 0.350 ms, the left arc root has crossed the gap and jumped from the arc runner onto the moving contact. Meanwhile, it is worth noting that at the location neighbouring the bottom of the leftmost splitter plates and the arc runner, a higher temperature zone appears and the local temperature is about 10,000 K. As mentioned above in this section, the vortices shown in Fig. 7.16

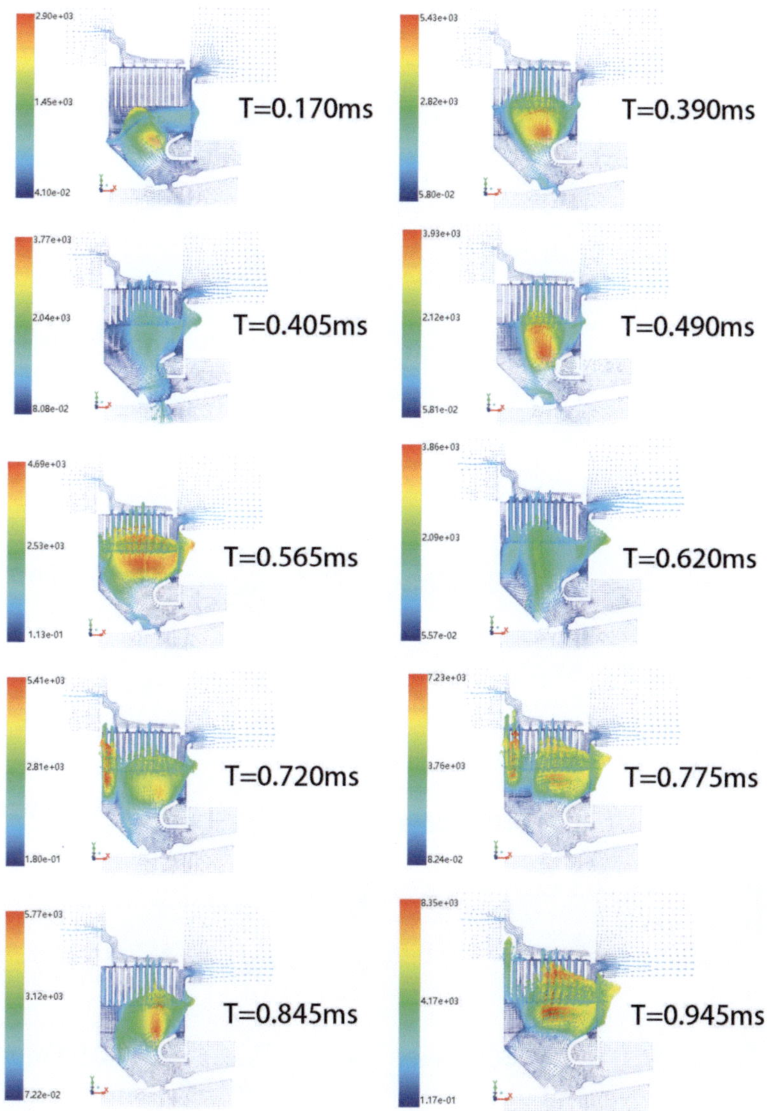

Fig. 7.10 Velocity contour from the simulation of the chosen MCB

bring the high-temperature gas in the arc column into this region and increases the electrical conductivity. This trend continues until a new arc root on the left arc runner is generated at t = 0.370 ms. By t = 0.385 ms, a new arc root on the moving contact has been formed.

At t = 0.385 ms, a driving force is formed to force the upper part of the arc column to move towards the inside of the arc chute, meanwhile another force to move the

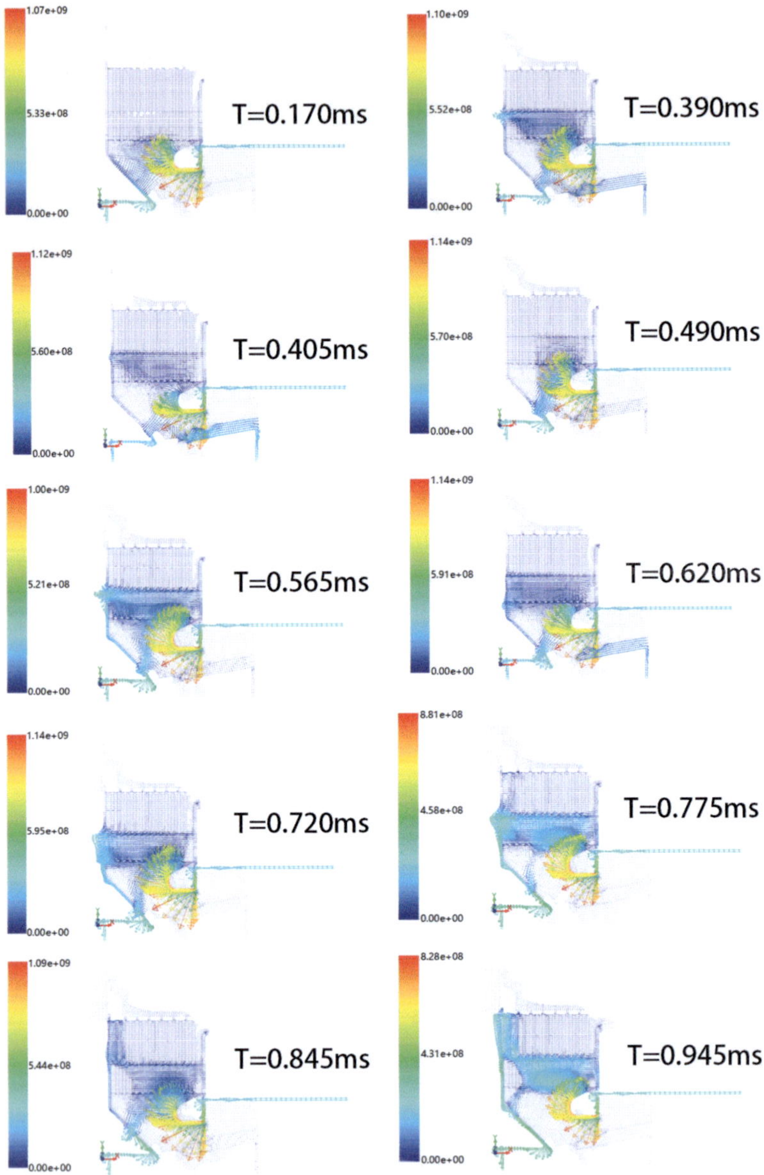

Fig. 7.11 Current density distribution from the simulation of the chosen MCB

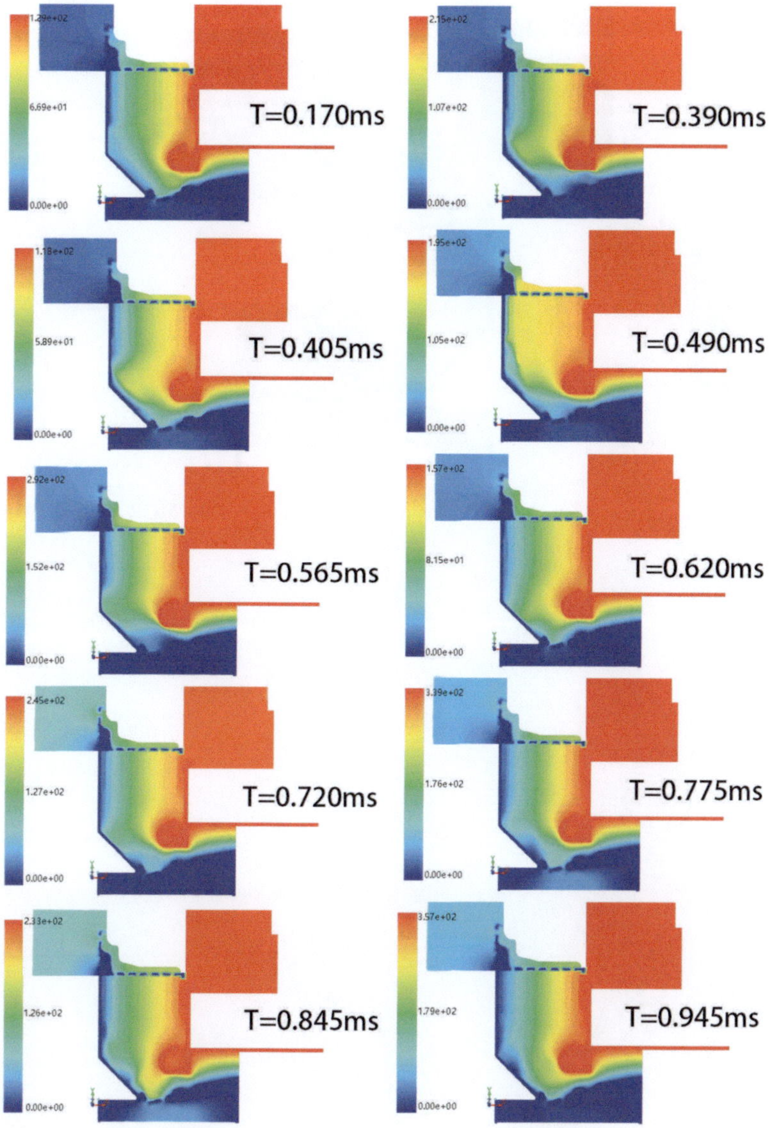

Fig. 7.12 Potential distribution from the simulation of the chosen MCB

lower part of the arc column and the left arc root to downloads. By t = 0.400 ms, it can be clearly seen that the arc column is almost split into two parts, and the temperature in the isolating strip gradually decreases. By t = 0.410 ms, the high-temperature region around the bottom of the arc chute has basically disappeared, and the main high-temperature region is concentrated between the moving and stationary contacts. At this time, a significant back-breakdown occurs, and it can be seen from the arc

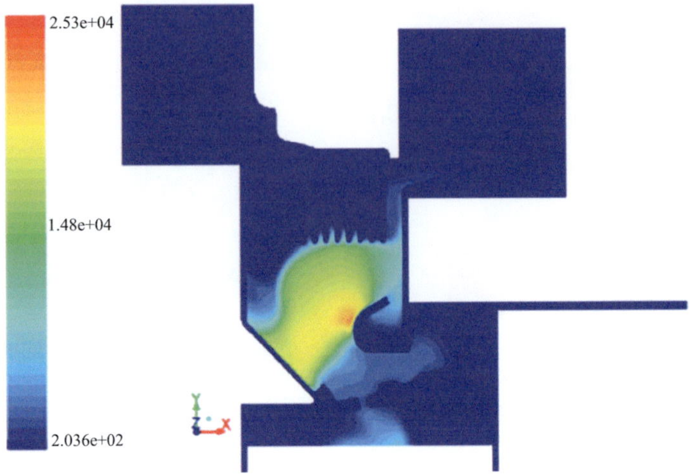

Fig. 7.13 Simulated temperature distribution in the arc chamber of the MCB at t = 0.195 ms

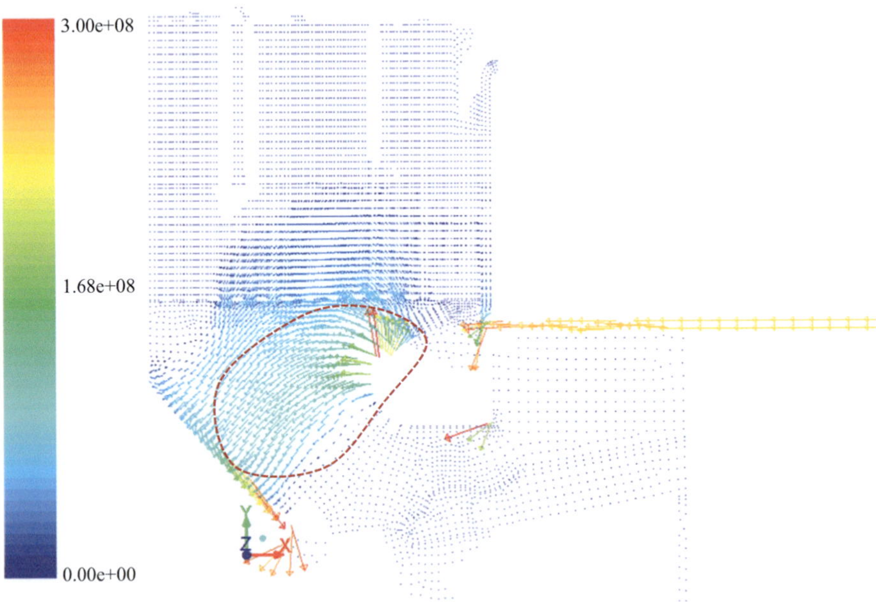

Fig. 7.14 Simulated current density distribution in the arc chamber of the MCB at t = 0.195 ms

current and voltage waveform in Fig. 7.6 that the arc voltage at this moment reaches a minimum value, approximately around 100 V. This is extremely detrimental for arc extinction. This back-breakdown phenomenon can be primarily attributed to factors relating to the flow field on the left side of the chamber.

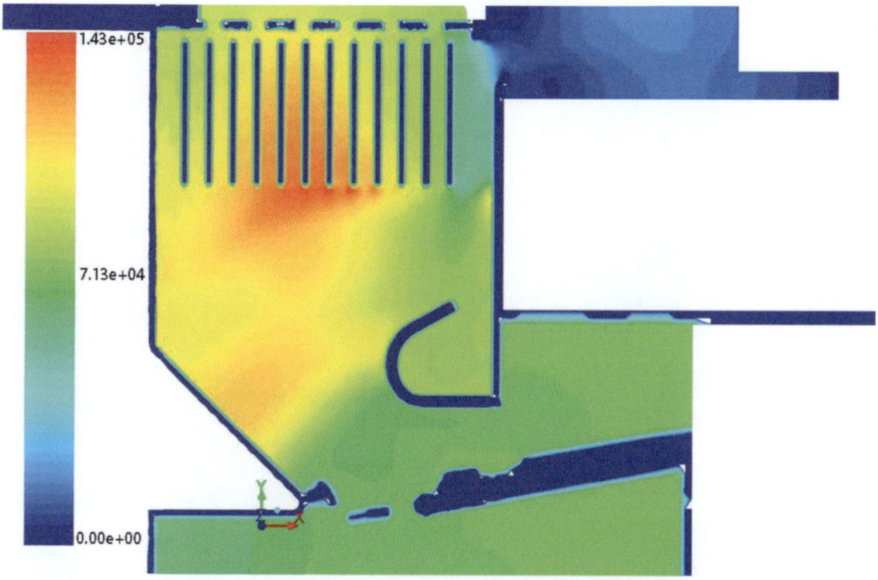

Fig. 7.15 Simulated pressure distribution in the arc chamber of the MCB at t = 0.195 ms

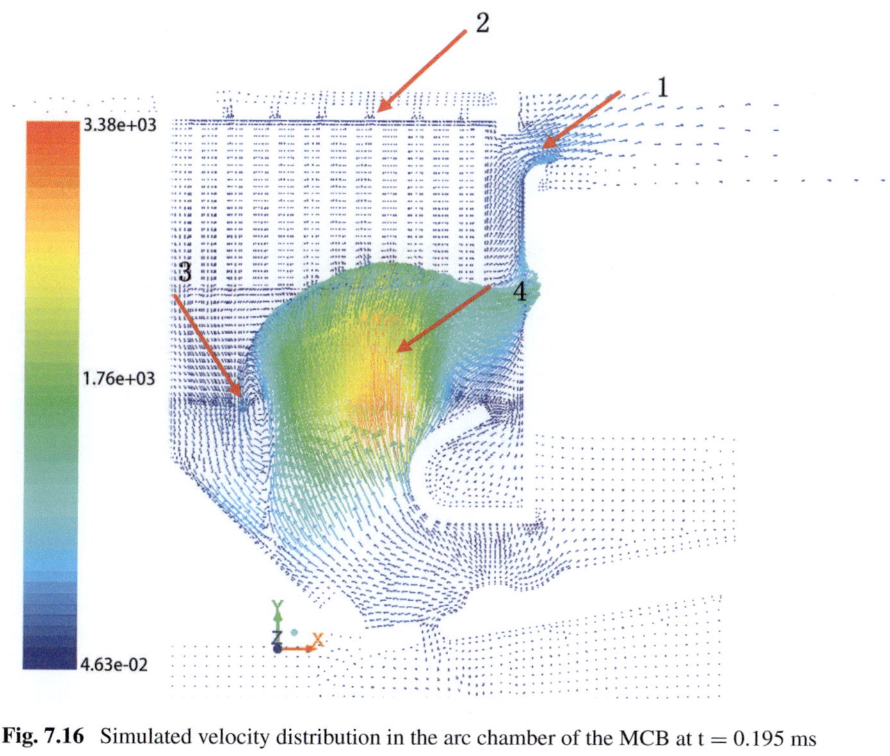

Fig. 7.16 Simulated velocity distribution in the arc chamber of the MCB at t = 0.195 ms

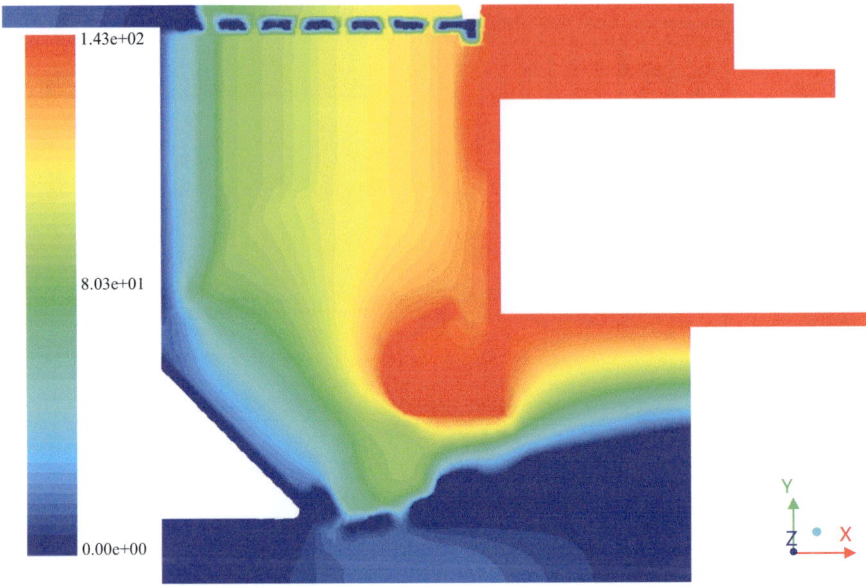

Fig. 7.17 Electric potential distribution in the arc chamber of the MCB at t = 0.195 ms

Subsequently, during the arc's movement towards the arc chute, two more distinct back-breakdowns occur. These phenomena correspond closely to the arc voltage and current waveforms.

By t = 0.900 ms, the whole arc column reaches the bottom of the arc chute again. At this time, the arc has two arc roots on the left side and one of them has entered the splitter space while the other is still on the arc runner. On the right side, there are also two arc roots. Overall, the arc continues to move upwards due to the aerodynamic effect and the Lorentz force. The arc root on the right side moves slower in comparison with the left side because it has to cross the large gap at the corner of the U-shaped arc runner. By t = 1.010 ms, most of the arc on the left side has entered the arc chute and been cut by the splitter plates. The lower arc root on the left side has almost disappeared, and the current path shifted almost entirely to the upper arc root. So at this point, the left side of the arc column can be considered to be fully within the splitter plates.

By t = 1.035 ms, the arc root on the right side has mostly crossed the gap between the U-shaped corner of the stationary contact and the arc runner, forming two arc roots on the U-shaped corner and the running arc. In addition, heavy arcing is visible on the back side of the splitter plates. This region is the gas exhaustion channel made of insulating plastics. So the arc causes severe thermal stress and damage to the structure. At t = 1.050 ms, all of the arc roots on the right side have reached the arc runner and there is stronger arcing at the back of the splitter plates. At t = 1.100 ms, the right arc roots have moved to the end of the arc runner and an arc is blowing outwards in the gas exhaustion channel, implying that there is severe

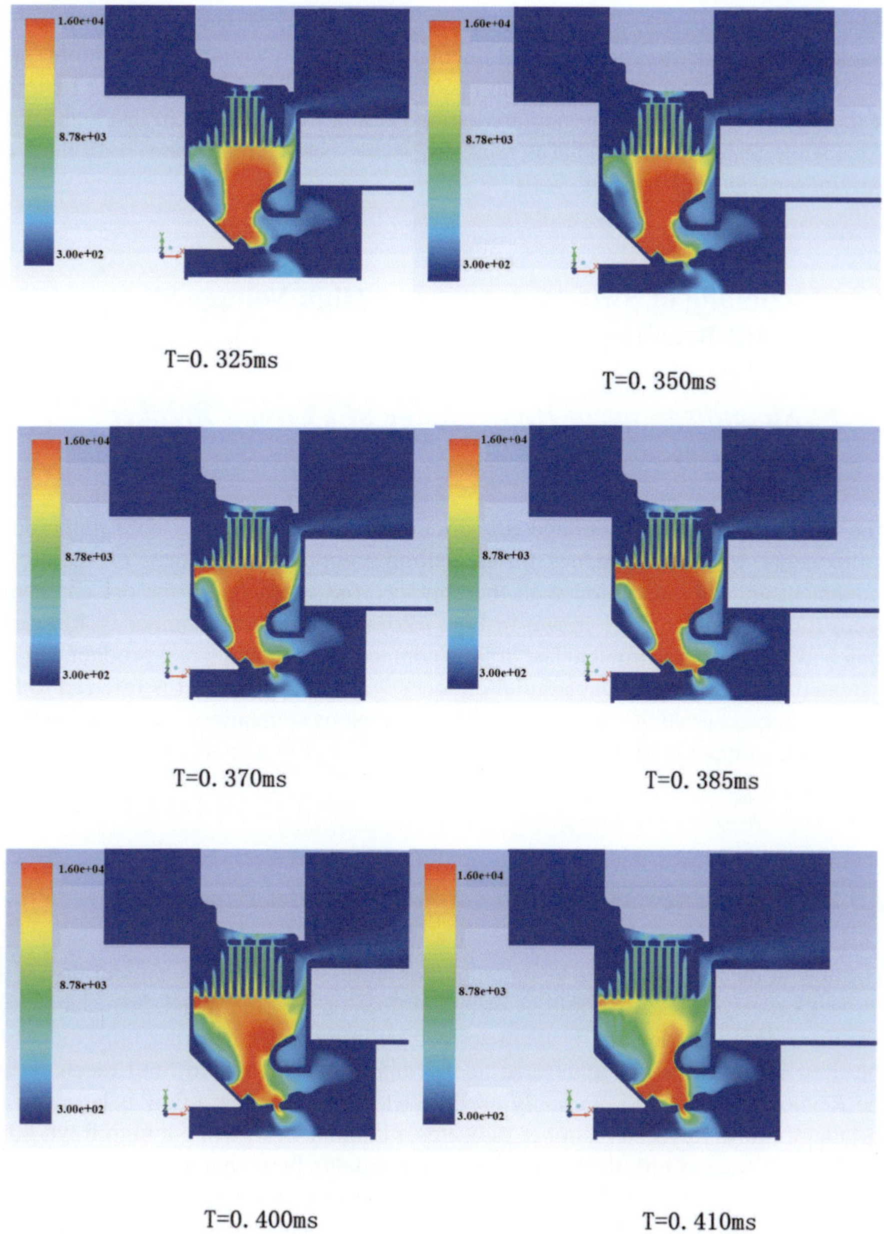

Fig. 7.18 An example of the back-breakdown phenomenon taking place during the arcing processs inside the MCB

ablation of the material. Also at this time, the arc column becomes stagnant overall inside the arc chute excepts localised fluctuation of the arc column. For example, at t = 1.250 ms, part of the arc column becomes fully connected around the bottom of the right side splitter plates and arc jetting becomes more apparent inside the gas exhaustion channel. The arc burns relatively stably between the plates until the arc extinguishes, as shown in Fig. 7.19.

7.3 Modelling of Switching Arcs in a High Voltage SF_6 Circuit Breaker

7.3.1 *Simplifications of High Voltage SF_6 Circuit Breaker Geometry*

The arc chamber in high voltage SF_6 circuit breakers is complex with numerous components. In order to reduce the modelling complexity and focus on the arc's movement and its interaction with the contacts, the geometry of the arc chamber needs to be simplified. As shown in Fig. 7.20, the entire arc chamber is 400 mm long and 150 mm in diameter. The hollow contact is moving in a real product, but is it treated as stationary in the simulation work. The hollow contact is referred to as the tulip contact or stationary contact in the context of simulation. The solid contact or the plug contact is moving in the simulation. The plug contact has a diameter of 30 mm.

7.3.2 *Turbulence Model and Choice of Model Parameters*

The flow of the arcing gas can be in laminar or turbulent regimes. A laminar flow is a smooth, orderly flow of the fluid. A turbulent flow, on the other hand, is in a chaotic, nonlinear flow in space and is characterized by intense and irregular movement of the fluid in the direction transverse to the main flow [7–11]. In fluid mechanics, the Reynolds number is commonly used to determine whether a flow is laminar or turbulent: a low Reynolds number indicates a laminar flow, while a high Reynolds number indicates a turbulent flow. Inside an annular flow channel, it is generally considered that a flow with a Reynolds number greater a certain value is turbulent [12, 13]. In SF_6 switching arcs, the Reynolds number is generally of the order of 10^6–10^9 and turbulent effects must be considered [14]. The k-ε or k-ω model are most convenient to use in the modelling of switching arcs in high voltage circuit breakers [15, 16].

The $k-\epsilon$ turbulence model introduced in Chap. 1 is suitable for arc simulation due to its ability to effectively capture the complex turbulent flow characteristics typical in such high-energy environments. This model provides a good balance between

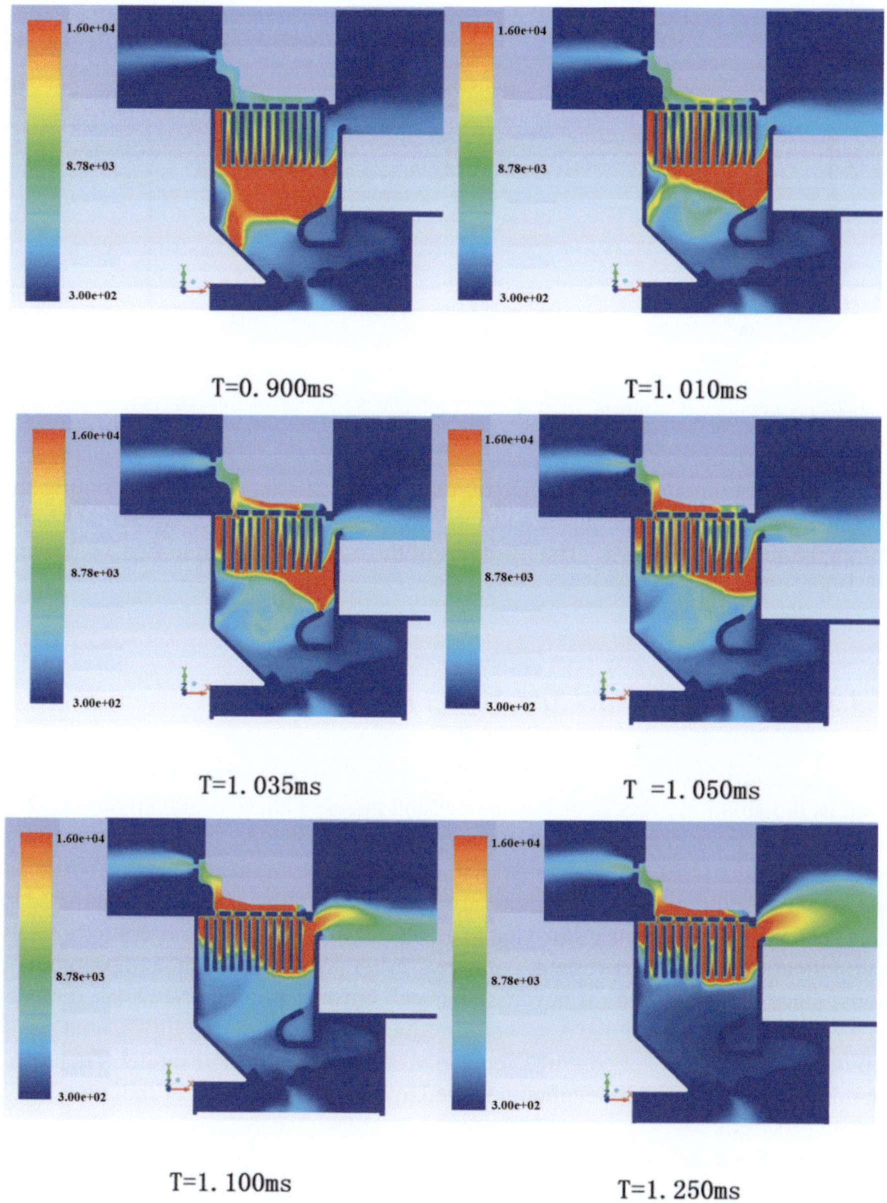

Fig. 7.19 Stable behaviour of the arc column inside the arc chute of the MCB

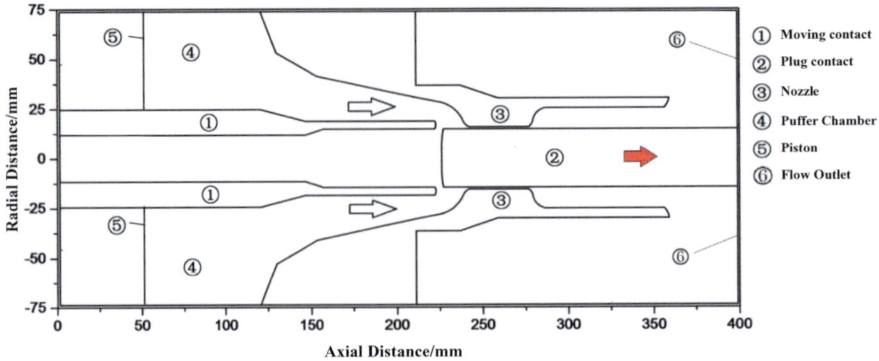

Fig. 7.20 Simplified structure of a part of a high voltage SF_6 circuit breaker selected for simulation

accuracy and computational efficiency, making it well-suited for representing the turbulent behaviour of the gas around the arc, including the high-speed and high-temperature effects [17–19]. The formulas of the model are given in Eqs. (1.30) and (1.31).

7.3.3 Radiation Model and Model Parameters

The radiation transfer process affects the cooling or heating of the arcing gas as well as the ablation process of the contact and nozzle materials and so on [20–22]. It is an important energy transport mechanism in switching arcs. The Net Emission Coefficient (NEC) based model and the P-1 radiation model are the two most commonly used models in switching arc simulation. The arc model established in this chapter is axisymmetric since the geometry of the arc chamber is approximately cylindrical. The NEC based model employs the concept of NEC in an approximate sense since the arc column is never isothermal. Furthermore, the NEC based model is a zero-equation model and does not introduce additional partial differential equations, resulting in a lower computational load compared to the P-1 model. Therefore, in this chapter the NEC based model is used to calculate the effect of radiation in the arcing process.

As discussed in Sect. 1.3, the NEC of a spherical or cylindrical plasma column of uniform temperature is a function of the plasma radius (emission radius), R [23]. According to [24] the central part of the arc column where the temperature is higher than $0.83T_{max}$ is defined as the net emission zone where T_{max} is the axis temperature in the arc. An equivalent emission radius can be defined based on experience. At the edge of the arc column, the arc temperature is decreasing rapidly and strong recombination/association of ions/atoms exists. A significant portion (80% in [25]) of the radiation flux from the net emission zone is absorbed in a region called the net absorption zone. In this example, the net absorption zone starts from the radial position of 4,000 K and ends at $0.83T_{max}$.

7.3.4 Material Properties for Simulation of Arcs in High Voltage SF_6 Circuit Breakers

The arc model in this example contains three materials, SF_6, contact material and nozzle material. Here is a brief introduction to each:

(a) SF_6 Real Gas Model. The material properties of SF_6 are critical in the arc simulation. The properties, such as the mass density ρ, specific heat capacity at constant pressure c_p, thermal conductivity λ, viscosity coefficient μ, and electrical conductivity σ do not stay constant. They are nonlinear functions of gas pressure and temperature.

(b) Contact and nozzle material properties. Contact erosion is not considered in this example due to the dual flow feature in the circuit breaker. The contacts do not participate in the flow simulation. The contact material mainly affects the conduction of current. In the area where the material is electrically insulating, then a small value of 10^{-6} S/m is used. Inside the contacts and associated conductors, the electrical conductivity is set to 10^6 S/m. For the nozzle and other insulating materials, the electrical conductivity is set to 10^{-6} S/m. Variation of the above values by a factor of 2 does not significantly affect the results.

7.3.5 Initial and Boundary Conditions for Simulation of Arcs in High Voltage SF_6 Circuit Breakers

Arc Simulation in this example involves solving the governing equations for the flow field, electric and magnetic fields, and radiation transfer. Each partial differential equation requires correct boundary conditions to ensure the convergence and accuracy of the solution process. The boundary conditions are described below with reference to Fig. 7.21.

1. Thermal boundary conditions

The circuit breaker chamber wall is treated as adiabatic boundary. One-dimensional thermal conduction is imposed on the nozzle and contact surfaces by:

$$Q = -k(T - T_0)/d \tag{7.5}$$

where k is the thermal conductivity of the chamber wall material, T the surface temperature of the chamber wall, and d the wall thickness. T_0 is the ambient temperature with a default value of 300 K. The boundary conditions settings are shown as Table 7.2.

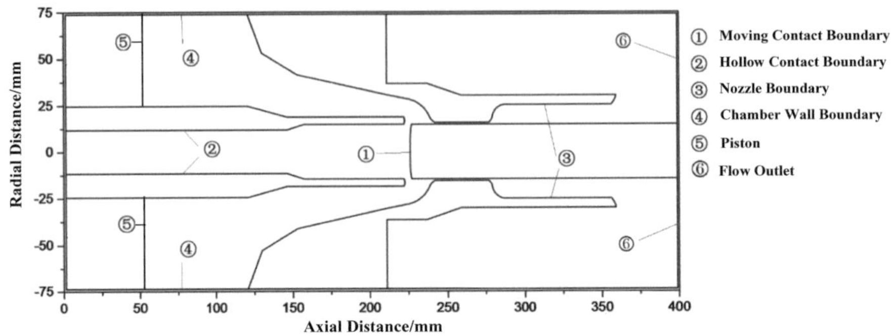

Fig. 7.21 Different boundaries used in the simulation for arc flow inside a high voltage SF_6 circuit breaker

Table 7.2 Thermal boundary conditions

Boundary	Boundary conditions
① Moving contact boundary	One-dimensional thermal conduction
② Hollow contact boundary	One-dimensional thermal conduction
③ Nozzle boundary	One-dimensional thermal conduction
④ Chamber wall boundary	Adiabatic
⑤ Piston	Adiabatic
⑥ Flow outlet	300 K (at constant pressure)

2. Flow boundary conditions

The computational domain under consideration is connected to its environment through two flow outlets. The pressure at the flow outlet is set to an absolute pressure of 6 atm for demonstration purposes. The other boundaries are shown in Table 7.3.

3. Electric and magnetic field boundary conditions

A hollow contact is used in high voltage circuit breakers to ensure sufficient electrical contact with the other contact, normally a plug contact. An advantage of this arrangement is that it minimises the possibility of the arc root staying at the same position on the contact surface and thus reduces the erosion of the contact materials. However, the hollow contact brings difficulties to the numerical simulation of the arc with a two-dimensional axisymmetric model since the arc column does not always stay on the axis of the domain. This is particularly a problem for low current arcs whose diameter is smaller than the bore diameter of the hollow contact. In fact in

Table 7.3 Flow boundary conditions

Boundary	Boundary conditions
① Flow outlet	Fixed pressure at 6 atm (absolute)

reality the arc root will move around and sometimes move towards the inner surface of the hollow contact under the influence of the flow field. In order to enable the solution process, a transparent contact is adopted [ref], that is, this transparent contact has high electrical conductivity (106 S/m) regardless the temperature of the gas at the location of the this transparent contact. The transparent contact fills the hollow contact. It does not pose any resistance to the gas flow but only helps collect the current to ensure current continuity.

In the simulation, the potential on the transparent contact is set to zero while the far end of the moving contact (in simulation) is imposed a current that changes with time. A power frequency current of 50 kA (RMS value) is used in this example. The normal gradient of the electric potential is set to zero at the surface of the insulating wall in the nozzle region (part 3). The current is imposed by:

$$j = I/S \tag{7.6}$$

where I is the prescribed current, S the cross-sectional area of the current inlet. The other boundary conditions are listed in Table 7.4.

4. Settings for moving mesh and initial conditions

In the high voltage SF_6 circuit breaker under simulation, the average contact opening speed is about 10 m/s, that is, the moving direction of component 2 in Fig. 7.20 is indicated by the arrow and its moving speed is 10 m/s. At the same time, the piston (component 5 of Fig. 7.20) moves in the same direction and at the same speed. A structured mesh system is used in the simulation and the layering mesh generation method in Fluent is chosen to enable contact and arc movement. As a commonly accepted rule, when the cell size in the direction of movement becomes more than 1.4 times or less than 0.2 times that of the original cell size, grid reconstruction will take place. The gas is initially at 6 atm (absolute) and room temperature, 300 K. A hot column is forced in part of the domain to start the arcing process. The settings of the initial conditions are shown in Fig. 7.22.

Table 7.4 Electric and magnetic field boundary conditions

Boundaries	Boundary conditions
① Nozzle boundary	Normal gradient od potential is zero
② Chamber wall boundary	Normal component of magnetic vector potential is zero
③ Piston	Zero potential
④ Flow outlet	Zero potential

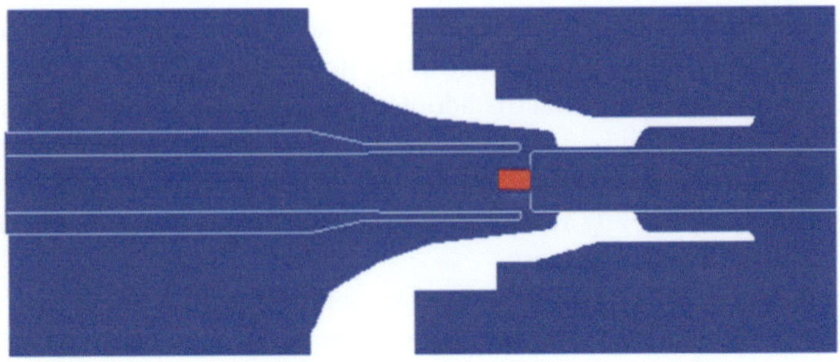

Fig. 7.22 Initial uniform gas temperature distribution in most of the computational domain that is selected for simulating the arcing process in a high voltage SF_6 circuit breaker. the red region indicates a hot and conducting column of gas

7.3.6 Results from Simulation

The results presented in this section are obtained based on the geometry shown in Fig. 7.20, with model parameters described in previous sections. Following mesh sensitivity studies, i.e. the influence of mesh on results) and a consideration of the accuracy and computing time, a cell number of 36,000 is adopted. The maximum time step is limited to 10^{-7} s and the duration is a half cycle (10 ms). Figure 7.23 shows the temperature distribution in the arc chamber during the arcing process of a 50 kA short-circuit current (RMS value). Figures 7.24 and 7.25 compare the experimental results with the simulated values of arc voltage and arc chamber pressure, respectively. Due to the thermal expansion of the high-temperature arcing gas and the blocking effect of the stationary contact, the pressure inside the compression chamber rises significantly.

There are discrepancies between the simulation prediction and measurements.

7.4 Summary

Two simulation examples are given in this chapter to illustrate the procedure to set up the arc simulation model. The choice of a proper computational domain based on knowledge and understanding of the physical phenomena and the operation of the circuit breakers is key to efficient (affordable computing time) and accurate (meeting the application need) simulation of the arcing process. Boundary conditions must be set up in a sensible manner for each partial different equation in the governing equations. Approximate boundary conditions can be used provided there are strong arguments or evidence that the level of approximation does not lead to significant errors in the solution. A sensitivity study may be needed for this purpose.

There are no universal turbulence models that are accurate to predict the effect of turbulence in the arcing flow. They need to be calibrated by test results before

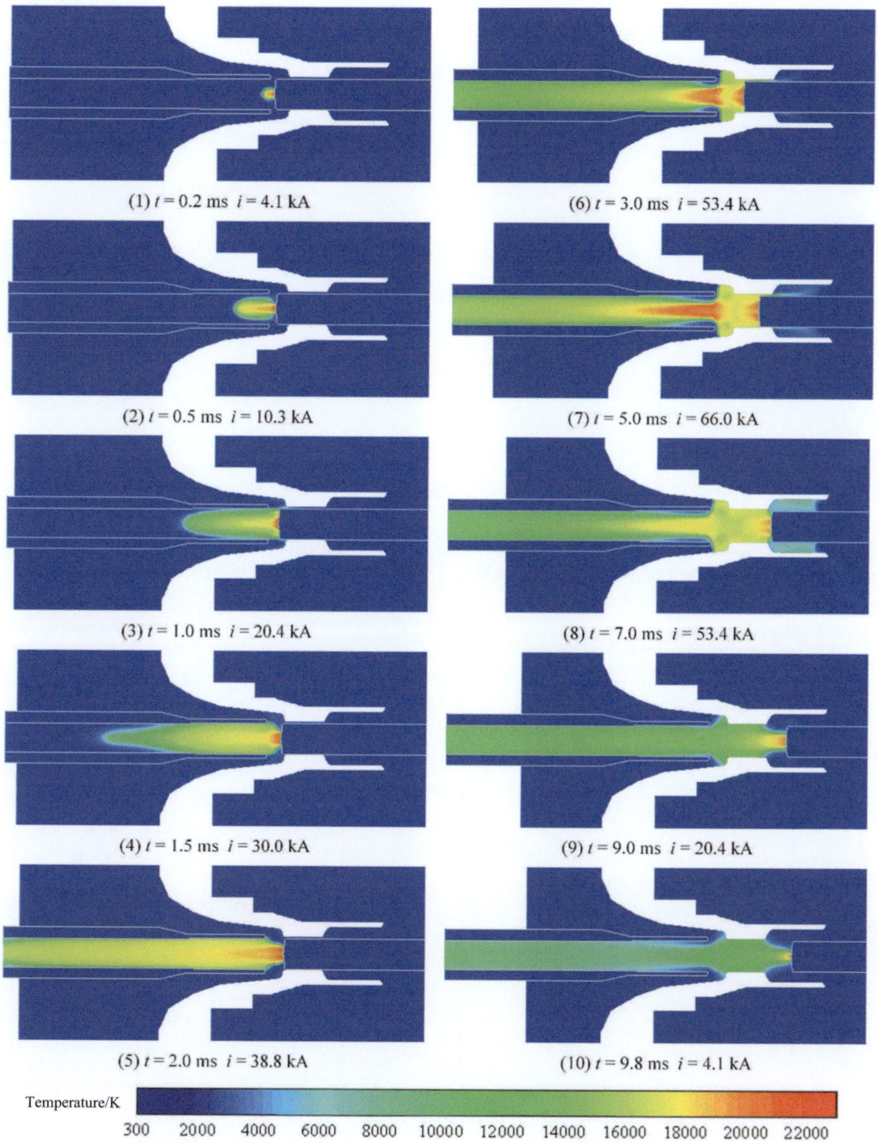

Fig. 7.23 Simulated temperature distribution at different instants of the arcing process with a power frequency current of 50 kA (RMS value). The filling gas for the high voltage circuit breaker is SF_6

confidence can be built up with the simulation results. The accuracy and applicability of radiation transfer calculation remains an issue to tackle.

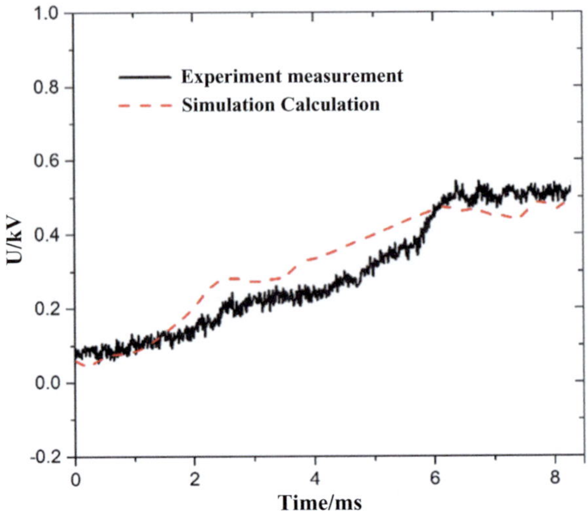

Fig. 7.24 Comparison between measurement and simulation prediction of the arc voltage for a power frequency current of 50 kA (RMS value). The filling gas for the high voltage circuit breaker is SF_6

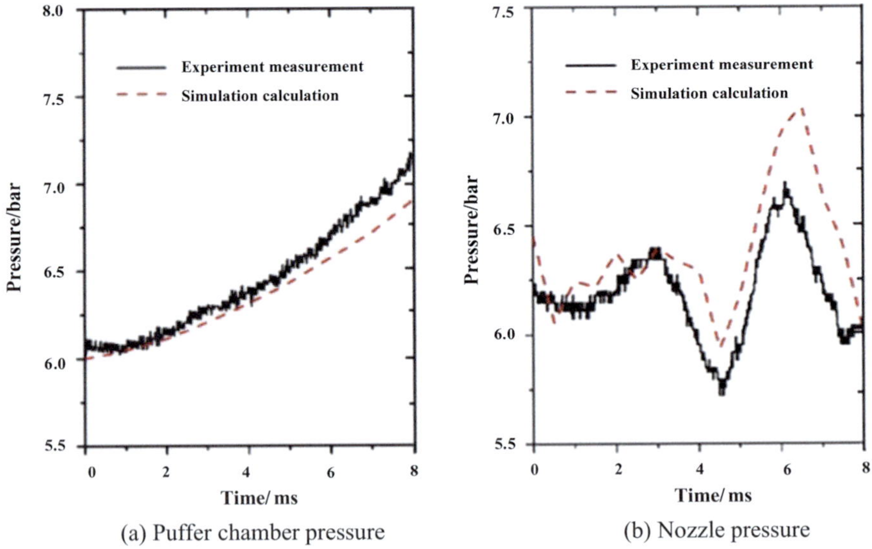

Fig. 7.25 Comparison between measurement and simulation prediction of the gas pressure variation inside the arc chamber of a high voltage SF_6 circuit breaker. The arcing current is 50 kA at power frequency

References

1. Yao J. Selection of miniature circuit breaker[J]. Electric World, 2000, 3: 1-3.
2. Yang F, Rong M, Wu Y, et al. Simulation and experimental study of grid cutting air arc considering grid ablation metal vapor[J]. Acta Physica Sinica, 2011, 60(5): 055208.
3. Pei J, Sun H, Liu Z, et al. Simulation study on the influence of metal vapor on arc process in low voltage circuit breaker[J]. Electrical Appliances and Energy Efficiency Management Technology, 2011, 3: 12-15.
4. Chen Z, Wu Y, Yang F, et al. Research on arc simulation of low voltage circuit breaker[J]. Electrical Appliances and Energy Efficiency Management Technology, 2014, 10: 10-17.
5. Li X, Chen D. Review of research on mathematical model and characteristics of air switch arc[J]. High Voltage Apparatus, 2007, 43(4): 269-273.
6. Wang W, Rong M, Wu Y, et al. Study on models of physical properties of equilibrium and non-equilibrium plasma[J]. High Voltage Apparatus, 2010, 46(7): 41-45.
7. Yan, J. D., K. I. Nuttall, and M. T. C. Fang. "A comparative study of turbulence models for SF6 arcs in a supersonic nozzle." Journal of Physics D: Applied Physics 32.12 (1999): 1401.
8. Zhang J F, Fang M T C, Newland D B. Theoretical investigation of a 2kA arc in a supersonic nozzle[J]. Journal of Physics D: Applied Physics, 1987, 20: 368-379.
9. Fang M T C, Zhuang Q, Guo X J. Current zero behaviour of a SF6 gas-blast arc. Part II: turbulent flow[J]. Journal of Physics D: Applied Physics, 1994, 27: 74–83.
10. Markatos N C. The mathematical modelling of turbulent flows[J]. Applied Mathematical Modelling, 1986, 10: 190-220.
11. Zhang J L. Computational investigation of arc behavior in an auto-expansion circuit breaker contaminated by ablated nozzle vapor[J]. IEEE Transactions on Plasma Science, 2002, 30(2): 706-719.
12. Günther A. Turbulent flow in a channel at a low Reynolds number[J]. Experiments in Fluids, 1998, 25(5): 503-511.
13. Trinh K T. On the critical Reynolds number for transition from laminar to turbulent flow[J]. arXiv preprint, 2010, arXiv:1007.0810.
14. Liu J. Analysis of the characteristics of DC nozzle arcs in air and guidance for the search of SF6 replacement gas[J]. Journal of Physics D: Applied Physics, 2016, 49(43): 435201.
15. Zhou Q. Comparative study of turbulence models on highly constricted plasma cutting arc[J]. Journal of Physics D: Applied Physics, 2008, 42(1): 015210.
16. Zhang Q, Yan J D, Fang M T C. The modelling of a SF6 arc in a supersonic nozzle: I. Cold flow features and dc arc characteristics[J]. Journal of Physics D: Applied Physics, 2014, 47(21): 215201.
17. Spalart P, Allmaras S. A one-equation turbulence model for aerodynamic flows[C]. 30th Aerospace Sciences Meeting and Exhibit, 1992.
18. Wilcox D C. Turbulence modeling for CFD[M]. La Canada, CA: DCW Industries, 1998.
19. Niyogi P. Introduction to computational fluid dynamics[M]. Pearson Education India, 2006.
20. Soloukhin R I. Handbook of radiative heat transfer in high-temperature gases[M]. CRC Press, 2000.
21. Chen W. Investigation and analysis of arc ablation on WCu electrical contact materials[J]. Journal of Materials Science: Materials in Electronics, 2016, 27: 5584-5591.
22. Ruchti C B, Niemeyer L. Ablation controlled arcs[J]. IEEE Transactions on Plasma Science, 1986, 14(4): 423-434.
23. Li M. Simulation of fault arc based on different radiation models in a closed tank[J]. Plasma Science and Technology, 2016, 18(5): 549.
24. Wu Y. Numerical analysis of three-dimensional air arc plasma using a six-band P-1 radiation model[J]. Acta Physica Sinica, 2008, 57(9): 5761-5767.
25. Sun Z. Three-dimensional numerical analysis with P-1 radiation model in low voltage switching arc[J]. IEICE Transactions on Electronics, 2007, 90(7): 1348-1355.

MIX
Papier aus verantwortungsvollen Quellen
Paper from responsible sources
FSC® C105338

If you have any concerns about our products,
you can contact us on
ProductSafety@springernature.com

In case Publisher is established outside the EU,
the EU authorized representative is:
Springer Nature Customer Service Center GmbH
Europaplatz 3, 69115 Heidelberg, Germany

Printed by Libri Plureos GmbH
in Hamburg, Germany